£12.05 TW/7

Library of
Davidson College
VOID

THE LOGICAL STRUCTURE OF MATHEMATICAL PHYSICS

SYNTHESE LIBRARY

MONOGRAPHS ON EPISTEMOLOGY,

LOGIC, METHODOLOGY, PHILOSOPHY OF SCIENCE,

SOCIOLOGY OF SCIENCE AND OF KNOWLEDGE,

AND ON THE MATHEMATICAL METHODS OF

SOCIAL AND BEHAVIORAL SCIENCES

Editors:

DONALD DAVIDSON, *Rockefeller University and Princeton University*

JAAKKO HINTIKKA, *Academy of Finland and Stanford University*

GABRIËL NUCHELMANS, *University of Leyden*

WESLEY C. SALMON, *Indiana University*

JOSEPH D. SNEED

THE LOGICAL STRUCTURE OF MATHEMATICAL PHYSICS

D. REIDEL PUBLISHING COMPANY / DORDRECHT-HOLLAND

Library of Congress Catalog Card Number 74–118136

ISBN 90 277 0166 0

All Rights Reserved
Copyright © 1971 by D. Reidel Publishing Company, Dordrecht, Holland
No part of this book may be reproduced in any form, by print, photoprint, microfilm,
or any other means, without written permission from the publisher

Printed in Great Britain

To my parents

PREFACE

This book is about scientific theories of a particular kind – theories of mathematical physics. Examples of such theories are classical and relativistic particle mechanics, classical electrodynamics, classical thermodynamics, statistical mechanics, hydrodynamics, and quantum mechanics. Roughly, these are theories in which a certain mathematical structure is employed to make statements about some fragment of the world. Most of the book is simply an elaboration of this rough characterization of theories of mathematical physics. It is argued that each theory of mathematical physics has associated with it a certain characteristic mathematical structure. This structure may be used in a variety of ways to make empirical claims about putative applications of the theory. Typically – though not necessarily – the way this structure is used in making such claims requires that certain elements in the structure play essentially different roles. Some play a "theoretical" role; others play a "non-theoretical" role. For example, in classical particle mechanics, mass and force play a theoretical role while position plays a non-theoretical role. Some attention is given to showing how this distinction can be drawn and describing precisely the way in which the theoretical and non-theoretical elements function in the claims of the theory. An attempt is made to say, rather precisely, what a theory of mathematical physics is and how you tell one such theory from another – what the identity conditions for these theories are. In connection with this, certain relations that might hold between theories of mathematical physics are discussed: the equivalence relation, exemplified by Newtonian and Lagrangian formulations of classical particle mechanics; and the reduction relation, exemplified by classical rigid body mechanics and classical particle mechanics. Finally, something is said about historical development of theories of mathematical physics – how people come to have them and come to give them up.

In Chapter I the notion of the logical structure of a scientific theory is discussed and something is said about what it means to exhibit the logical structure – provide a logical reconstruction – of an existing scientific theory. The role of various sorts of axiomatizations of theories in illuminating their logical structure is considered. Particular attention is given to axiomatized deductive theories in first order logic and definitions of set-theoretic predicates. It is argued that there actually exist axiomatizations of theories of mathematical physics that are essentially definitions of set-theoretic predicates. Yet, the relevance of these axiomatizations to illuminating the logical structure of these theories is not apparent. An answer to the question of how these set-theoretic axiomatizations illuminate logical structure is suggested. It is roughly this. The set-theoretic predicate defined by the axiomatization characterizes the formal, mathematical structure associated with the theory. This predicate is used to make the empirical statements of the theory. Several ways in which such predicates might be used to make empirical statements are considered in the subsequent four chapters.

In Chapter II the simplest proposal for using a set-theoretic predicate to make empirical claims is considered. The set-theoretic predicate is predicated of a singular term. Some questions are considered which arise in connection with any proposal for using set-theoretic predicates involving quantitative notions to make empirical claims. In this connection, the theory of measurement and the question of the "approximate" nature of claims in mathematical physics are discussed briefly. The proposal considered here is shown to be essentially equivalent to a common view of the way in which the statements of a theory are obtained from the axiomatized deductive theory in first order logic which axiomatizes the theory. Considerable effort is devoted to characterizing a theory-relative notion of theoretical term, independently of any sweeping epistemological assumptions. This notion is then employed to describe a difficulty that could be encountered in attempting to employ the sentence form proposed here in logical reconstruction. An attempt is made to relate this difficulty to more traditional discussions of the problem of theoretical terms.

A second proposal for using a set-theoretic predicate to make empirical claims is considered in Chapter III. Essentially the sentence-form is this. Given a model for the non-theoretical part of the predicate, there is some way of adding a theoretical part to this model to yield a model for the entire predicate. This proposal is quite similar to the method of dealing with theoretical terms proposed by Frank Ramsey [35]. The relation between these two proposals and some formal questions raised by taking this line, are examined. Among these questions is the possibility of getting on without theoretical terms. An attempt is made to define a concept of eliminability of theoretical terms that is applicable to situations in which a set-theoretic predicate is being used to illuminate the logical structure of a theory. Finally, the question of the adequacy of this proposal to the task of logical reconstruction is considered. Though the proposal appears to solve the problem of theoretical terms it is found to be inadequate because it fails to provide a means of accounting for the way measured values of theoretical functions appropriate to one application of the theory can be employed in other applications of the same theory.

Chapter IV considers an emendation of the Ramsey view. It employs existential quantification over theoretical-function places to avoid the problem of theoretical terms. In addition, it proposes that we add to our rendering of the theory's empirical content the claim that the theoretical functions that satisfy the existential claim for various applications, taken all together, satisfy certain *constraints* that operate across applications. It is argued that the notion of constraints on the array of theoretical functions appearing in different applications of the theory allows us to account for non-trivial measurements of theoretical-function values. In addition, regarding the empirical content of a theory to be a statement of this form provides some insight into certain "holistic" views of theories such as those held by Duhem [10] and Kuhn [21].

In Chapter V one final proposal for using a set-theoretic predicate to make empirical claims is considered. The motivation for considering this proposal is provided by examples of situations in which we apparently "postulate" or "hypothesize" that theoretical functions have some special form in certain applications of a theory. It appears that a logical reconstruction of theories with such claims will require something more than sentences of the form considered in Chapter IV. The proposal for dealing with such theories is roughly this. A number of predicates are employed, all of them defined by restrictions of the definition of the same basic predicate. These predicates are used to construct a sentence which says that there are theoretical functions which make all intended applications models for the basic predicate, make some designated sub-sets of intended applications models for the "restrictions" of this basic predicate, *and* satisfy certain constraints. The "restrictions" of the basic predicate are to characterize various special forms that the theoretical function is hypothesized to have. This claim is regarded as the central empirical claim of the theory.

In Chapters II–V some simple examples of set-theoretic predicates are used to illustrate the alternative ways such a predicate could be used to elucidate the logical structure of a theory of mathematical physics. In Chapter VI these alternatives are again illustrated in attempting to provide a logical reconstruction of a real theory of mathematical physics – the Newtonian formulation of classical particle mechanics. This serves to illustrate, in a more concrete way, the difficulties with some of these alternatives and ultimately provides a *sketch* of an adequate logical reconstruction of this theory. This sketch, together with the notion of eliminability of theoretical terms developed in Chapter III, provides a means of treating in a systematic and perspicious way some frequently raised questions about the epistemological status of the concepts of mass and force. It also illuminates questions about the measurability of masses and forces and the status of specific force laws in measuring forces.

The account of the logical structure of empirical claims in theories of mathematical physics developed in the first five chapters is exploited in Chapter VII to clarify some other questions about these theories. An attempt is made to say precisely what a theory of mathematical physics *is*. Roughly, such a theory is identified as an ordered pair consisting of its characteristic mathematical structure – including "essential" constraints on its theoretical functions – and its characteristic range of intended applications. This account is then employed to investigate the properties of two relations – equivalence and reduction – that hold between some theories of mathematical physics. The Lagrangian and Hamiltonian formulations of classical particle mechanics are considered as examples of theories that are, in some sense, equivalent to the Newtonian formulation of classical particle mechanics. Classical rigid body mechanics is considered as an example of a theory which reduces to classical particle mechanics.

In Chapter VIII the previously developed static account of the logical structure of theories of mathematical physics is brought to bear on the dynamic aspect of theorizing – the way theories grow and develop in time. Considerable attention is given to the question of what it means to say a person *has* a theory of mathematical physics. This leads to an account of the characteristic ways a person's beliefs may change while he still has the same theory – how he may expand the same theory to encompass a wider range of phenomena or say more about the same range of phenomena. Finally, something is said about how people come to have theories, how they cease to have them, and how the conceptual apparatus in theories once held is related to the conceptual apparatus in theories held subsequent to them.

Having mentioned some of the things that are attempted in this book, perhaps it is also appropriate to mention some that are not.

What is described in this book is a certain way of theorizing. It involves using mathematical structures in certain ways to make claims about parts of the world. Theories of mathematical physics are examples of this way of theorizing. They are perhaps the most familiar examples and, historically, the first examples of theorizing in this way. No claim is made in this book that physical theories must, or do in fact, provide the only successful examples of this way of theorizing. Whether or not there is, or could be, theorizing in the behavorial sciences conforming to this model is a question I do not attempt to answer.

On several occasions in this book, I find it necessary to explicitly disclaim any intention of saying what a physical system is. For this reason, among others, the account of what a theory of mathematical physics is that appears in Chapter VII is admittedly deficient. I do not avoid this question because I believe it to be unimportant. I simply have nothing new to say about it. However, I do believe that a good many true and interesting things can be said about theories of mathematical physics without answering this question. I hope I have said some of these things.

In view of the pervasive role of probability concepts in contemporary mathematical physics, it must appear rather peculiar that a philosophical treatise on mathematical physics entirely avoids mentioning these concepts. This is no accident. There appear to be no insurmountable difficulties in describing precisely the mathematical structure associated with theories like statistical mechanics and quantum mechanics. Yet, a precise understanding of the empirical statements made with these structures is very elusive. Essentially, this understanding eludes us because we do not know how to interpret the probability measures appearing in these theories. Most physicists appear to believe that something like a relative frequency interpretation is appropriate. Yet, I know of no attempt to elucidate the empirical content of probabilistic theories of mathematical physics that avoids all the well known difficulties with this interpretation of probability. On the other hand, a thoroughgoing subjectivistic interpretation of the probabilities appears to require that we expand the "domain" of the theory's applications to include a human "observer" to whom the subjective probabilities belong. Even if this approach could be reconciled with our "unreconstructed" understanding of these theories, it remains to be shown that it can be carried out in detail. It is for these reasons that I avoid considering examples of probabilistic theories of mathematical physics. Whether ultimate clarification of the content of these theories will reveal that they too conform to my account of the logical structure of theories of mathematical physics, I can not say. However, I can see no compelling reason to think they will not.

The intellectual foundation for the ideas presented in this book clearly lies in the work of two philosophers: Frank Ramsey and Patrick Suppes. It has been my good fortune to be associated with the latter, both as student and colleague. His criticism and encouragement have been invaluable. Less obvious, but no less significant, is my debt to Donald Davidson. His conception of the philosophical enterprise has constantly influenced my approach to the problems treated in this book.

I have profited from the criticism and advice of many people in connection with this work. Extensive and detailed discussions with Professor John Wallace and Dr. Carole Ganz have contributed so much to shaping my thoughts on these matters that it is impossible to acknowledge their contributions in detail. Professor Carl Hempel graciously consented to read and comment upon an early draft of Chapters I–IV. I am much indebted to Professor Herbert Simon and Dr. Raimo Tuomela for their assistance in clarifying the notion of Ramsey eliminability (Chapter III). Professor Robert Causey and Mr. Gary Bower shared with me their insights into the theory of measurement. The early stages of my thinking about the historical development of scientific concepts (Chapter VIII) were influenced by discussion with Dr. Renate Bartsch and Dr. Lothar Schäfer. In ways too numerous to list, I have profited from the careful and insightful criticism of Professor Zoltan Domotor, Professor Stig Kanger, Mr. David Miller, Mr. Peter Oppenheimer and Dr. Krister Segerberg. To many others – students and colleagues – with whom I have discussed parts of this work, I am also indebted. I regret that space does not permit me to acknowledge explicitly these debts.

Though I have been the fortunate recipient of much good counsel, I have not always heeded it. For the errors and deficiencies remaining in the work, I alone am responsible.

The main features of the view of mathematical physics presented in Chapters I–VI were outlined in a course of lectures given at the University of Michigan in the winter of 1966. This view was developed further in seminars at Stanford during 1967 and 1968. Some of the material in Chapter III was presented at the Third International Congress for Logic, Methodology and the Philosophy of Science at Amsterdam in 1968. I am indebted to Stanford University and the National Science Foundation for freeing my time during the fall and winter of 1969 to complete work on Chapters VII and VIII. Much of this work was done at the Philosophy Department, Uppsala University, Sweden to which I am indebted for providing office space and clerical assistance. Without the patient assistance of the secretarial staff in the Stanford Philosophy Department this work would never have appeared.

Stanford, California, May 13, 1970

CONTENTS

PREFACE	VII
I. LOGICAL STRUCTURE AND AXIOMATIZATION	1
II. THE TRADITIONAL VIEW	15
III. THE RAMSEY VIEW	41
IV. THE RAMSEY VIEW EMENDED	65
V. THEORETICAL FUNCTIONS WITH SPECIAL FORMS	96
VI. CLASSICAL PARTICLE MECHANICS	110
VII. IDENTITY, EQUIVALENCE AND REDUCTION	154
VIII. THE DYNAMICS OF THEORIES	249
BIBLIOGRAPHY	308
INDEX	310

CHAPTER I

LOGICAL STRUCTURE AND AXIOMATIZATION

In this chapter the stage will be set for the subsequent discussion of the logical structure of theories of mathematical physics. The notion of the logical structure of a scientific theory will be discussed briefly and the notion of providing a logical reconstruction of an existing scientific theory introduced. Three methods of providing logical reconstructions will be considered. It will be noted that one of these methods has met with some apparent success when applied to theories of mathematical physics. However, the nature of this success is not altogether intelligible. A precise formulation of the philosophical question raised by this method will be suggested, and a sketch of a proposed answer given. Out of the elaboration of this answer in later chapters, will eventually come the promised characterization of the logical structure of theories of mathematical physics.

A widely accepted claim about scientific theories is this.

(A)   Scientific theories are sets of statements; some of which are empirically true or false.

This is a general claim about the sort of thing scientific theories are. It is a very plausible claim. Even those who believe it to be false or misleading do not deny this. Presumably, what one finds in the textbooks and journals germane to a particular scientific discipline are statements – the sort of thing one may properly claim to be true or false. At least some of these statements are claimed to be true or false because they are, or are not, supported by experimental data. And, presumably also, it is, at least in part, scientific theories that are being expounded in these textbooks and journals.

Though (A) is perhaps more widely accepted than most philosophical theses, its acceptance is by no means universal. There are those who claim to see beyond its superficial and vulgar plausibility and propound a more sophisticated view of scientific theories (e.g. the so-called 'instrumentalist view' propounded by Toulmin [45]). For the moment, I should like to be allowed to ignore the alternatives to (A) without giving my reasons for believing this to be justified. In Chapter VIII I shall say what I can about these alternatives.

Though (A) is usually regarded as plausible, even by proponents of more sophisticated accounts of scientific theories, it is not usually regarded as interesting, even by those who believe it to be true. It does not really say anything very explicit about what a scientific theory is like. The set of my random remarks to a companion about the surf, sand, and birds as we walk along the beach is certainly not a scientific theory. At best, (A) is a very weak necessary condition for a scientific theory.

What is needed to characterize more explicitly scientific theories? Philosophers who accept (A) generally believe that sets of statements which *could* constitute a scientific theory may be distinguished from those which *could not* by their logical structure. Statements may stand in various sorts of relations with one another which we call "logical relations". Examples of logical relations are: 'is entailed by', 'is consistent with', 'is confirmed by'. The first two are deductive logical relations; the third is an inductive logical relation. The logical structure of a set of statements is, roughly speaking, the logical relations (both inductive and deductive) holding among members of the set. Thus, it is alleged that certain logical relations must hold among members of a set of statements if this set is to be a scientific theory.

A great part of the activity of philosophers of science has been directed toward describing the sort of logical relations that must hold among the members of a set of statements if it is to be a scientific theory ([6], [15], [29]). The aim is to provide a stronger necessary condition for a scientific theory than (A). Clearly, since it is a necessary condition that is sought, a working assumption of this endeavor is that *all* scientific theories have the *same* logical structure. That is, it is assumed that there are some logical relations among the members which are common to all sets of statements which are scientific theories. The task is then to characterize or describe this common logical structure.

Related to the enterprise just described is a second enterprise – that of clarifying or elucidating the logical structure of a particular scientific theory. This enterprise has also been the concern of philosophers of science and of philosophically oriented scientists as well. It is related to the first enterprise – characterizing the common logical structure of all scientific theories – in at least two ways. First, successful completion of the first task would provide a pattern or prototype that *might be* helpful in dealing with a particular theory. The hedging is important here. It could turn out that whatever we could say *in general* about the logical structure of scientific theories was *so* general as to be of no help at all when we came to trying to clarify the structure of some particular theory. Second, successful completion of the second task, for some theory, would provide an example from which to build or to check a general account of the logical structure of scientific theories.

A clear appreciation of the second way these enterprises might be related is important to understanding the significance of the results to be presented in this book. Most of this book will be devoted to a discussion of the various possibilities afforded by a certain formal device – a set-theoretic predicate – for clarifying the logical structure of scientific theories. It will then be argued that *one* of these possibilities appears to give a satisfactory account of the logical structure of a real scientific theory – classical particle mechanics. It will be suggested that this theory is not atypical in this respect. The claim will be made that there is a wide class of scientific theories whose logical structure may be exhibited in this way. This class includes, it will be argued, all the theories we commonly recognize as theories of mathematical physics. It may contain other theories as well. There is, however, no claim made that it contains *all* scientific theories. There is no claim made about the logical structure of scientific theories in general. However, there is a claim made about the logical structure of a significant class of scientific theories and that claim is supported primarily (though not exclusively) by exhibiting a member of the class which has this logical structure.

Since the notion of clarifying the logical structure of a scientific theory is central to most of the discussion in this book, it is important to understand it clearly. My conception of the enterprise is roughly this. We are presented with an existing scientific theory as it is expounded in textbooks and technical literature (and perhaps in the unrecorded colloquies of scientists working with the theory). We have reasonably clear intuitions about what the empirical claims of the theory are and what the logical relations among them are. Here 'reasonably clear' means that in most specific cases we can confidently claim, for example, that such-and-such is, or is not, an empirical claim of the theory. 'Intuitive' means that we do not, in fact, appeal to explicit criteria in justifying such claims. With this as our starting point, we would like to produce some comprehensive and perspicuous form for exhibiting the claims of this theory and their logical relations. Let us call this '*a logical reconstruction* of the theory', and the activity of attempting to produce it 'logical reconstruction'.

Certainly we should demand that the logical reconstruction be, in some sense, compatible with our intuitive ideas about the structure of the theory. Beyond this, it should provide a tractable and systematic way of codifying these intuitions which could be appealed to as justification for specific claims about the theory. It should provide a means of answering questions about the theory on which our intuitions seem to be hazy or conflicting. It may be that, in the process of attempting to provide such a logical reconstruction, we come to believe that some of our intuitive conceptions about the claims of the theory are confused or even incompatible. We might be forced to make a choice to preserve some intuitions at the price of giving up others. In this sense, the enterprise of logical reconstruction is a normative one. But, in overall outlook, it is descriptive. We presume the practicing scientists' conception of what he is doing to be roughly correct, until proven otherwise.

Thus, the task of providing a logical reconstruction, as I conceive it, is one of codifying and systematizing an existing scientific theory, applying no external standards of judgment beyond simple clarity and logical consistency. In particular, I do not conceive the enterprise to be that of providing an epistemological critique of the concepts employed in the existing theory from the viewpoint of some epistemological credo (e.g. operationalism, logical empiricism, etc.). If a scientist claims that he observes certain things in his laboratory, in my view, we are committed to at least attempt to deal justly with this claim in our logical reconstruction, rather than denounce it as conceptually confused from the point of view of some external criterion.

It is evident that the aim of logical reconstruction, as I conceive it, is to provide what might be called a '*static* account of a scientific theory'. It aims only to provide a clear and accurate picture of a particular theory at a particular time – an account of what the theory claims, at this time, about the way the world is and how these claims are related logically. But, one might contend, scientific theories are the sort of things which change with time. Certainly, conceived as a body of empirical claims, quantum mechanics now encompasses a larger body of such claims than it did 35 years ago. For example, since that time we have incorporated into the theory a body of claims about electrical conductivity in metals. In the light of this fact about scientific theories, it might be held that my notion of a logical reconstruction is too restrictive to be interesting. The really interesting questions about a scientific theory are *dynamic* ones – questions about how theories change, grow, come to be accepted and rejected. It might even be hoped that if we could answer these questions for some particular theories, we might generalize the results and arrive at what might be called 'a general theory of *scientific methodology*'. For those philosophers of science who see their task to be one of providing such a general theory of scientific methodology, the enterprise of logical reconstruction might appear to be so modest in scope as to be uninteresting. Though the enterprise is admittedly modest in its scope, I think that the results of the subsequent investigation into the means of providing a static account of a scientific theory do in fact shed some light on dynamic questions. The implications of these results for a dynamic account of a scientific theory will be discussed in Chapter VIII.

If we are interested in clarifying the logical structure of some scientific theory – in producing a logical reconstruction of the theory – how should we proceed? A second widely accepted claim about scientific theories is this.

(B)   The logical relations among the statements of a scientific theory may be exhibited by an axiomatic system.

This claim can be most plausibly maintained for deductive logical relations. However, it is reasonably clear that any account of inductive logical relations will presuppose an account of deductive relations. Thus, the subsequent discussion may be regarded as, strictly speaking, confined to the deductive relations. But, it should be remembered that this is a necessary prelude to any serious discussion of the inductive relations. We shall not, however, completely avoid talk about inductive relations such as confirmation. But such talk will be on a naive, intuitive level. No explicit account of these inductive relations and their connection with deductive relations will be given. These difficulties aside, it is clear that (B) is not much help as an answer to our question until we understand what is meant by 'an axiomatic system'.

In discussions of the logical structure of scientific theories the term 'axiomatic system' has been used to denote different things. Among these are: (i) a certain kind of set of statements; (ii) an axiomatized deductive theory in some formal language; (iii) a definition of a set-theoretic predicate.

The traditional meaning of 'axiomatic system' is (i). The kind of set of statements which is an axiomatic system is characterized in the following way. There is some finite subset – the axioms – of which all other statements in the set are logical consequences. The relation of being a logical consequence is an informal one applicable to statements in ordinary discourse. The paradigm of this sort of axiomatic system is Euclid's axiomatization of geometry. For this sort of axiomatic system, the relation between the axiomatic system and the scientific theory whose logical structure it exhibits is obvious. The two sets of statements are usually held to be coextensive.

In order to understand the second meaning of 'axiomatic system' one must first understand what a formal language is and how it may be related to sets of statements.

A formal language (e.g. the first order predicate calculus) is a set of symbols called 'sentences'. For most interesting examples of formal languages this set of symbols is defined recursively by listing a finite set of basic symbols and stating rules for constructing sentences from these basic symbols. Defining the set of sentences in this way is often regarded as analogous to stipulating the rules of grammar or syntax for so-called 'natural languages' like English.

The sentences of a formal language may be related to statements of ordinary discourse by providing an *interpretation* of the formal language. Formally, an interpretation of a formal language is a function which maps the set of sentences of the formal language onto a set of two objects $\{T, F\}$. For the first order predicate calculus (without identity and operation symbols) this function is specified in two steps. First, certain basic symbols – the predicate symbols – are assigned to subsets of a set D – the *domain* of the interpretation – and its powers D^n. Intuitively, this corresponds to an assignment of meanings (references) to the non-logical symbols (predicates). Next, rules are provided for employing this assignment to determine whether T or F is to be assigned to any given sentence in the formal language. Intuitively, this corresponds to an assignment of meanings to the logical symbols (sentential connectives and quantifiers). If we assign meanings to the logical symbols in such a way that they correspond roughly to the meanings of logical symbols in a natural language, then we may regard the interpretation as establishing a correspondence between sentences in the formal language and statements. The corresponding statement is true or false depending upon whether the sentence of the formal language is assigned T or F by the interpretation.

Let us fix the assignment of meanings to the logical symbols of the formal language to correspond to the meanings of logical symbols in a natural language. Then each different assignment of meanings to the non-logical (e.g. predicate) symbols determines a different function from the sentences of the formal language into the set of statements. Let us say that each interpretation of the non-logical symbols determines a function from the set of sentences of the formal language into the set of statements. It is important to note that the entities which an interpretation assigns to the non-logical symbols of the formal language are set-theoretic entities. In particular, they are subsets of the domain of the interpretation and its powers.

Having fixed the assignments of meanings to the logical symbols we can then define the concept of *logical consequence* for sentences of the formal language. Roughly, α is a logical consequence of β if and only if α is mapped onto T by all interpretations of the non-logical symbols of the formal language by which β is mapped onto T. The intuitive content of this notion is such that the logical consequence relation on sentences of the formal language corresponds to the usual logical consequence relation among statements in ordinary discourse. This is to say that, if α is a logical consequence of β and $I(\alpha)$ and $I(\beta)$ are the statements corresponding to α and β under the interpretation I of the non-logical symbols, then $I(\alpha)$ is a logical consequence (in the intuitive sense) of $I(\beta)$.

The notion of logical consequence allows us to define an *axiomatized deductive theory* in a formal language. An axiomatized deductive theory in formal language L is a set of sentences $B \subset L$ having the following property. There is a finite proper subset $A \subset B$ such that every member of $B - A$ is a logical consequence of the conjunction of the members of A. Any interpretation of the non-logical symbols of L by which all the members of A are assigned T (and consequently all members of B are assigned T) is customarily called a *model* for B.[1]

The relation between a scientific theory θ and an axiomatized deductive theory B in a formal language which allegedly exhibits the logical structure of θ is usually described in the following way. There is some interpretation I of the non-logical symbols of the formal language such that $I(B)$ – the set of statements corresponding to members of B under I – is a proper subset of θ, i.e. $I(B) \subset \theta$. The set of statements $I(B)$ is clearly an axiomatic system in sense (i) with $I(A)$ as its axioms. The statements describing the interpretation I (sometimes called 'coordinating definitions') are usually also counted to be statements of the theory θ. Thus B exhibits the logical structure of only a part of θ. The logical structure of the statements describing I is not given by B though, of course, it might be given in part, by some other axiomatized deductive theory in the same formal language.

Most of the recent discussion by philosophers of science about the logical structure of scientific theories in general is carried on under the assumption that the logical structure of any theory can be exhibited by some axiomatized deductive theory in a formal language. Usually the formal language is some close relative of the first order predicate calculus.[2]

This might appear surprising to the philosophically naive, given the scarcity of examples of substantial scientific theories axiomatized in the first order predicate calculus. The only example, known to me, of a significant physical theory's yielding to this approach is a first order axiomatization of classical particle mechanics due to Montague [28].

On the other hand, there do exist various examples of axiomatizations of physical theories – specifically those in the province of mathematical physics – for example: classical thermodynamics by Cathéodory [5]; and Giles [12]; classical particle mechanics by Hamel [14]; Simon [37]; and Suppes *et al.* [26]; continuum mechanics by Truesdell [46]; rigid body mechanics by Adams [1]; quantum mechanics by Mackey [24]; and many others. The standards of logical rigor and the intuitive acceptability of these axiomatizations differ widely. But there is general agreement that these axiomatizations do not miss the point entirely. They are at least relevant to elucidating the logical structure of the theories they axiomatize, though they may not be fully adequate to the job. It is at least possible to raise the question of their adequacy and to consider ways of making them more adequate. This is more than can be said for any axiomatization of these theories in the first order predicate calculus.

The usual view among philosophers of science about these existing axiomatizations of physical theories is that they are "informal" axiom systems – i.e. axiom systems in sense (i). They contain statements about the world and these could be generated by interpreting the sentences of some axiomatized deductive theory in a formal language – if only one was clever and patient enough to discover how to do this. However, this is not the only way to view these existing axiomatizations.

It is well known that mathematical theories – e.g. group theory, lattice theory etc. – may be axiomatized informally, as well as by employing the first order predicate calculus. Recently attention has been given to the logical criteria of acceptability of these informal axiomatizations and there is available a well understood body of doctrine on this subject (see, for example [41] Chapter 12). The essence of this doctrine is that all such informal axiomatizations of mathematical theories may be regarded as, more or less, adequate definitions of set-theoretic predicates – that is roughly, predicates definable with the conceptual apparatus of set theory. The doctrine proceeds further to elaborate the criteria of logical adequacy for such definitions. Suppes *et al.* have attempted, with some success, to apply this same method to axiomatizing classical particle mechanics and relativistic kinematics. Adams has axiomatized rigid body mechanics in this way. These successes suggest that perhaps all existing axiomatizations of theories of mathematical physics *may be* fruitfully regarded as, more or less, adequate attempts at defining a set-theoretic predicate.

Two questions naturally arise from the preceding discussion. First, why are we convinced that at least some of these existing informal axiomatizations of physical theories are adequate – in some sense enlightening about the logical structure of the theory? Second, why are we convinced that they may all be regarded as, more or less, successful or logically precise attempts at defining a set-theoretic predicate? Most of the remainder of this book is concerned with attempting to answer the first question. The answer to the second question may be given now. So far as I know, it is only by assuming that these informal axiomatizations are essentially definitions of set-theoretic predicates that a satisfactory answer to the first question may be provided. The justification for this answer to the second question remains to be provided.

To understand more precisely the nature of axiomatization by definition of a set theoretic predicate it is convenient to have an example at hand. Suppes ([41], pp. 249–259) considers at length the example of group theory. This example will also suffice here as an introductory illustration. An adequate axiomatization of group theory by defining a set theoretic predicate is the following:

(DG) x is a group if and only if there exists a D, $\circ$ such that:

(1) $x = \langle D, \circ \rangle$;
(2) D is a non-empty set;
(3) $\circ$ is a function whose domain is $D \times D$ and whose range is a subset of D;

(4) for all $a, b, c \in D$
$$a \circ (b \circ c) = (a \circ b) \circ c;$$
(5) for all $a, b \in D$, there is an $e \in D$ such that $a = b \circ e$;
(6) for all $a, b \in D$, there is an $e \in D$ such that $a = e \circ b$.

The definition (DG) defines the predicate 'is a group'. The individuals of which this predicate is true are set-theoretic entities – i.e. sets. In particular, 'is a group' is true of ordered pairs having the properties (DG.2–DG.6). Another way of saying this is that (DG) defines a class of set-theoretic entities – groups. The predicate 'is a group' is true of all and only members of this class. It should be noted that particular members of this class – particular groups may have a set D composed of physical objects like bits of clay, or of non-physical objects like rotations.

Particular members of the class of entities for which 'is a group' is true, are sometimes said to be *models* for group theory, or models for (DG). The sense of 'model' employed here is related to the sense in which 'model' is employed to refer to interpretations of a formal language which assign T to all sentences of an axiomatized deductive theory. To see this relation, consider an axiomatized deductive theory in the first order predicate calculus with identity which axiomatizes group theory. The axioms of one such theory are:

(AG)   (1) $(x)(y)(z)(x \circ (y \circ z) = (x \circ y) \circ z)$
            (2) $(x)(y)(\exists z)(x = y \circ z)$
            (3) $(x)(y)(\exists z)(x = z \circ y)$.

where '∘' is a binary operation symbol – the only non-logical symbol in (AG). Models for the axiomatized deductive theory (AG) consist essentially of interpretations with a domain D and a function $\bar{\circ}$ from $D \times D$ into D, having the properties (DG.4–DG.6), which is assigned by the interpretation to '∘'. Thus, roughly, models for (AG) are interpretations in which entities comprising models for (DG) are assigned to the non-logical symbols of (AG). This relation between axiomatizations of group theory by definition of a set-theoretic predicate and axiomatizations of group theory in the first order predicate calculus with identity, is sometimes described by saying that the two axiomatizations have the same class of models.

In more complicated examples of mathematical theories, for example probability theory, it may be relatively easy to produce an axiomatization of the theory by defining a set-theoretic predicate, and fairly difficult and tedious to provide an axiomatized deductive theory in the first order predicate calculus which axiomatizes the same theory. Nevertheless, the relation between the models of these two axiomatizations should be roughly the same as the relation between the two axiomatizations of group theory we have just seen. I say 'roughly' for this reason. In more complicated cases the axiomatized deductive theory in the first order predicate calculus must include a formalization of theories like set theory and real number theory. Thus its models will not, in general, have the same set-theoretic structure as the models for the set-theoretic predicate which axiomatizes the theory. However, it is usually possible to describe a natural way in which the models for the axiomatized deductive theory correspond to models for the set-theoretic predicate. Thus, speaking elliptically, we can say that the class of models for the two theories are coextensive. Indeed, that this relation holds might be taken as the *criterion* which determines that we really have two axiomatizations of the same theory. From this point of view, what we are trying to do in axiomatizing a mathematical theory is to determine a class of set-theoretic entities. It matters not whether we do this by defining a set-theoretic predicate or listing some sentences (axioms) in the first order predicate calculus. The two approaches are simply two different ways of skinning the same cat.

Having seen a simple example of a mathematical theory axiomatized by defining a set-theoretic predicate, it would now seem appropriate to consider an example of a physical theory axiomatized in this way. However, for reasons of simplicity and clarity of exposition, it is expedient to defer consideration of a substantive example of a physical theory until later (Chapter VI). For the moment, let us consider a definition of a set-theoretic predicate which is quite similar in structure to the McKinsey–Sugar–Suppes axiomatization of classical particle mechanics, and yet somewhat less complex.

(D1) x is an S if and only if, there exists a D, t and n such that:

(1) $x = \langle D, t, n \rangle$;
(2) D is a finite, non-empty set;
(3) n and t are functions from D into the real numbers;
(4) $t(y) > 0$, for all $y \in D$;
(5) $\sum_{[y \in D]} n(y)t(y) = 0$.

So far as I know, (D1) has never been proposed as an axiomatization of any real scientific theory – physical or otherwise. However, this fact does not preclude its being a useful example of how such definitions function as axiomatizations.

Let us suppose that (D1) is offered as an axiomatization of some scientific theory – the claim being that it is, in some way, enlightening about the logical structure of the statements of this theory. The most obvious question to ask is this: How is (D1) related to the statements of this theory? Of course (D1) is itself a statement. It could be a statement of the theory. But clearly it is not an empirical statement. On any plausible account of the nature of empirical statements, definitions would not be included. Thus our question might be reformulated. How is (D1) related to the empirical statements of the theory?

A possible answer to this question is suggested by the relation between the axiomatizations of mathematical theories produced respectively by defining a set-theoretic predicate and by producing an axiomatized deductive theory in the first order predicate calculus. The answer is this: Suppose that we could produce an axiomatized deductive theory in the first order predicate calculus, AS, which had the same class of models as (D1). By providing an interpretation I for AS, we may produce a set of statements $I(AS)$. We might then identify $I(AS) \cup$ {the set of statements describing I} as the statements of the scientific theory whose logical structure is supposedly illuminated by (D1).

This is the answer one might expect from a philosopher of science who believes that employing an axiomatized deductive theory in some formal language is the most fruitful way to exhibit the logical structure of a scientific theory. Subsequently, I will maintain that this is not the *only* plausible answer to this question and that, in some cases at least, it is clearly not the correct answer. The view I will maintain is this. The empirical statements of any scientific theory axiomatized by (D1) are, roughly speaking, statements about the entities which satisfy 'is an S'. The answer sketched above will be seen to correspond to one type of statement about entities which satisfy 'is an S'. I will argue, however, that statements of this type are clearly not the only interesting ones to be made about the entities which satisfy 'is an S', and further that statements of these other types are very likely candidates for the empirical claims of certain scientific theories.

It is important to understand at the outset that nothing that I have to say counts against the feasibility, or even the fruitfulness of attempting to exhibit the logical structure of a theory by an axiomatized deductive theory in the first order predicate calculus. Assuming that we agree that the predicate 'is an S' is related to the statements of the theory in one of the ways alternative to that sketched above, then what I say will count decisively against using AS to exhibit the logical structure of the theory. But it will not count against using *some other* axiomatized deductive theory in the first predicate calculus for this purpose. Indeed, my suggestions can be formulated as a proposal about what the interpretation of the first order predicate calculus must be like – suggesting a domain and interpretation of certain predicate letters – for such an attempt to be successful. This will be elaborated later. The question of whether anything further in the way of insight into the logical structure of the theory is to be gained by a formalization in the first order predicate calculus along these lines, is not one with which I shall be concerned.

The substance of the discussion thus far is this. If you believe that scientific theories are sets of statements, some of which are empirical, then the question of the logical structure of these statements is significant. In particular, there is good reason to be concerned with the enterprise of logical reconstruction for particular scientific theories. The device of defining a set-theoretic predicate appears to be relevant to the enterprise of logical reconstruction – especially to the logical reconstruction of physical theories. However, the precise nature of this relevance is not readily apparent. Just exactly how the set-theoretic predicate, which allegedly axiomatizes a physical theory, illuminates the logical structure of the empirical claims of the theory remains unclear. Until this question is clarified, we have little understanding of what is to be gained by providing an axiomatization of this sort, or why one example of such is to be preferred to another. Much of the remainder of this book is devoted to considering the question of how definition of a set-theoretic predicate could illuminate the logical structure of a scientific theory.

The general nature of the answer I will offer to this question may be given immediately. I will suggest that the definition (D1) is relevant to illuminating the logical structure of a scientific theory if some of the empirical claims of the theory may be rendered as statements made with sentences in which the predicate 'is an S' or closely related predicates appear. There are, of course, many ways sentences can be constructed with a given predicate. Some of these sentence forms may be more useful than others in rendering the empirical claims of scientific theories.

In the following four chapters, I will consider four sentence forms containing the predicate 'is an S' and discuss, in some detail, the prospects for using each of these for making empirical statements. The purpose of these exhaustive discussions of sentences containing 'is an S' – rather than some realistic example – is twofold. First, I want to emphasize the general applicability of the discussion. Attention to familiar examples tends to obscure this. Second, many of the conceptual points I want to make come through more clearly by using an example which minimizes the need for tedious technical manipulation and avoids troublesome, and essentially irrelevant, mathematical questions. However, the discussion is brought down to earth in Chapter VI. In that chapter a detailed discussion of the logical structure of classical particle mechanics is offered, employing the ideas developed in the preceding four chapters.

NOTES

[1] This sketchy account of the syntax and semantics of formal languages may be filled out by referring to any of a number of standard texts. Perhaps the most accessible treatment which contains all the essential information is in [25].
[2] For examples of this approach see [6], [15].

CHAPTER II

THE TRADITIONAL VIEW

In this chapter one proposal – the simplest – for using a set-theoretic predicate to make an empirical claim is considered. Some attention is given to questions which arise in connection with this proposal, and proposals subsequently considered, as well. In this connection, the theory of measurement and the question of the "approximate" nature of claims in mathematical physics are discussed briefly. The proposal considered here is shown to be essentially equivalent to the view of how the theory's statements are obtained from the set-theoretic predicate that was described in Chapter I. Considerable effort is devoted to characterizing a theory-relative notion of theoretical term, independently of sweeping epistemological assumptions. This notion is then employed to describe a difficulty that could be encountered in attempting to employ the sentence form proposed here in logical reconstruction. Finally, an attempt is made to relate this difficulty to more traditional discussions of the problem of theoretical terms.

The simplest way the predicate 'is an S' can be used to construct a sentence is to predicate 'is an S' of some singular term. Let us begin by examining this possibility. Consider a sentence of the form:

(1) Q is an S.

where 'Q' is a name or definite description. Is it plausible to expect that sentences of this form might be used to make empirical statements in a scientific theory?

The meaning of a sentence of form (1) may be explained in the following way. The predicate 'is an S' characterizes (is true of all and only) entities which have a certain set-theoretic structure. One might say that 'is an S' characterizes a certain formal, mathematical structure. The force of the adjective 'formal' here is roughly that of 'abstract'. It means simply that a great many different individuals – indeed individuals of radically different kinds – may have this structure. The force of 'mathematical' is that the structure is composed in part of mathematical objects – numbers. The singular term 'Q' *may* name or describe some entity which is composed in part of concrete physical objects. Of course, it may also refer to completely abstract entities. The claim of (1) is that the entity referred to by 'Q' has the formal, mathematical structure characterized by 'is an S'.

One of the central claims to be developed in this book is very roughly this. The essential, distinguishing feature of theories of mathematical physics is that each has associated with it a formal, mathematical structure. This structure forms, so to speak, the core of the theory, or the mathematical formalism characteristic of the theory. It is this formal, mathematical core which is precisely described by the definition of a set-theoretic predicate which "axiomatizes" the theory. This predicate is then used to make the empirical claims of the theory. In sentences of form (1), is one possible way which this predicate could be used.

For example, *suppose* 'is an S' characterized the mathematical formalism of Newtonian mechanics (clearly, it does not) and 'Q' referred to some entity including the set of bodies in the solar system – roughly, this set of bodies, together with their positions, masses and the forces acting on them. Then (1) asserts that this entity has the formal, mathematical structure of Newtonian mechanics. One might say roughly that (1) asserts that the mathematical formalism of Newtonian mechanics is *applicable* to the bodies in the solar system. In general, for physical theories, 'Q' will refer to a physical system and (1) makes the claim that the mathematical formalism of this theory applies to that physical system.

Let us now begin to consider in more detail the properties a sentence of form (1) must have to be a plausible candidate for an empirical claim of some scientific theory.

First, it is intuitively clear that the entities which it "makes sense" for some person to claim to be S's are just those which are believed by him to have a certain minimal set-theoretic structure namely, the structure characterized by (D1.1–D1.3). The reason for this is evident. For most, if not all, entities it will, as a matter of fact, be known (to the claimant) whether or not the entity has this minimal structure. If it does not, then (1) is trivially false and (presumably) believed to be so. Generally (at least in making scientific claims) it does not "make sense" to assert what one believes to be false. For this reason, we may confine our attention to 'Q's which refer to entities which have this minimal structure. Others do not provide plausible candidates for empirical claims of a scientific theory.

We can be more precise about this and this precision will prove useful later. Consider:

(D0) x is an S_0 if and only if there exists a D, n, and t such that:
(1) $x = \langle D, t, n \rangle$;
(2) D is a finite, non-empty set;
(3) n and t are functions from D into the real numbers.

The entities which are S_0's are just those entities that it "makes sense" to claim to be S's. In line with our previous usage of 'model' in which anything which is an S is a model for S, let us call anything which is an S_0 'a *possible model* for S'. If 'Q' refers to a possible model for S, then (1) says that this possible model for S is actually a model for S. Subsequently we shall consider only 'Q's which refer to possible models for S.

Though this consideration above does not rule out the possibility that 'Q' is a proper name, it does seem to suggest that it should be ruled out. It is hard to see how a proper name for a possible model for S could function intelligibly in a sentence of form (1), unless a gloss of the name in terms of definite descriptions were readily available. For this reason let us restrict our attention to 'Q's which are definite descriptions of possible models for S.

If we restrict 'Q' in this way, then a statement made with a sentence of form (1) is true if and only if Q satisfies (D1.4) and (D1.5). This sentence may be used to make an empirical claim only if the question of whether Q satisfies (D1.4) and (D1.5) is one to be answered by empirical investigation. A necessary condition for this is that the descriptions 'Q' provides of the set D, and functions n and t are not *all* simply lists of their members. If they were all lists, then it would be purely a matter of calculation – not empirical investigation – to determine whether Q satisfied (D1.4) and (D1.5).

How might one describe a possible model for S without simply listing the members of all the sets involved? For the set D, the answer is simple. This set may be described by naming some property possessed by all and only members of D. If we want to know whether a particular individual is in D, instead of looking for a name of this individual in a list, we check to see if it has this property.

For the functions n and t from D into the real numbers, the answer is somewhat more complex, but reasonably well understood. The answer is essentially this: Relations of a certain kind (i.e. having certain logical properties) on D suffice to determine classes of functions from D into the real numbers. By further specifying some values of this function (considerably less than all its values) a particular member of this class may be singled out. To describe a function from D into the real numbers one has only to name the appropriate sort of relational properties and specify a few values of the function. Discussions of this way of describing functions are usually carried on under the heading 'theory of measurement' and the problem sometimes described as 'obtaining quantitative data (numerical functions) from qualitative data (facts about relations)'. There is so vast a literature on this subject (see, for example [4], [17], [44]) that another discussion – with nothing new to add – is hardly excusable. However, the subsequent discussion proceeds much more smoothly by referring occasionally a concrete example of this way of describing functions.

To provide such an example, we begin by defining an extensive system and a numerical extensive system.[1]

(DE) x is an *extensive system* if and only if there exists an A, R and $\circ$ such that:
     (1) $x = \langle A, R, \circ \rangle$;
     (2) A is a non-empty set;
     (3) R is a binary relation on A;
     (4) $\circ$ is a function from $A \times A$ into A;
     For all a, b and $c \in A$,
     (5) If $a\,R\,b$ and $b\,R\,c$ then $a\,R\,c$;
     (6) $(a \circ b) \circ c \; R \; a \circ (b \circ c)$;
     (7) If $a\,R\,b$ then $a \circ c \; R \; c \circ b$;
     (8) If not $a\,R\,b$, then there is a $c \in A$ such that $a\,R\,b \circ c$ and $b \circ c \,R\, a$;
     (9) Not $a \circ b \,R\, a$
     (10) If not $b\,R\,a$ then there is a number n such that $b\,R\,n\,a$, where the notation na is defined recursively as follows: $1a = a$ and $na = (n-1)a \circ a$.

(NE) x is a *numerical extensive system* if and only if there exists an N, $\leq$ and $+$ such that:
     (1) $x = \langle N, \leq, + \rangle$;
     (2) N is a non-empty set of positive real numbers, closed under addition and subtraction of smaller numbers from larger numbers;

(3) ≤ is the numerical less than or equal relation;
(4) + is the numerical addition operation restricted to N.

Two important facts about extensive systems and numerical extensive systems are the following:

(V)   Every extensive system $\langle A, R, \circ \rangle$ is homomorphic to some numerical extensive system $\langle N, \leq, + \rangle$; the homomorphism h being such that, for all $a, b \in A$
$h(a) = h(b)$ if and only if $a\,R\,b$ and $b\,R\,a$.

(U)   If $\langle A, R, \circ \rangle$ is an extensive system homomorphic to the numerical extensive systems $\langle N, \leq, + \rangle$ and $\langle N', \leq, + \rangle$ under the homomorphisms h and h', satisfying V then there exists a positive real number α such that for all $a \in A$, $h(a) = \alpha h'(a)$.

One way of describing the content of facts (V) and (U) is this. Every extensive system $\langle A, R, \circ \rangle$ determines a class of functions from A into the real numbers. To every numerical extensive system satisfying (V) there corresponds a function in this class, namely the homomorphism h. To determine a particular member of this class, h^*, we have only to specify the value of h^* for some *one* member $a \in A$. This is because (U) does not allow that two different members of this class of functions have the same value for a. Thus, if we want to describe a function n from D into the real numbers, one way to do it is to describe an extensive system $\langle A, R, \circ \rangle$ such that $D \subseteq A$ and specify one value of the function n. (Note that the value specified may be for a member of $A - D$.) Thus the problem of describing a numerical function has been replaced by the problem of describing an extensive system. This we may do by naming properties – the defining property of A – and the relational properties determining R (a binary relational property) and $\circ$ (a tertiary relational property).

The relation between extensive systems and numerical extensive systems is an example of what is called 'fundamental measurement'. There is another, well understood, way of using empirical relation systems (an example of which is an extensive system) to determine classes of functions from "empirical objects" (of which physical objects are an example) into the real numbers. The basic idea is to define a function f on A as some numerical function (e.g. the product function) of the values of the functions g and h on A which are defined directly by empirical relation systems. The traditional example of this is the density function which is defined as the quotient of the mass and volume functions. This way of determining a numerical function is called 'derived measurement'.

It must be emphasized that there are other kinds of fundamental measurement besides that provided by extensive systems. In discussing the description of functions appearing in theories of mathematical physics, I do not consider these other kinds. The reason for this is simply that I do not believe the differences between these kinds of fundamental measurement are relevant to the points I want to make. The reason is not that I believe only extensive systems do (or can) appear in theories of mathematical physics. Some philosophers (e.g. Campbell [4] and Hempel [17]) have maintained or suggested that this is the case. Nothing that I say entails that this is or is not the case.

Suppose we have a description of an extensive system $\langle A, R, \circ \rangle$, plus a specification of one value of a function n from A into the real numbers; say $n(a)$, $a \in A$. How do we discover the values on $D \subseteq A$ of the function denoted by this description? In particular, how do we discover a particular value of this function; say $n(b)$, $b \in D$? The facts (V) and (U) above do not tell us how to do this, beyond saying that an examination of the situation of b and other members of A with respect to the R relation and the $\circ$ operator is relevant. But, in fact, the proof of (V) contains, implicitly, instructions for determining such values. Rather than describe this in detail, a sketch of a concrete example is more to our purposes.

A common, frequently discussed (see [11], [44]) example of an extensive system is this. The set A is taken to be a set of "weights" – presumably, medium size physical objects. This is a bit peculiar since (DE.4), (DE.5) and (DE.9) entail that A is an infinite set. But, let us ignore this for a moment. The relation R is taken to be the relation 'weighs less than, or just as much as'. The most obvious interpretation of $\circ$ is physical concatenation – "grouping together". Intuitively, this operation must have the idempotence property, $a \circ a = a$. But, since $h(a) > 0$ and $h(a \circ a) = h(a) + h(a)$, $\circ$ clearly cannot be idempotent. Idempotence is also seen to contradict (DE.9) and (DE.10). Clearly, what we want, as an interpretation of $\circ$, is some operation on physical objects such that $a \circ a$ is a physical object which, intuitively, weighs twice as much as a. One way to provide this is to let $a \circ b$ be the physical object that you get by pouring enough sand into a bottle on one pan of a balance to balance a and b placed on the other pan. It is at least conceivable to suppose that A is closed under this operation, as is not the case for physical concatenation. Of course we require an infinite supply of sand and this seems a bit implausible.

Let w be a member of the class of functions from A into the real numbers determined by this extensive system, and let $w(a)=1$, $a \in A$. It is easy to see intuitively and *roughly* how $w(b)$ is discovered. If a weighs less than b, then concatenate with a just as many weights which weigh the same as a as are required to make this concatenation weigh the same as b. The number of "equivalents" of a needed is the value of $w(b)$. A more detailed account could be given, but it is not essential for our purposes. It is only important to note that the value of $w(b)$ is determined by noting how b is related to various members of A and their physical concatenations by the 'weighs less than or just as much as' relation.

It is important to note here that one can use the extensive system just mentioned to determine functions on subsets of A which do not necessarily meet the strong requirement (D.8) and (D.10), nor have an infinite number of members. If one is interested in describing the "weight" function on a small, finite set D, one need only "embed" D into the domain A of an extensive system of the sort just mentioned. To discover values of the functions thus determined, one must discover how the members of D are related to some members of A and their concatenation relations by the 'weighs less than or just as much as' relation. This is essentially what we do when we weigh an object by comparing it to a set of standard weights.

Another example of an extensive system, relevant to the subsequent discussion of classical particle mechanics is this. Let A be a set of "lengths", say rigid rods, R, the 'is shorter than, or just as long as' relation, and ∘ be something like the 'laid end-to-end in a straight line with' operation. Again this interpretation of ∘ is too simple. We need some physical operation on rigid rods which produces a new rod just as long as a and b laid end-to-end. It is easy to see, intuitively, how to provide this using an infinite supply of rod-making material and a marking crayon. An extensive system of this sort determines a class of functions whose values are "lengths" of rigid rods in A.

From the preceding discussion, and from a naive perusal of the literature on measurement, one might get the impression that these methods of describing classes of functions may be employed *only* when the domain of the function is a part of some empirical relation system, like an extensive system. This limitation is both implausible and unnecessary.

Consider a set of small physical objects P – particles. How might we describe a function from $P \times P$ into the real numbers which we might identify as the distance function? If we try to use an extensive system, we seem to be stymied because there is no obvious way to define a concatenation relation ∘ on ordered pairs of particles. But, suppose we have an extensive system whose domain A is a set of rigid-rods. Clearly, it is not implausible to think that ordered pairs of particles might stand in the 'is shorter than, or just as short as' relation with members of A. If this is so, then we can say that

$$d(\langle p_1, p_2 \rangle) = h(a)$$

if and only if $\langle p_1, p_2 \rangle\, Ra$ and $aR\, \langle p_1, p_2 \rangle$. That is we can identify the value of the distance between two particles p_1 and p_2 as the length of a rigid rod which is just as long as the pair $\langle p_1, p_2 \rangle$. This is roughly what we do when we use measuring rods to measure distances between things. A precise mathematical account of this could obviously be given by extending the notion of an extensive system to allow situations in which ∘ is defined only on some sub-set of A.

From this point of view, there is no objection to allowing certain "non-physical" objects (for example, "lengths" which are not rigid-rods, but what you *would* get if you laid down a real rod end-to-end *with itself* a finite number of times) in the domain of an extensive system. One would only have to be certain that a satisfactory account of the R-relation between these non-physical members of A to its physical members could be given. Traditional ways of rendering counterfactuals as general laws might provide this. Such an account would remove the implausibility that (DE.8) and (DE.10) may have for sets of physical objects. So far as I know, no attempt has been made at producing a satisfactory "Platonistic" interpretation of (DE).

One additional point needs to be made about this way of describing functions from D into the real numbers. One *need not* assume that naming a relation on some set $A \supseteq D$ gives *all* the information needed to discover whether or not that relation holds between any two members of A. That is, adopting this way of describing functions *need not* commit one to giving something like "operational definitions" of the relations employed. For example, one *may* gloss the 'is longer than or just as long as' relation on rigid rods in terms of what happens when you place them side-by-side with their ends resting on a flat surface. But, so far as this method of description is concerned, one may admit that there are, in fact, many accepted ways of discovering the truth value of statements involving this relation in different situations. The relations between these methods – for example, whether they are all "derivative" from one basic method – is an interesting (and usually difficult) question. To answer it is essentially to provide a logical reconstruction of our theory of the 'is longer than, or just as long as' relation. This may not be a trivial task.

Many people would contend that a thorough logical reconstruction of any scientific theory requires that a satisfactory logical reconstruction of all the "sub-theories" underlying the descriptions of the functions in the original theory be given. There is no doubt a strong motivation behind such thoroughness. One might plausibly contend that without such an account the truth conditions of the empirical claims of the original theory have not been made explicit. However, it does seem to be possible to say much that is non-trivial about the empirical claims of some theories without providing a logical reconstruction that is complete, in this sense. At the very least, it seems possible to separate questions of the logical structure of the original theory from questions about the logical structure of theories underlying it. This is the line to be taken in Chapter VI in discussing classical particle mechanics. Attention will be focused on the logical structure of the empirical claims of this theory, ignoring questions about the logical structure of theories of geometry and time which underlie it. This is not because I think the latter questions are unimportant, but rather because I have nothing new to offer as answers to them.

Let us return now to our consideration of sentences of form (1). Suppose that 'Q' describes a possible model for S, and that it does so by providing descriptions of D, n, and t which are not all simply lists of their members. Clearly, (1) will still not be an empirical claim unless, roughly speaking, it is an empirical question as to whether or not a particular individual is in D and what the value of the n and t function for that individual is. For D, this simply means that statements attributing the defining property of D to individuals must be empirical statements. For the functions on D, this means that statements attributing the relational properties, used to describe these functions to members of D (and perhaps to other individuals), must be empirical statements. For example, it must be an empirical question as to whether or not two individuals stand in the 'weighs less than, or just as much as' relation.

I employ the term 'empirical statement' here without offering any account of its meaning. It is, if you like, a primitive term in my account of the logical structure of a certain kind of theory. I assume that we can, in some cases at least, recognize that particular statements are, or are not, empirical statements. I assume that we can do this even though we lack an acceptable general account of the meaning of 'empirical statement'. I do not believe that anything I say subsequently commits me to any particular view about what this account must be like.

In particular, I do not believe that anything I say commits me to either side on the question of whether or not the notion of an empirical statement can be explicated in terms of notions like "directly observable" or "operationally definable". As will be made more explicit shortly, I simply leave open the question of whether relations appearing in empirical statements may be "defined" in terms of what we take as acceptable ways of obtaining evidence for them. I should, however, emphasize the following point. As should be clear from the preceding discussion, I take "empirical statement" broadly enough to include statements in scientific theories, other than those which some philosophers might call 'observation reports'. The description 'Q' may contain terms which refer to quite high level, theoretical concepts of another theory, related to "observation reports" only by a complex network of physical laws and definitions.

Subject to the restrictions we have placed on the description 'Q', it now seems clear that sentences of form (1) *could be* used to make empirical statements, and that these statements *might be* part of some scientific theory. However, there is at least one strong reason to think that this is not the case. It might be argued that one simply cannot expect to find "real" physical models for S, or anything like it. If we limit ourselves to domains consisting of physical objects or other non-abstract individuals (e.g. sounds or events) which we can, in some way, perceive, and define functions in terms of "observable" relations among these objects, then (1) will always (or almost always) be false. Speaking platonistically, S is an ideal of which real objects are, at best, imperfect copies. There is some obscurity about this claim. Most apparent is the question of its epistemological status. Is it offered as a conceptual truth, or simply as a substantive fact about existing theories and attempts to provide logical reconstructions of them? If we construe it in the latter way, then there appears to be a substantial body of evidence supporting the view. It is very unlikely that any set of physical objects, together with a specification of their masses and the forces acting on them was ever seriously claimed to be *exactly* a model for any plausible axiomatization of Newtonian mechanics.

There appear to be at least two distinct lines one might take to deal with this problem. First, we can retain 'Q' as it is and modify (1) to read:

(1') Q is approximately an S.

We must then undertake explication of this use of 'approximately' so that the truth conditions of sentences of form (1') can be precisely stated. One way to do this might be to employ some predicate 'is an S^+', related to 'is an S' in a certain way and such that Q *is* an S^+. This seems to be something like what we do when we justify claiming that freely following bodies approximately satisfy Galileo's law by claiming that they exactly satisfy *some* law which takes into account air friction, the change in the distance between the earth and the body, etc. and that, in some appropriate sense, this law reduces to Galileo's law. How to carry out in detail this approach to dealing with (1'), or whether it can be carried out, is not clear to me.

Alternatively, we can retain (1) and change 'Q' so that it describes some "ideal" entity which corresponds in some way to the "real world". We are then, of course, obliged to explicate the nature of this correspondence. For example, consider the case of planetary motions. The actual data that we have about this phenomenon are a quite complex welter of descriptions of particular observations of relative angular positions at particular instants. Let us suppose we are committed to accounting for these data within the framework of Newtonian mechanics and have some precise characterization of the mathematical formalism of this theory. In some way we pass from the actual descriptions of what happens in our laboratory (or observatory) to descriptions of entities which we want to claim have *exactly* the structure of this mathematical formalism. If our claim that Newtonian mechanics accounts for this data is not to be vacuous, then the way we make the transition from data to the entity we claim to be here, the structure of Newtonian mechanics cannot be completely arbitrary. We must provide some general account of it. Again, I have no suggestions about how such an account is to be provided (however, see [43]).

It seems apparent that a comprehensive account of the logical structure of any significant scientific theory must await a satisfactory solution to this problem. However, the questions of logical structure I wish to examine can, I believe, be considered independently of this matter. For convenience of exposition, I shall continue to refer to sentences of form (1), and subsequently to other sentences open to the same objection. I recognize that a more complete logical reconstruction of the theory *might* dictate some modification of these sentences along the lines of the first alternative above. I hope it will be obvious that my account would not have to be essentially changed to accommodate these modifications.

We have seen that sentences of form (1) might be used to make empirical statements included in a scientific theory. The view that this is the *only* way a set-theoretic predicate may be used to make empirical claims, I call 'the traditional view'. The appropriateness of this name is made clear by the following considerations.

In Chapter I we consider one way of generating a set of statements from a definition of a set-theoretic predicate. In the case of (D1), it consisted of constructing an axiomatized deductive theory in the first order predicate calculus, AS, which had, roughly speaking, the same class of models as (D1). Providing an interpretation I for the first order predicate calculus, makes the members of AS correspond to statements – AS yields a set of statements $I(AS)$. If we assert all the statements of $I(AS)$, *together with* all the statements describing I, this simply amounts to asserting that I is a model for AS. But since the class of models of AS is coextensive with the class of models for S, this amounts to asserting that I is a model for S. Thus, if we let 'Q' denote the interpretation of the first order predicate calculus, I, then a sentence of form (1) may be used to make a statement equivalent to the conjunction of the statements in $I(AS) \cup$ {the set of statements describing I}. But, on the view sketched in Chapter I, this set of statements is the scientific theory whose logical structure is illuminated by (D1). Hence, the view that a sentence of form (1) is the only way to use 'is an S' to make an empirical claim is equivalent to the view sketched in Chapter I.

I believe that the view sketched in Chapter I is, at least implicitly, the view of most contemporary philosophers of science. I say 'implicitly' because I know of no philosopher who considers this question explicitly. However, most of these philosophers are committed to a discussion of the logical structure of theories under the assumption that they can (in some sense) be axiomatized in the first order predicate calculus. It would be very tempting for them to regard definitions of set-theoretic predicates which allegedly axiomatize theories as simply "half-way houses" on the road to an axiomatized deductive theory in the first order predicate calculus. That is, once you had an acceptable axiomatization of a theory, given by defining a set-theoretic predicate, it would be only a matter of time, perseverance, and logical cleverness until the desired axiomatization in the first order predicate calculus was forthcoming. No further insight into the nature of the theory's claims would be needed – only technical skill. This is essentially what the view sketched in Chapter I says. Any view which allowed for a more subtle relation between an existing, intuitively acceptable axiomatization given by defining a set-theoretic predicate, and the yet-to-be-produced axiomatization in the first order predicate calculus would be, *prima facie*, less congenial to a philosopher committed to this way of discussing theories. Less congenial, because it would not lend such direct and compelling support to his assumption that all theories can be axiomatized in the first order predicate calculus.

What reasons are there for believing that sentences of form (1) may not be adequate for making all the empirical claims of a theory axiomatized by (D1)? One may answer this question by characterizing, in a rather general way, a difficulty that might be encountered in attempting to use sentences of this form in logical reconstruction.

Consider a situation in which we are attempting to use the predicate 'is an S' to elucidate the logical structure of some existing scientific theory. Presumably, one reason for thinking we might be successful in this is that functions which have properties something like n and t – particularly (D1.4) and (D4.5) – appear in the literature of the theory. In most "real life" situations (D1) will be easily recognized as simply a "precisification" of the mathematical apparatus expounded in textbooks germane to the theory. Suppose we think that the central empirical claim (or claims) of the theory can be made with sentences of form (1). How do we proceed to discover what these are?

It is, I think, characteristic of theories of mathematical physics that we speak of applications of these theories. We say that the BCS theory of super conductivity is a successful application of quantum mechanics to explain the phenomenon of super conductivity; that classical thermodynamics may be applied to explain chemical equilibria in gaseous reactions; that classical particle mechanics may be applied to explain the motion of bodies in the solar system. I want to suggest that it is claims of this sort – claims that such-and-such is a successful application of the theory – that are to be taken as the empirical claims of this theory. I am aware that this distinction between theory and application may seem, at first glance, a difficult one to draw with any precision. I am also aware that it runs counter to one feature of the common usage of 'theory'. We frequently speak of the *theory* of gaseous reactions, and of celestial mechanics as a scientific *theory*, in itself. I am not particularly concerned about doing violence to ordinary usage, but I do take seriously the question of whether this distinction may be drawn precisely. I shall have more to say on this point later. For the moment, let us simply assume that the empirical content of a theory of mathematical physics lies in claims that it applies to certain phenomena.

If this is so, then a first step in providing a logical reconstruction of the empirical claims of theories of mathematical physics is to explain what we mean by saying the theory applies to a certain situation – i.e. what empirical claim is made in saying this. A step in the right direction is to say we mean that the formal, mathematical structure – the core of the theory – applies to this situation. This might be a particular, or a general claim. One might claim that the mathematical structure applies to a particular physical system, or that it applies to all physical systems of a certain kind. We have already suggested that particular claims of this sort may be made with sentences of form (1), provided 'is an S' characterizes the mathematical core of the theory. Let us suppose that the theory we are considering makes only particular claims. The difficulty pointed out subsequently for singular claims of form (1) can easily be seen to arise for conjunctions of such claims, and thus for simple generalizations of the form that every possible model of S, of a certain kind, is an S.

If we have such a theory, and our account of the application claims is correct, then we must proceed in the following way. Examine the literature of the theory in which its applications are described – articles in technical journals, examples in textbooks – and try to extract, from the account of each application, a description of a possible model for S. We then try to render the empirical content of the claim that this is a successful application of the theory as a statement of form (1), with 'Q' denoting this possible model for S.

Now suppose that, in examining the various applications of the theory, the following situation arises. For each application i, it is possible to settle upon an empirical description of a domain D_i which we are convinced must be involved in any adequate reconstruction of the empirical claims of the theory. This is to say, we are confident that we clearly understand what individuals the theory makes claims about in each of its applications. Note that the domains D_i might overlap. That is, some individuals might appear in more than one application of the theory.

It is now important to distinguish two kinds of situations that might arise when we examine the existing discussion of an application i trying to settle upon the appropriate description of a function from D_i into the real numbers to include in our description of the possible model for S. For example, suppose we are trying to decide upon the appropriate description of the n-function, n_i.

When we examine closely the existing discussions of application i we might expect to encounter the following situation. There are descriptions of various ways that values of n_i are (or may be) obtained for different members of D_i. For some members of D_i, more than one way might be described; for others, no way. One might think that the totality of these descriptions of ways of obtaining n_i-values would somehow constitute the description of the n-function appropriate to this application. But this need not be the case. To be sure, these ways of determining n_i-values are going to be related to the appropriate description of n_i, but this relation *may be* only contingent.

To see this, let us suppose, for the moment, that we have already hit upon a description of n_i that we intuitively recognize as appropriate. Let us further assume that it is provided by describing an extensive system $\langle A, R, \circ \rangle$ ($D_i \subseteq A$). Then, the existing discussion's descriptions of ways of obtaining values of n_i may be accounted for in the following two ways.

(1) They involve determining "directly" whether or not members of D_i stand in the R-relation with other members of A.

(2) They involve determining the status of members of D_i and other members of A with respect to some relations which are connected by accepted physical laws to R.

The adjective 'directly' in (1) means, roughly, that our evidence for the statement that $a\,R\,b$ does not depend upon (presuppose) the truth of any physical laws. More precisely, our evidence for $a\,R\,b$ is not of the form: '$a\,R\,b$ follows from $a\,S\,b$ and physical law L; there is evidence for $a\,S\,b$ and for L.' This does not rule out the possibility that our evidence for $a\,R\,b$ is that it follows from $a\,S\,b$ for which there is evidence, *not* of the sort just mentioned, and some non-empirical truth. Relations which admit of such direct evidential support are called by some philosophers 'directly observable relations' or 'protocol relations'. Whether there are, in fact, any such relations is a much debated question (see, for example [34]). According to the views of some philosophers, category (1) may always be vacuous.

If category (1) is vacuous, then *all* the ways of determining the values of

n_i are only contingently related to the description which determines n_i. That is they are acceptable ways of determining values of n_i, *provided that* certain physical laws are true. If this is so, then one cannot expect to provide anything better than a conditional "definition" of n_i in terms of the ways its values are determined – the conditioning proposition being the physical laws in question.

The question of whether category (1) is vacuous is one I should like, if possible, to avoid. I attempt to do this in the following way. Even if some ways of determining values of n_i are non-contingently related to the description of n_i, this clearly does not require that all are. The distinction *I* am driving at is made in terms of ways of making measurements of n_i-values which are only contingently related to the description of n_i. But, it is drawn in such a way that it does not entail that category (1) (nor, indeed category (2)) is, or is not, vacuous. I subsequently cite examples which have the force of showing that category (2) is not vacuous. But I do not believe that I say anything which entails that category (1) is, or is not, vacuous.

Let us then focus our attention on category (2). In this case, we might say intuitively that the ways of measuring *n*-function values offered in the existing discussion are "justified" by appealing to certain physical laws. These are the laws which connect other relations to the *R*-relation. They allow us to infer statements involving the *R*-relation from statements involving these other relations. The logical structure of these laws may be quite complex. There may be an entire theory into which the *R*-relation is incorporated, together with other relations. This theory may be tacit and informal, or it may be explicit and formal. It may even itself be a theory of mathematical physics. A part, but not necessarily all its logical structure, may be exhibited by $\langle A, R, \circ \rangle$. Indeed, the relations which this theory connects with R may themselves be imbedded in still other theories. It may be that citing *all* the laws relevant to justifying a way of measuring *n*-function values is a very difficult task. It may even be, in some sense, an impossible task. However, these difficulties need not prevent us from drawing a very important distinction in terms of the laws appealed to in justifying ways of measuring *n*-function values.

Thus far, this discussion has been carried on under the assumption that we have already satisfactorily identified the appropriate *description* of the n_i-function. The purpose of this assumption was solely expository. We may replace it by one more congenial to a situation in which one faces a raw, completely unreconstructed theory. It is only necessary to assume that in the existing exposition of the theory we can recognize descriptions of acceptable ways of measuring n_i-values, *and* that we can recognize the physical laws (if any) appealed to in arguing for their acceptability. We *need not* suppose that we can successfully depict these laws as connecting all these ways of determining n_i-values to some single description of n_i. Indeed, since attempting to account for ways of making measurements in this fashion is intimately connected with attempting to render all claims of a theory in form (1), it is my view that, at least in some cases, an attempt at such an account would be a misguided effort (see Chapter IV). At any rate, I propose now to draw a distinction based on the ways of measuring n_i-values and the physical laws appealed to in justifying them, as they actually appear in existing expositions of a theory.

The distinction is roughly this. It may, or may not, be that in the existing discussion, the laws appealed to in justifying a way of measuring n-function values, *in the particular application i*, are statements of the very same theory whose logical structure we are attempting to clarify. We can make this a bit clearer by supposing, tentatively, that this theory can, in fact, be reconstructed along the line we are pursuing. That is, we assume that all its empirical claims – all its empirical laws – can be rendered in form (1).[2] On this assumption, the distinction may be formulated as follows. A way of measuring values of the function n_i in application i may, or may not, be justified by some statement of the theory which is expressed by 'Q_j is an S'. Alternatively, a way measuring n-function values employed in one application of a theory may or may not presuppose that some (other) application of the *same* theory is successful.

We may introduce a bit of terminology here to describe this distinction. Assuming that the formal, mathematical structure of theory θ is characterized (in part) by (D0):

(C)   The function n_i is *measured in a θ-dependent way*, if and only if there is some individual $x \in D_i$ such that the existing exposition of application i of theory θ contains no description of a method of measuring $n_i(x)$ which does not presuppose that some application of θ is successful; n_i is *measured in a θ-independent way* if and only if it is not measured in a θ-dependent way.

For brevity, I shall sometimes speak of n_i's being θ-dependent and θ-independent, as well as of n_i being measured in θ-dependent and θ-independent ways.

It should be evident now that, in the case (if there is one) where the means of determining n_i-values involves determining "directly" whether or not members of A stand in the R-relation, no physical laws are presupposed and *a fortiori* no laws of θ. Hence, if *all* means of measuring n_i-values are of this sort, n_i is θ-independent.

An example of a θ-independent function is the position function of the bodies in the solar system. Here θ is classical particle mechanics and it is being applied to account for the motion of the bodies in the solar system. It seems intuitively clear that we *can*, and do, describe means of determining the positions of these bodies whose acceptability does not presuppose that classical particle mechanics applies to any physical system. Of course, these means of determining positions may presuppose a vast and intricate structure of physical laws and theories, e.g. geometrical optics, but they do not presuppose any laws of classical particle mechanics. It must be understood that I do not contend that *all* means of determining the positions of these bodies are independent of applications of classical particle mechanics – only that *some* are. It may be that positions *can* be determined by a telescope with an automatic tracking device whose use presupposes mechanical principles, but they *can* also be determined without this piece of apparatus.

An example of a θ-dependent function is the mass function in an application of classical particle mechanics to a projectile problem. In this case we typically determine the mass of the projectile by "comparing" it to some standard body with a device like an analytical balance or an Atwood's machine. What we are doing, albeit indirectly, is determining the ratio of the mass of the projectile and the mass of an arbitrarily chosen standard. But, the *only* reason we believe that these comparison procedures yield mass-ratios, and not just numbers, completely unrelated to classical particle mechanics, is that we believe classical particle mechanics applies (at least approximately) to the physical systems used to make the comparisons. If someone asks why the number $(a/g-1 \mid a/g+1)$, calculated from the acceleration observed in an Atwood's machine experiment, is the mass-ratio of the two bodies involved, we reply by deriving it from the application of classical particle mechanics to this system. I maintain that examination of *any* acceptable account of how the mass of a projectile might be determined would reveal the same sort of dependence on an assumption that classical particle mechanics applied to the physical system used in making the mass determination. This point will be discussed further in Chapter VI.

It must be understood that the examples offered in the preceding two paragraphs have the status of substantive claims about the nature of a particular scientific theory – classical particle mechanics – as it is actually expounded. They may be substantiated, or refuted, by a detailed examination of existing expositions of this theory. I have not offered much in the way of support for them here. This is because I believe their truth will be obvious, after a moments reflection, to anyone who has endured an elementary college physics course. But, I *may* be wrong. They may not be obvious, nor even true. However, only a close examination, actual expositions of these applications of classical particle mechanics will show this. "*A priori*" claims, about the epistemological status of scientific concepts (e.g. mass) in general, are not relevant to evaluating empirical claims about the actual, existing expositions of a given theory.

Having made a distinction between θ-dependent and θ-independent descriptions of a function that appears in a *single* application of the theory θ, we can employ this to draw a further distinction which is not tied to a specific application of the theory. This distinction may be drawn in the following way. Assuming that the formal, mathematical structure of theory θ is characterized (in part) by (D0):

(D) The function n is *theoretical with respect to* θ if and only if there is no application i of θ in which n_i is θ-independent; n is *non-theoretical with respect to* θ if and only if there is at least one application i of θ in which n_i is θ-independent.

For brevity, I will speak of θ-theoretical and θ-non-theoretical functions.

It should be noted that both (C) and (D) are conditional definitions. The condition in both – that the formal, mathematical structure of theory θ is characterized (in part) by (D0) – means, roughly, that we can reconstruct θ in terms of only the functions n and t appearing in its existing expositions. Once we have picked out the functions we propose to use in logical reconstruction, we can then decide whether they are θ-theoretical or θ-non-theoretical. This condition commits us to talking about possible models for S in our reconstruction, but it does not commit us to using the predicate 'is an S', nor to using sentences of any particular logical form (e.g. form (1)) in talking about possible models for S. Thus we may draw our distinction between θ-theoretical and θ-non-theoretical functions without assuming anything very explicit about the logical structure of θ.

Examples of theoretical functions with respect to classical particle mechanics are mass and force. An example of a non-theoretical function with respect to classical particle mechanics is the position function. I claim that there is no application of classical particle mechanics in which the means of measuring masses and forces employed in the application do not presuppose that some application of classical particle mechanics is successful. In contrast, there are *at least some* applications in which the means of measuring position do not presuppose this. Other examples are provided by classical thermodynamics. Mechanical functions (pressure, volume, etc.) can be measured, in some applications, without assuming that the measuring apparatus obeys the laws of thermodynamics, while thermodynamic functions (heat, entropy, thermodynamic temperature) cannot. Again, these examples are offered as substantive claims about the existing expositions of these theories.

It is important to understand that a function may be θ-non-theoretical even though there are *some* applications of θ in which the values of the function are only obtained via other applications of θ. Thus, the possibility that there are *some* applications of classical particle mechanics in which the values of the position function are calculated by appealing to other applications of classical particle mechanics does not rule out position as a non-theoretical function with respect to classical particle mechanics. Position would be a theoretical function with respect to classical particle mechanics only if this were the situation in *all* applications.

The distinction between θ-theoretical and θ-non-theoretical functions is offered in an attempt to illuminate the same feature of scientific theories that other philosophers have tried to illuminate by drawing a distinction between theoretical and observational predicates, or between a theoretical and observational vocabulary. I attempt to formulate this distinction in a way which sidesteps epistemological questions surrounding the notion of direct evidence. Instead, I attempt to draw the distinction in terms, very roughly, of different ways functions appearing in a theory are *used* in applying the theory. The suggestion that such an attempt might be fruitful is due to Putnam:

A theoretical term, properly so-called, is one which comes from a scientific *theory* and the almost untouched problem, in thirty years of writing about 'theoretical terms' is what is *really* distinctive about such terms. ([34], p. 243.)

I have attempted to say what is "really distinctive" about theoretical terms, as they appear in theories of mathematical physics.

The distinction that I have drawn is, in one crucial respect, much more modest than that sometimes called upon by philosophers of science. I make no attempt to draw a sweeping distinction between predicates, or functions, which holds good in *all* contexts in which they might appear. Rather, I draw the distinction theoretical versus non-theoretical *relative to a given theory*. It may be that a certain function, say pressure, is non-theoretical with respect to one theory – thermodynamics – and yet, theoretical with respect to another theory – classical particle mechanics. Despite its limited scope, I believe this distinction is adequate to deal with many, if not all, the problems for which some philosophers have called upon the more comprehensive distinction. It appears to me to be adequate, at least, to deal with these problems in the form in which they arise in theories of mathematical physics.

Though my distinction between θ-theoretical and θ-non-theoretical can be drawn without appealing to the controversial epistemological distinction between direct and inferential evidence, it cannot be drawn without appealing to *any* other distinctions. Since I have spoken rather superciliously about the epistemological presuppositions underlying a sweeping distinction between theoretical and observational vocabulary, it is only fair that I lay my own cards on the table. At the very least, my distinction presupposes that we are very clear about the identity criteria for theories. We must be able to say with confidence that a particular law (used in justifying a measurement procedure) is a part of this theory, rather than that theory. Moreover, it presupposes that we can make a clear distinction between various different applications of the same theory – roughly, that we may partition the empirical statements of the theory into sub-sets corresponding to different applications. In addition to this, it is presupposed that, in the existing exposition of a theory, one can unequivocally identify the methods it prescribes for determining values of functions and the physical laws these methods presuppose.

I do not claim that these assumptions, required to make my distinction intelligible, are true of all scientific theories. I do, however, believe that they can be successfully maintained for theories of mathematical physics. My reasons for believing this are partly deductive and partly inductive. I have a theory – yet to be fully expounded – about the logical structure of theories of mathematical physics. From this theory one may derive these assumptions. But my evidence for this theory is that its logical consequences – among which are these assumptions – appear to be in accord with what we recognize, on independent grounds, to be truths about theories of mathematical physics.

Although my theory about the logical structure of theories of mathematical physics has not been completely stated, enough of it is available to see roughly how it supports the first two of these assumptions. In my view we may distinguish two components in any theory of mathematical physics: its formal mathematical structure – characterized by a set-theoretic predicate; and a set of physical systems to which the theory claims this formal structure "applies". Statements of the theory θ may be recognized by noting that the formal structure characteristic of θ is used in making these statements *and* that they are statements "about" physical systems in the set of intended applications characteristic of θ. One way to make such statements is by using different singular claims of form (1). There are others. We can recognize different applications of the same theory by noting that the same mathematical formalism is used to make claims about different physical systems in the set of intended applications. A more detailed account of the identity criteria for theories of mathematical physics is given in Chapter VII. As to the third assumption, Chapter IV provides an account of how physical laws, used in justifying measurements, appear in logical reconstructions of theories.

As to the reasons for accepting these presuppositions that are independent of my account of the logical structure of mathematical physics, I rest my case, for the moment, on the examples already mentioned. Further support will be provided by the more detailed consideration of classical particle mechanics in Chapter VI.

Having the distinction (D) between θ-theoretical and θ-non-theoretical functions, we can now describe a difficulty one might encounter in attempting to use *only* sentences of form (1) in a logical reconstruction. Suppose, as before, that we are attempting to use 'is an S' to elucidate the logical structure of theory θ. Suppose further, that after an examination of the existing expositions of this theory, we become convinced that the t-function is θ-theoretical, while the n-function is θ-non-theoretical. Consider one particular application i of θ. We propose to render the claim that θ applies to i as a sentence of the form 'Q_i is an S' where 'Q_i' denotes a possible model for S.

Assume now that we are successful in this endeavor. We actually have, for each i, a sentence of the form 'Q_i is an S' which is an intuitively satisfactory rendering of 'θ applies to i'. What happens when we set about trying to discover whether or not these statements are true? If we want to know whether it is true that Q_i is an S, we must presumably produce values of the n_i and t_i functions, described by 'Q_i', and check to see if they satisfy (D1.4) and (D1.5). How might we obtain values of these functions? In general, any method of producing such values presupposes the truth of some physical laws – except *perhaps* in the case of the θ-non-theoretical function n which *might* be described in terms of a "directly observable" relation. In the case of the θ-theoretical-function t, there is reason to expect that any method will presuppose the truth of some law of θ. This is because any method appearing in the existing exposition does so. But, on our assumption, all laws of θ may be rendered in form (1). Thus, we cannot obtain numbers which we have any good reason to believe are values of the t_i-function described by 'Q_i', unless we have good reason to believe that some claim of the form 'Q_j is an S' is true. Consequently, we cannot have evidence that Q_i is an S (or that it is not) unless we have evidence that Q_j is an S. The makings of a vicious circle, or infinite regress (depending on whether there are a finite or infinite number of intended applications) are evident.

The situation I have described appears to be intolerable. However, at least two ways it might be avoided are apparent. First, for some i, there might be ways of obtaining evidence that Q_i is an S without producing values of the t_i-function. For example, one might deduce that Q_i is an S from some other theory for which there was evidence. Second, it might be that the t_i-function, as described by 'Q_i', has properties that are not fully revealed in the existing exposition. It may turn out that once we describe the t_i-function clearly, we see ways of determining its values which do not presuppose laws of θ. Both of these "ways out" appear to entail that the t-function appears in some theory besides θ. My definition of θ-theoretical-functions does not rule this possibility out, nor should it. I am quite ready to admit that one is not *necessarily* led into the difficulty I have described in attempting to reconstruct, exclusively with claims of form (1), a theory containing functions that are theoretical with respect to this theory. It is sufficient, for motivation of the rest of my discussion, only to demonstrate that it is *possible* to encounter such a difficulty.

However, were this an unrealized possibility, one might justifiably question the utility of discussing it further, I do not believe that it is unrealized. If one employs an axiomatization of classical particle mechanics (such as that of McKinsey, Sugar, and Suppes), which takes both mass and force, as well as position, as primitives, and attempts to formulate the claims made in various applications of this theory using sentences of form (1), one seems to encounter just the situation I have described. We cannot discover the values of the mass and force functions to use in checking the claim for *any* application, unless we presuppose that the claim for some (perhaps not even 'some other') application is true.

The difficulty just described I will call 'the *problem of theoretical terms*'. It is important to understand that, at least so far as theories of mathematical physics are concerned, it is not occasioned *solely* by the presence of θ-theoretical-functions. Equally responsible for its occurrence is the commitment to employ the mathematical formalism of the theory, in one particular way, to make the empirical claims of the theory. This may be emphasized by looking at the relation between the problem of theoretical terms, as I have formulated it, and the way it is formulated by philosophers who discuss scientific theories in terms of axiomatized deductive theories in the first order predicate calculus.[3]

If one encounters the problem of theoretical terms in attempting to reconstruct a theory using 'is an S' in sentences of form (1), two alternatives are available. One may give up 'is an S', or one may give up using claims of form (1). If, for some reason one is committed to using claims of form (1), the only alternative is to try some predicate other than 'is an S'. We want a predicate that, in some sense, exhibits the relations (D1.4) and (D1.5). Otherwise, we could not expect the predicate to be adequate for reconstruction of a theory whose existing exposition involves these relations. But, we want to avoid having to give a description of the t-function in sentences of form (1). Clearly, what we want is some predicate whose possible models can be described by describing only the n-function, and which nevertheless expresses the content of (D1.4) and (D1.5). That is, we want a predicate whose possible models are structures of the form $\langle D, n \rangle$ and such that a t-function, satisfying (D1.4) and (D1.5), may be introduced by explicit definition in terms of the n-function.

It might, of course, be the case that such a predicate lies near at hand. Perhaps (D1) is such that a sentence explicitly defining t, in terms of n, can be derived from it. Then, one might think that our difficulty is occasioned only by an infelicitous formulation of the formal, mathematical structure of the theory. It is only a matter of a little mathematical cleverness to come up with a predicate of the sort we want; we haven't missed connections between t and n that are essential. But in fact, a trivial application of Padoa's principle shows that a sentence explicitly defining t, in terms of n, cannot be derived from (D1). If we are committed to defining t, in terms of n, this suggests that (D1) leaves out something important; that there must lie buried somewhere in the existing expositions of the theory we are trying to axiomatize by (D1), connections between t and n that do not appear in (D1) – *and* that these connections are properly a part of the formal, mathematical structure of the theory.

I have sketched earlier (p. 26) the source of the traditional view's commitment to claims of form (1). Indeed, as outlined there, it seemed that this view was committed to using *one* claim of form (1) to render the empirical content of a theory. This is the case if one is committed to the view that a *single* interpretation of the first order predicate calculus provides all the statements of the theory. It is difficult to tell, in the absence of specific examples, whether, and with what tenacity, traditional philosophies of science are committed to this. Let us suppose that they are committed to using a *single* interpretation, and see what happens when the theory contains θ-theoretical functions.

If (D1) appears to be a suitable characterization of the formal, mathematical structure of θ, then the traditional view is committed to providing a single interpretation I of AS which produces all the statements of θ. If t is a θ-theoretical function, it seems reasonable to expect that AS will have the following property. There will be some two-place predicate letter ρ in AS to which I assigns the relation R, appearing in the extensive system determining the t-function. The R-relation will be such that one can have (observational) evidence that $a\,R\,b$ is (or is not) true, only if one has evidence that some statement of θ is true. This is not obviously unacceptable, unless we show that this presupposed statement must also contain a statement involving R, say $c\,R\,d$. In the absence of a full account of the relation between (D1) and AS, I see no way to show that this must be the case for all θ-theoretical functions. But it seems intuitively that this might be so. If it is, then serious doubts are raised about the acceptability of R as an interpretation of ρ. If one identifies the meaning of an R-statement with ways of obtaining evidence for it, then these doubts become compelling.

At this point one might try to find some alternative interpretation to give to ρ or one might try to avoid having to interpret ρ at all. The first approach may be expected to be ruled out by constraints of fidelity to the existing exposition of the theory. At any rate, the second seems to have been the one most thoroughly discussed by philosophers of science. The obvious way to avoid interpreting ρ is to replace AS by some axiomatized deductive theory in which ρ does not appear as a primitive predicate. This is precisely analogous to the search for a definition of a set-theoretic predicate in which t may be defined explicitly in terms of n. The point to be made here is that this approach to the problem of theoretical terms, as described here, is only one among several lines open. It is characterized by a commitment to employ only one sentence of form (1) to render the empirical claims of a theory.

It might be that, by employing more than one claim of form (1), i.e. more than one interpretation of the first order predicate calculus, an alternative approach to this problem might be found within the scope of the traditional view of the relation between (D1) and statements of θ. I know of no attempt to provide such an approach.

In the next three chapters, alternatives which avoid the use of claims of form (1) are considered.

NOTES

[1] For a detailed treatment of this see [44], p. 42 ff.

[2] It may seem a bit bizarre to take such singular claims as empirical laws. It seems less so when one realizes that general statements about the individuals in D_i can be derived from 'Q_i is an S'.

[3] It should be noted that I make no claim that what I call here 'the problem of theoretical terms' includes *all* problems philosophers have discussed under this and similar names. In particular, I have nothing to say, at this point, about so-called 'theoretical entities' or 'unobservable individuals'. I do, however, think that I am dealing with a significant portion of the traditionally discussed problems.

CHAPTER III

THE RAMSEY VIEW

In this chapter a second proposal for using a set-theoretic predicate to make empirical claims is considered. This proposal is quite similar to the method of dealing with theoretical terms proposed by Frank Ramsey [34]. The relation between these two proposals, as well as some formal questions raised by taking this line, will be examined. Among these questions is the possibility of getting on without theoretical functions. An attempt is made to define a concept of eliminability of theoretical functions which is applicable to situations in which set-theoretical predicates are being used to illuminate the logical structure of a theory. Finally the question of the adequacy of this proposal to the task of logical reconstruction will be considered. It will be seen that this proposal is inadequate because it fails to provide a means of accounting for the ways measured values of theoretical functions appear to be used in existing expositions of theories.

To discuss the other ways in which the predicate 'is an S' might be used to make empirical claims, some additional concepts are required.

Consider the following definition:

(D2) x is a P_0 if and only if, there exists a D and n such that:
   (1) $x = \langle D, n \rangle$;
   (2) D is a finite, non-empty set;
   (3) n is a function from D into the real numbers.

If x is a P_0, we shall say that x is a *possible partial model* for S.

If we suppose that n is a θ-non-theoretical function, then, intuitively, possible partial models for S are just the "fragments" of possible models for S that can be described in θ-non-theoretical terms. Roughly, a configuration of *paths* for bodies in the solar system would be a possible partial model for an axiomatization of classical particle mechanics in which mass and force, as well as position, appeared as primitives. In the subsequent discussion we shall continue to employ illustrations in which n is assumed to be non-theoretical and t theoretical, relative to a theory θ

whose logical structure we are considering. The definitions given do not depend on this assumption, but they are obviously motivated by it.

For any possible partial model of S, $\langle D_0, n_0 \rangle$ we may define the predicate 'is an $E_{\langle D_0, n_0 \rangle}$ in the following way:

(D3)   If $\langle D_0, n_0 \rangle$ is a P_0, then x is an $E_{\langle D_0, n_0 \rangle}$ *if and only if there* exists a t such that:
(1) $x = \langle D_0, t, n_0 \rangle$;
(2) t is a function from D_0 into the real numbers.

If x is an E_y we shall call x 'an *extension* of y'.

Intuitively, an extension of the possible partial model of S, $\langle D_0, n_0 \rangle$, is a possible model of S ((D0), Chapter I) obtained from $\langle D_0, n_0 \rangle$ by simply "adding" a t-function. The same possible partial model may be extended, or "filled-out" in many different ways to produce different possible models for S.

Consider now a sentence of the form:

(2) $(\exists x)(x$ is an $E_Q \wedge x$ is an $S)$

where Q is a possible partial model for S. Intuitively, (2) says that there is some extension of Q which is a model for S. If we regard 'Q' as a description of the observed facts about a situation, then (2) says that this description of the observed facts can be "filled-out" – in some way – to produce a model for S. That is, one can produce a description of a t-function which, when combined with Q, yields a model for S.

Another way of looking at (2) is the following: Consider the sets:

$U_{P_0} = \{x \mid x$ is a $P_0\}$
$V_S = \{x \mid x$ is a $P_0 \wedge (\exists y)(y$ is an $E_x \wedge y$ is an $S)\}$.

V_S is the set of all possible partial models of S which are extendible to models for S. Clearly $V_S \subseteq U_{P_0}$. Intuitively, U_{P_0} is the set of all possible states of affairs that can be described in θ-non-theoretical terms – i.e. all possible n-functions on a set of individuals. V_S is the set of all such "observable" states of affairs that can be filled-out, by adding a θ-theoretical function t, to produce a model for S. The sentence (2) claims that $Q \in V_S$, i.e. that Q is one of those "observable" states of affairs that can be made into a model for S by adding some t-function.

The first question to ask about sentences of form (2) is this. Is it reasonable to expect that sentences of form (2) might be used to make empirical claims? Before considering this question, it is instructive to see what (2) entails about the values of the n-function for domains containing small numbers of individuals. Let $Q = \langle D_0, n_0 \rangle$ and consider the following examples.

(E1)   (a) $D_0 = \{x\}$. Claim (2) then reduces to this.

Given the value of n_0, $n_0(x)$, there is at least one number, call it $t(x)$, greater than zero, such that:

$$t(x)n_0(x) = 0.$$

This entails that $n_0(x) = 0$, and clearly $n_0(x) = 0$ entails (2), so that (2) is equivalent to a statement about the values of n_0.

(b) $D_0 = \{x, y\}$. Claim (2) then reduces to this.

Given the values of n_0, $n_0(x)$, $n_0(y)$, there is at least one pair of numbers, call it $\{t(x), t(y)\}$, both greater than zero, such that

$$n_0(x)t(x) + n_0(y)t(y) = 0.$$

This entails that either $n_0(x) = n_0(y) = 0$ or $n_0(x) > 0$ if and only if $n_0(y) < 0$, i.e. either all values of n_0 are zero or not all have the same sign. Here again (2) is equivalent to a statement about the values of n_0. It should also be noted that (2) entails that

$$\frac{t(x)}{t(y)} = -\frac{n_0(y)}{n_0(x)}.$$

The ratio of the t-function values required to satisfy (2) is uniquely determined by the n-function values.

(c) $D_0 = \{x, y, z\}$. Claim (2) then reduces to this.

Given the values of n_0 for x, y, and z, there exists at least one triple of numbers, call it $\{t(x), t(y), t(z)\}$ such that

$$n_0(x)t(x) + n_0(y)t(y) + n_0(z)t(z) = 0.$$

Clearly (2) is equivalent to the claim that either all values of n_0 are zero, or not all values have the same sign. In this case, the ratios of t-function values which satisfy (2) are not uniquely determined by n-function values, though certain ratios can obviously be ruled out by n-function values.

From these examples it should now be clear that, for any D_0, claim (2) is equivalent to the claim that either $n_0(x) = 0$, for all $x \in D_0$, or not all values of n_0 have the same sign.

Let us suppose now that the description 'Q' of $\langle D_0, n_0 \rangle$ appearing in (2) describes both D_0 and n_0 in ways which are not simply lists of their members. Further, let us suppose that it is clearly an empirical question as to whether or not a certain individual is in D_0 and what the value of n_0 for that individual is. It is evident that the truth value of (2) would then be a question to be settled by empirical investigation. Since (2) is equivalent to certain conditions on the values of the function n_0, one would simply determine these values empirically and check to see if they satisfy these conditions.

Of course the situation may be different for claims of form (2) made with other predicates besides 'is an S'. For some predicates, it might be that one could prove that the corresponding claim of form (2) was true for *any* Q which was a P_0, or for *no* Q. In these cases (2) clearly could not be used to make an empirical claim.

Alternatively, for some predicate S', it might be that (2) was not equivalent to some statement about the values of n_0. In this case it still appears that it would be an empirical question as to whether or not Q was extendible to a model for S. Empirical investigation would determine all, or some, of the members of D_0 and their values of n_0. Then we would employ our mathematical skills to try to discover whether there was a t-function which could be adjoined to these to yield a model for S'. A predicate which occasions a situation of this kind will be considered shortly.

In the light of this, it does not seem implausible to expect that one might depict a part of the activity of an empirical science as being directed toward testing empirically claims of form (2). 'Testing' is understood here to mean simply gathering data about n-functions which might show the claim of form (2) to be false. Typically, one might expect scientists to search for individuals having the defining property of D_0, and then employ the description of n_0 to determine the values of n_0 for these individuals. The object of their search would be to discover n_0-values which definitely preclude filling out the possible partial model $\langle D_0, n_0 \rangle$ to a model for S. The failure to produce such n_0-values would constitute what one might call '*prima facie* confirmation' for (2). It should be noted, however, that empirical investigation of this sort to discover the truth value of (2) does not involve anything that could be identified as observing or measuring t-function values. Roughly speaking, the t-function only comes to be considered after all the empirical work has been done. It is, if you like, a calculational device. We will return later to a detailed consideration of this point.

It also appears that claims of form (2) can be employed to justify predictions. If one has good reason to believe (2) for some particular $\langle D_0, n_0 \rangle$, this might justify at least two types of predictions.

(i) Let D_0 be described by naming a property possessed by all and only members of D_0. If all known members of D_0 have values for n_0 which do not allow $\langle D_0, n_0 \rangle$ to be extended to a model for S, then one can predict that there are, as yet undiscovered, members of D_0.

An example of this sort of prediction is provided by celestial mechanics. The paths of all the planets known prior to 1846 could not be extended to a model for Newtonian mechanics with inverse square forces. Yet, at this time, there was reason to believe that a claim something like (2) was true, where $D_0 =$ the bodies in the solar system. This led astronomers to predict the presence (and indeed the path) of a previously unobserved planet, Neptune.

(ii) Suppose one has good reason to believe that D_0 has exactly k members and the n_0-values for $k-l$ members are known. Then one might be able to predict some facts about the n_0-values for the remaining l members.

Predictions of this sort are typical of certain applications of classical particle mechanics. In these applications the initial positions of particles correspond to the members of D_0 whose n_0-values are known and their subsequent positions correspond to the unknown n_0-values.

Suppose now that we are in the situation, described in Chapter I, of attempting to use the predicate 'is an S' to elucidate the logical structure of the existing theory θ. The preceding remarks suggest that it is not implausible to expect that some of the empirical claims of this theory may be rendered as claims of form (2). In particular, we might hope to render the claim 'θ applies to i' as a claim of form (2) with $Q_i = \langle D_i, n_i \rangle$. Is there any reason to expect that an attempt to use sentences of form (2) to express these claims may succeed, where attempts to use sentences of form (1) fail?

It appears that there is some reason to expect this to be so. In a situation where the t-function appearing in existing expositions of the theory is θ-theoretical, while n is θ-non-theoretical, the use of sentences of form (2) offers a possible solution to the problem of theoretical terms. Consider a claim of form (2) made for $\langle D_i, n_i \rangle$. To check the truth value of this claim, we have only to produce values of the n-function on D_i. There is no need to produce any values of the t-function. But, since n is θ-non-theoretical, we can find methods of determining n_i-values which do not presuppose any laws of θ. Thus, we can expect to be able to obtain evidence that a claim of form (2) for $\langle D_i, n_i \rangle$ is true without having good reason to believe that any other claim, 'θ applies to j', is true. Claims of form (2), in contrast to claims of form (1), may be confirmed or disconfirmed without presupposing the truth of any other claim of the theory. This means that there is a possibility of rendering *all* claims that θ applies to i as claims of form (2), even in the cases where the t-function occurs in no theory other than θ. The circularity or regress which arises when one attempts to do this with claims of form (1) does not arise.

The suggestion that sentences of form (2) might be used to make the empirical statements of theories containing theoretical terms, is closely related to a view of the logical functioning of theoretical terms first suggested by Ramsey [35]. Though Ramsey's view is formulated for theories axiomatized in the first order predicate calculus, it is easily seen to be equivalent, in a sense to be made clear shortly, to the suggestion being considered here. For this reason, I shall call this suggestion 'the *Ramsey solution* to the problem of theoretical terms'. Before considering the adequacy of sentences of form (2) to the task of logical reconstruction, it is enlightening to examine two formal questions which arise in connection with the Ramsey solution. In this connection we will have occasion to examine the differences between Ramsey's formulation of the suggestion and the formulation offered here.

To provide the framework for raising one of these questions, consider the possibility of constructing set-theoretic predicates similar to 'is an S_0' and 'is a P_0' by, roughly speaking, keeping the same basic set-theoretic structure, but placing additional restrictions on the n-function and t-function. By doing this for 'is an S_0' we can define various mathematical structures similar to S, differing only in the nature of the n-function and t-function and the relations between them. For example, we might replace (D1.5) by:

(D1'.5) $\sum_{[y \in D]} t(y)n(y) = \prod_{[y \in D]} (t(y)+n(y)).$

to obtain (D1′), a definition of 'is an S''. Similarly, we might add to (D2):

(D2′.4) Either,
  (i) for all $y \in D$, $n(y) = 0$, or
  (ii) for all $y \in D$, if $n(y) > (<)0$ then there exists a $z \in D$ such that $n(z) < (>)0$.

to obtain a definition (D2′) of the predicate 'is a P''.

We can try to be a bit more precise about this. Let a *restriction* of (D0) or (D2) be a definition of a set-theoretic predicate obtained from (D0) or (D2) by adjoining to these definitions additional axioms containing, besides logical and mathematical symbols, only symbols occurring in the brackets $\langle \ \rangle$ in the first numbered clause of the original definition. For (D0), these symbols are 'D', 'n', and 't', for (D2), they are 'D' and 'n'. Both (D1) and (D1′) are restrictions of (D0). Definition (D2′) is a restriction of (D2). For an example of something that is not a restriction of (D2), consider adding the following axiom to (D2).

(D2″.4) There exists a function t, from D into the real numbers such that:
  (i) for all $y \in d$, $t(y) > 0$;
  (ii) $\sum_{[y \in D]} t(y) n(y) = 0$.

Intuitively, definition (D2″) is not a restriction of (D2) because the added axiom (D2″.4), contains the symbol 't'.

The notion of a restriction of a definition of a set-theoretic predicate is not as precise as one would like it to be. The question of whether or not an axiom we add to (D2) contains symbols other than 'D' and 'n' may not always be an easy one to answer. For example, consider:

(D2″.4) There is some set of positive numbers, N, such that
  $\sum_{[y \in D]} n(y) a = 0, \quad a \in N$.

Does (D2″.4) contain symbols other than 'D' and 'n'? One might say that the presence of 'N' requires us to say that it does. But 'N' appears as a bound variable, just as do 'y' and 'z' in (D2′.4). If we count 'N' as an additional symbol, why do we not count 'y' and 'z' too? On the other hand, if we refuse to count variables as new symbols then 't' does not appear in (D2″.4). But (D2″.4) is, intuitively, just the kind of axiom we are trying to rule out.

48   THE LOGICAL STRUCTURE OF MATHEMATICAL PHYSICS

The intuitive idea behind the notion of a restriction of a definition of a set-theoretic predicate seems reasonably apparent. A restriction is simply a further specifying, or saying more about, the entities mentioned in the initial definition *without introducing any additional conceptual apparatus*. That we have difficulty making this notion precise should not be surprising. It is notoriously difficult to be precise about "linguistic" notions such as primitive symbols and definitions of one symbol in terms of others without resorting to a formal language. The notion of a restriction seems to be of this sort. So long as we avoid formalizing our definitions of set-theoretic predicates in some formal language, these notions seem destined to elude precise formulation. It would, however, be a mistake to think that the vagueness of such notions, in the absence of formalization, renders them useless. It seems intuitively clear that (D2′) is a restriction of (D2) and that (D2″) is not. Whatever insight, if any, this fact affords into the logical structure of theories should not be denied us simply because we cannot make the notion of a restriction precise. I will argue below that an interesting issue can be raised, in terms of this notion, for definitions of set-theoretic predicates. To allay, insofar as possible, any suspicions about this issue, occasioned by the notion of a restriction, I will then consider the analogous issue for axiomatized deductive theories in the first order predicate calculus.

To raise this issue, let 'is an S^i' be defined by some restriction of (D0). Let V_{S^i} be the set of all possible partial models for S^i which are extendible to models for S^i. Note that the possible partial models are the same for all restrictions of (D0). They are, in every case, just the models for P_0. The notion of one of these possible partial models being extendible to a model for S^i is exactly parallel to the notion of its being extendible to a model for S.

Now, consider this question. Under what circumstances, i.e. for what S^i's does there exist a restriction of (D2), defining 'is a P^i' such that U_{P^i}, the set of all models for P^i, is identical with V_{S^i}? The question can be put in another way. The claim of a sentence of form (2) (made with 'is an S^i' (2^i)) is that some particular member of U_{P_0} (the set of all models for P_0, or possible partial models for S^i) is also a member of V_{S^i}. Can we make an equivalent claim using a predicate 'is a P^i', defined by some restriction of (D2), in a sentence of the form:

(1^i) $\langle D_0, n_0 \rangle$ is a P^i?

Roughly speaking, can we determine the same sub-set of possible partial models for S^i (models for P_0) as is determined by (2^i) without using a predicate which employs the t-function? If we regard U_{P_0} as the set of all possible states of affairs that can be described in θ-non-theoretical terms, and V_{S^i} as the sub-set of these determined by (2^i), then we are asking whether this same sub-set can be determined without resorting to predicates which contain θ-theoretical functions. That is, can we characterize the set V_{S^i} using only θ-non-theoretical terms?

Let us regard

(2^i) $(\exists x)(x$ is an $E_y \wedge x$ is an $S^i)$

as a sentential function in y, i.e. as a one place predicate whose argument is y. If we carry out the program outlined above, replacing 'is an S^{i}' by 'is a P^{i}' which does the same job of picking out a sub-set of "observable" states of affairs, then we might naturally say that the t-function has been eliminated from the "predicate" (2^i). One might say loosely that it has been eliminated from sentences of form (2^i). 'Eliminated' in the sense that we have produced a predicate 'is a P^{i}', extensionally equivalent to (2^i), which does not "contain" the t-function. If this program can be carried out, we shall say that the t-function is *Ramsey eliminable* from sentences of form (2^i).[1] It is important to note that the notion of Ramsey eliminability is defined relative to particular predicates like 'is an S^{i}'. If sentences of form (2), constructed from different predicates like 'is an S^{i}', were used to make all the empirical claims of a theory, and if the t-function were Ramsey eliminable from all of them, then one might say that the t-function was Ramsey eliminable from the theory. But it might be the case that the t-function was Ramsey eliminable from some claims of a theory and not from others. In the case there was one single restriction of (D2) which served to eliminate the t-function from *all* claims of the theory, one might want to say that the t-function was *Ramsey eliminable in the strong sense* from this theory.

In the example we are considering, the t-function is clearly Ramsey eliminable from sentences of form (2), constructed with 'is an S'. The conditions for $\langle D_0, n_0 \rangle$ to be extendible to a model for S can obviously be stated entirely in terms of the values of n_0. They are simply that, either all values of n_0 are zero, or not all have the same sign. It is easy to see that $U_{P'}$ – the set of all models of P' – is coextensive with V_S – the set of all possible partial models for S that are extendible to models for S. Thus (D2′) is a restriction of (D2) which defines, without using the t-function, a predicate determining the same sub-set of possible partial models for S as is determined by the sentence of form (2), constructed with 'is an S'. Since the same restriction serves to eliminate the t-function from all claims of form (2) constructed with 'is an S' the t-function is Ramsey eliminable in the strong sense from the theory whose mathematical structure is characterized by 'is an S'.

It should be noted that this example shows that a function may be Ramsey eliminable, without being explicitly definable in terms of the other primitive symbols. A trivial application of Padoa's principle ([41], p. 169) shows that the t-function is not explicitly definable in terms of the n-function. It is easy to find models for S in which D and n are the same and yet t is different. The n-function on D does not, in general, uniquely determine the t-function on D needed to produce a model for S. Of course, if t is explicitly definable, then it is certainly Ramsey eliminable. One would simply replace all occurrences of t by its definitional equivalent to obtain the requisite restriction of (D2). The important point to note is that this "piecemeal" elimination of t is not the only way to get rid of it. We can "in one stroke" replace 'is an S' by a new predicate, not containing t, which does the same work that 'is an S' does in sentence (2).

In the case of a sentence of form (2) using 'is an S'' (2'), the situation is less clear. There appears to be no simple, direct way, like (D2'.4), of describing the class of n-functions such that there exists a t-function satisfying (D1'.4) (the same as (D1.4)) and (D1'.5). For one-element domains it is obviously:

$$\frac{n(y)}{n(y)-1} > 0.$$

For two-element domains, one obtains a second degree equation in $t(y_1)$ and $t(y_2)$. There are well known methods for describing, in terms of their coefficients, the conditions under which such an equation has positive solutions. In general, for n-element domains one obtains n-degree equations in the t-function values. For $n>2$, there is no simple way of describing conditions under which the equation has positive solutions. Of course, one could describe an algorithm for determining, for any putative n-function, whether or not the requisite t-function exists. But, having done this, it is unclear how one would decide whether or not the description of the algorithm "involved the t-function". Intuitively, one would suspect such algorithms to proceed by locating one solution to the n-degree equation and constructing others by "trial and error" from this. This seems to involve the t-function, but it is not obvious that, by clever circumlocution, this could not be avoided. In view of the difficulty one might expect in giving a single algorithm suitable for all finite D's, it appears that we might want to say that t was not Ramsey eliminable in the strong sense from a theory whose structure is mathematical characterized by 'is an S''.

It is important to understand that, even in this case where we can only with great difficulty, if at all, describe explicitly the constraints that (2') puts on n-function values, it is nevertheless an empirical question as to whether (2') is true. Whether the values of n_0 are consistent with the existence of a t-function satisfying (D1'.4) and (D1'.5) is decided, *if* it is decided, by finding members of D_0 and discovering their n-function values. Of course, even if we knew *all* the values of n_0, we might simply not be clever enough mathematicians to discover whether the requisite t-function exists.

That the concept of Ramsey eliminability and the notion of a restriction we use to formulate it are useful tools in the analysis of actual physical theories, will be supported by the treatment of classical particle mechanics in Chapter VI. At this point, I will mention briefly a point discussed in some detail in that chapter (p. 140 ff.). Consider a set-theoretic axiomatization of Newtonian mechanics – classical particle mechanics with third-law, central forces, and a claim of form (2) made with this set-theoretic predicate. This claim says roughly that the paths of certain physical objects can be augmented with mass and force functions to produce a model for Newtonian mechanics. It turns out that this is a non-trivial claim. It is not true of all possible configurations of paths. A claim of this sort appears to have been offered by various people (e.g. Pendse [31], [32], [33] and Simon [37]) as expressing at least a part of the empirical content of classical celestial mechanics. In considering the question of Ramsey eliminability of mass and force from this claim, it turns out that force is clearly Ramsey eliminable. This fact is essentially equivalent to noting that linear and angular momentum are conserved in systems with third-law central forces. The situation with the mass function is less clear, but it appears not to be Ramsey eliminable.

Though examples of the sort just mentioned might convince us of the utility of the concept of Ramsey eliminability, they are not of much help in clarifying the vagueness associated with it. Although I see no direct way to remove this vagueness, it seems enlightening to consider the way one would formulate the concept of Ramsey eliminability for axiomatized deductive theories in formal languages. This will also allow us to see more clearly how the view being considered in this chapter is related to Ramsey's views.

We can, without losing any essential features of the situation, consider an axiomatized deductive theory with an underlying logic L – the first order predicate calculus with identity – and one extra-logical axiom $F(\bar{T}, \bar{O})$, where $\bar{T}$ and $\bar{O}$ are both two place predicates.[2]

Let D be a given domain and let

$$M_D = \{x \mid x = \langle D, O \rangle \text{ and } O \subseteq D \times D\}$$

i.e. M_D is the set of all binary relations on D. We can then introduce the following definition:

(D4) For all $m = \langle D, O \rangle \in M_D$, m is D-extendible to a model for $\bar{F}(\bar{T}, \bar{O})$ if and only if there exists a $T \subseteq D \times D$ such that $\bar{F}(\bar{T}, \bar{O})$ is true in $m^* = \langle D, T, O \rangle$.

We then formulate the claim that '$\bar{T}$' is Ramsey eliminable from the sentence $F(\bar{T}, \bar{O})$ when this sentence is used to make statements about members of D.

(R1) '$\bar{T}$' is D-Ramsey-eliminable from $F(\bar{T}, \bar{O})$ if and only if there exists a sentence $H(\bar{O})$ such that, for all $m \in M_D$, $H(\bar{O})$ is true in m if and only if m is D-extendible to a model for $F(\bar{T}, \bar{O})$.

Intuitively, (R1) says that '$\bar{T}$' is Ramsey eliminable from a statement about members of D made with $F(\bar{T}, \bar{O})$ if and only if the sub-set of M_D whose members are D-extendible to models for $F(\bar{T}, \bar{O})$ is coextensive with the sub-set of M_D whose members are models for some sentence $H(\bar{O})$ which does not contain '$\bar{T}$'. This is to say, exactly the same observable circumstances in D are consistent with the theory having $H(\bar{O})$ as its one extra-logical axiom as can be "filled out" by adding a theoretical relation so as to be consistent with the theory having $F(\bar{T}, \bar{O})$ as its one extra-logical axiom.

Consider now the claim:

(R2) Given D, for all $F(\bar{T}, \bar{O})$, '$\bar{T}$' is D-Ramsey eliminable from $F(\bar{T}, \bar{O})$.

A result of Craig and Vaught [9] entails that this claim is true for any finite D. Examples of non-finite D's and $F(\bar{T}, \bar{O})$'s for which it is not true can be given (see below).

It is important to understand the difference between (R2) and a proposal, made by Ramsey, about how we are to "understand" axiomatized deductive theories which contain theoretical terms. Ramsey's view was essentially this ([35], p. 231). We are to understand the empirical claim of our axiomatized deductive theory with $F(\bar{T}, \bar{O})$ as its single extra-logical axiom to be made by a sentence of the form, $(\exists t)F(t, \bar{O})$, in the second order predicate calculus. Adopting this view is sometimes thought of as a way of eliminating theoretical terms. However, it is much more natural to regard this as a proposal for obtaining statements – things that can be true or false – from sentences in the first order predicate calculus without providing an interpretation of the theoretical terms. Hempel has noted this ([16], p. 81). Viewed in this way, Ramsey's proposal is analogous to our proposal that sentences of form (2) be used to make the empirical claims of a theory axiomatized by (D1). The Ramsey sentence $(\exists t)F(t, \bar{O})$ is true in just those $m \in M_D$ which are extendible (in the sense of (D4)) to models for $F(\bar{T}, \bar{O})$. The claim of (R2) however, is that we can *always* find a sentence $H(\bar{O})$, in the first order predicate calculus, which is extensionally equivalent – determines the same sub-set of M_D – to $(\exists t)F(t, \bar{O})$.

There is some reason to believe that (R2) is not a very "interesting" eliminability claim. First, note that, for a fixed $F(\bar{T}, \bar{O})$, different $H(\bar{O})$ may be required to eliminate '$\bar{T}$' when $F(\bar{T}, \bar{O})$ is used to make statements about different finite D's. Next, note that in many theories the domain is described intentionally in a way that gives us no information about its cardinality. We describe D by naming a property, e.g. being a body in the solar system, without knowing how many things have that property. This suggests that one might like to have a single $H(\bar{O})$ that serves to eliminate '$\bar{T}$' from $F(\bar{T}, \bar{O})$ in any domain, i.e. we might want a notion of Ramsey eliminability that is independent of D.

(R3)   '$\bar{T}$' *is Ramsey eliminable from* $F(\bar{T}, \bar{O})$ if and only if there exists an $H(\bar{O})$ such that for all D and $m \in M_D$ $H(\bar{O})$ is true in m if and only if m is D-extendible to a model for $F(\bar{T}, \bar{O})$.

For this stronger sense of Ramsey eliminability we have the following eliminability claim analogous to (R2).

(R4)   For all $F(\bar{T}, \bar{O})$, '$\bar{T}$' is Ramsey eliminable from $F(\bar{T}, \bar{O})$.

Note that (R4) is analogous to the claim, for a set-theoretic axiomatization, that the theoretical function is Ramsey eliminable in the strong sense. In our example, the claim is that there is *one* restriction of (D2) which eliminates the t-function from all applications of the theory – all claims of form (2). The weaker claim is that, for every application of the theory, there is some restriction of (D2) which eliminates the t-function in that application. Later we shall see some further reasons for believing that the strong sense of Ramsey eliminability is the interesting one.

Ramsey considers some eliminability claim and believes it to be true of the $F(\overline{T}, \overline{O})$ in his example.

> Can we say anything in the language of this theory that we could not say without it? Obviously not; for we can easily eliminate the functions of the second system and so say in the primary system all that the theory gives us ([35], p. 219).

Whether he is considering (R2) or (R4) is unclear, as well as whether he believes the claim is generally true or only true for his example. Subsequent remarks indicate that Ramsey clearly distinguished eliminability in the sense of (R2) or (R4) from explicit definability of theoretical terms.

Let us consider (R4). If it is the case that Ramsey believed (R4) to be true then he was mistaken. It is possible to find examples of sentences $F(\overline{T}, \overline{O})$ for which no sentence $H(\overline{O})$ exists which, for all D, is true in all and only members of M_D that are extendible to models for $F(\overline{T}, \overline{O})$. These examples also suffice to show that (R2) is false for infinite D's. A simple example is the following[3]:

(E2) $F(\overline{T}, \overline{O}) =$ (i) $(x)(y)(\overline{T}xy \to (\exists z)\overline{O}xz \land \neg(\exists z)\overline{O}yz) \land$
 (ii) $(x)((\exists z)\overline{O}xz \to (\exists y)[\overline{T}xy \land (w)(\overline{T}xw \to w=y)]) \land$
 (iii) $(x)(\neg(\exists z)\overline{O}xz \to (\exists y)[\overline{T}yx \land (w)(\overline{T}wx \to w=y)])$

Sentences (i), (ii), and (iii) are true exactly in models $\langle D, T, O \rangle$ in which there is a one–one correspondence between individuals which stand in the first place in the O-relation and those which do not. Yet it can be shown that there is no sentence, containing only the predicate $\overline{O}$ and identity, which is true exactly in models $\langle D, O \rangle$ in which there is such a one–one correspondence.[4]

One might raise a question as to whether the fact that there is a one–one correspondence between individuals which stand in the first place of the O-relation and those which do not, is an "observable fact". If one takes 'observable fact' to mean 'fact expressible in the observation vocabulary', then clearly it is not. This is just the force of the example. On the other hand, it is obviously a fact about the O-relation. It is just the same sort of fact about the O-relation as that O is transitive. If one, in some sense, discovers by observation whether or not individuals stand in the O-relation, then there is no difference in the way one would check to see whether some particular O-relation had either of these properties. The point is that there are some facts about O-relations that cannot be expressed by sentences containing only the $\bar{O}$-predicate.

A somewhat more complex, yet more familiar, example is provided by an axiomatization of the theory of ordered fields in the first-order predicate calculus with identity. Let the ordering relation correspond to O, and the operations + (addition) and · (multiplication) correspond to T. For any sentence containing only the order-relation predicate and identity, there will always be models which cannot be extended to produce models for the full theory. This is to say, intuitively, the notion of an ordered field cannot be fully characterized by sentences in the first-order predicate calculus with identity, containing (besides identity) only the order-relation predicate. Any attempt to do this will always let in models that we want to exclude.

Having noted that (R4) is false, it is instructive to compare it with Craig's result about the eliminability of auxiliary expressions [7], [8]. At first glance, both Craig's result and (R4) seem to be getting at the same thing – a duplication of the empirical content of a theory containing theoretical terms without the use of theoretical terms. To investigate this more closely, we must formulate Craig's result for this special case.

Let,

(4) $C(y) = \{x \mid x$ is a sentence of L and x is a logical consequence of $y.\}$
i.e. $C(y)$ is the set of all logical consequences of y. And let,

(5) $\delta(\bar{T}) = \{x \mid x$ is a sentence of L and $\bar{T}$ does not occur in $x.\}$
$\delta(\bar{T})$ is the set of sentences of L which do not contain the theoretical predicate $\bar{T}$. Craig's result may be stated roughly as:

(C1)   For any sentence $F(\bar{T}, \bar{O})$, there exists a sentence $H(\bar{O})$ such that
$C(F(\bar{T}, \bar{O})) \cap \delta(\bar{T}) = C(H(\bar{O}))$

More precisely, $H(\bar{O})$ may not be a sentence, but an infinite set of sentences. The real interest of Craig's result is that it provides a decidable method of constructing this set. This refinement of (C1) is not essential to a comparison of Ramsey and Craig eliminability. Intuitively, however, it is important to note that $H(\bar{O})$ is not merely the conjunction of members of $C(F(\bar{T}, \bar{O})) \cap \delta(\bar{T})$. Since L is complete, we can give an equivalent semantical formulation of (C1).

(C1′)   For any sentences $F(\bar{T}, \bar{O})$, there exists a sentence $H(\bar{O})$ such that, for all D and $m \in M_D$, all members of $C(F(\bar{T}, \bar{O})) \cap \delta(\bar{T})$ are true in m if and only if $H(\bar{O})$ is true in m.

Intuitively, what (C1′) tells us is that we can always find a sentence $H(\bar{O})$ that determines the same class of models as is determined by the class of all "observational consequences" of $F(\bar{T}, \bar{O})$. Clearly, this would not be very interesting if $H(\bar{O})$ were nothing but the infinite conjunction of these observational consequences. The interest in Craig's result lies in the possibility of finitely specifying $H(\bar{O})$, even though it may, in fact, be an infinite set of sentences, rather than a single sentence. From our point of view, it is not particularly important how the observational consequences are characterized, whether $H(\bar{O})$ is a single sentence of an infinite set of sentences. What is important is to note that the class of all models for the observational consequences of $F(\bar{T}, \bar{O})$ is not necessarily coextensive with the class of all models that are extendible to models for $F(\bar{T}, \bar{O})$. The $H(\bar{O})$ provided by Craig's result does not necessarily serve as the $H(\bar{O})$ which (R4) asserts to exist.

To see that there is always some D such that

$$U_0 = \{x \mid x \in M_D \text{ and all members of } C(F(\bar{T}, \bar{O})) \cap \delta(\bar{T}) \text{ are true in } x\}$$

and

$$V_F = \{x \mid x \in M_D \text{ and } x \text{ is extendible to a model for } F(\bar{T}, \bar{O})\}$$

need not be coextensive, one has only to consider our example (E2). It was noted that there is no sentence containing only $\bar{O}$ and identity which is true only in models where there is a one–one correspondence between the individuals which do and do not occupy the first place of the O-relation. In a like manner, it can be shown that we could do no better by taking infinite sets of such sentences. By using only $\bar{O}$ and identity, we could never rule out models in which there was no such correspondence.

It is obvious that if m is extendible to a model for $F(\bar{T}, \bar{O})$ then m is a model for $C(F(\bar{T}, \bar{O})) \cap \delta(\bar{T})$ so that we have the situation as depicted in Figure 1.

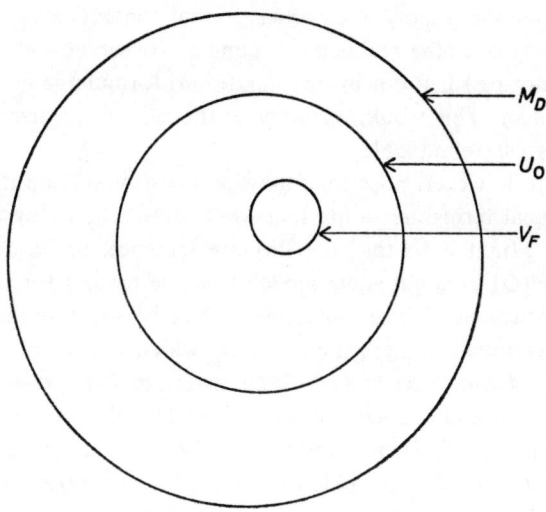

Fig. 1

One conclusion to draw from this is that the notions of Ramsey and Craig eliminability are not equivalent. The Craig result, which assures us that Craig eliminability is always (in situations like the one described here) possible, does not assure us of the possibility of Ramsey eliminability. More important than this, however, is the insight that these facts afford us into the functioning of theoretical terms. It can be shown that the set of all observational consequences of the Ramsey sentence $(\exists t)F(t, \bar{O})$ is the same as the set of all observational consequences of $F(\bar{T}, \bar{O})$ (i.e. $C(F(\bar{T}, \bar{O})) \cap \delta(\bar{T})$). However, the class of models for the Ramsey sentence $(\exists t)F(t, \bar{O})$ is not necessarily coextensive with the class of models for its observational consequences. The Ramsey sentence is, in general, stronger than the conjunction of all the observational consequences of the theory. It is not merely stronger in a trivial sense of entailing some sentences not in the set of observational consequences. It is stronger in the sense that one could discover (empirically) some situations (members of M_D) that were consistent with all the observable consequences of the theory, and yet the Ramsey sentence would not be true of them. That is, they could not be filled out to produce a model for $F(\bar{T}, \bar{O})$.

If we accept the Ramsey proposal that the empirical claim to be made with the axiomatized deductive theory $F(\bar{T}, \bar{O})$ is made with a sentence of the form $(\exists t)F(t, \bar{O})$, then the "empirical import" of the theory may be more than just the totality of its observational consequences. The theoretical terms may be doing real work – ruling out observable states of affairs that could not be ruled out by any conditions formulable in the observation vocabulary. This would certainly be the case if the theoretical terms were not Ramsey eliminable.

One might, however, raise the following question. Could it be the case that theoretical terms are doing real work, even when they are Ramsey eliminable? Might it be that the Ramsey sentence, or its observational equivalent $H(\bar{O})$, rule out some models that are models for the observational consequences? To see that this cannot be so, note that, when '$\bar{T}$' is Ramsey eliminable, the sentence $H(\bar{O})$ which is true in all and only members of M which are extendible to models of $F(\bar{T}, \bar{O})$ is an observational consequence of $F(\bar{T}, \bar{O})$. Thus any model for the set of observational consequences of $F(\bar{T}, \bar{O})$ is a model for $H(\bar{O})$. On the other hand, any member of M which is extendible to a model for $F(\bar{T}, \bar{O})$ is a model for the set of observational consequences of $F(\bar{T}, \bar{O})$. Thus, any model for $H(\bar{O})$ is a model for the set of observational consequences. This shows us that, when '$\bar{T}$' is Ramsey eliminable, the sentence $H(\bar{O})$ which Ramsey eliminates it is logically equivalent to the set of observational consequences of $F(\bar{T}, \bar{O})$. Thus, the theoretical term '$\bar{T}$' is not doing any work in ruling out observable states of affairs that could not be done with the observational vocabulary, albeit perhaps in a less "economical" way.

This digression to consider the concept of Ramsey eliminability for axiomatized deductive theories in the first-order predicate calculus with identity should serve to mitigate skepticism about the significance of the concept for definitions of set-theoretic predicates. Clearly, it does not provide any direct means of making this concept more precise in the latter context, but it does show that, if we are willing to employ the apparatus of a formalized language, the concept can be made precise. Moreover, it can be made precise in such a way that clear-cut examples of theoretical terms which are not Ramsey eliminable can be given. This suggests that our intuition that 't' is not Ramsey eliminable from (2') may not be entirely baseless. We can see that our feeling that there is something like a minimal amount of conceptual apparatus needed to characterize a certain class of set-theoretic structures is borne out in those cases where we have strict syntactical criteria for determining what conceptual apparatus is in fact being used. This leads us to be more trusting about our intuitions in the cases where clear-cut syntactic criteria are lacking. It seems reasonable to compare the role of the t-function in claim (2') to the role of the T-relation in (E2). They both serve to help us rule out certain possible observable states of affairs – to say that these states of affairs do not occur – and moreover there is no other way to rule out *just* these states of affairs by talking about only observable properties. It thus appears that 'Ramsey eliminability' might prove to be a useful tool in understanding precisely what the θ-theoretical terms are doing for us in the theory θ.

The second formal question raised by the Ramsey solution to the problem of theoretical terms is this. Suppose that claim (2) is true for a particular $Q - \langle D_0, n_0 \rangle$. Is there a *unique* t-function which makes $\langle D_0, n_0, t \rangle$ a model for S? The answer to this question will in general depend not only on what S is but also on what Q is like. In none of our examples (E1–a, b, c) is the t-function uniquely determined by the values of n_0. But in (E1–b) the ratio of the t-function values is uniquely determined.

We shall see later that the answer to this question is relevant to accounting for talk about measuring values of θ-theoretical functions in a logical reconstruction of θ. For the time being, it is important to note the relation between the t-function's being uniquely determined by n-function values and the t-function's being explicitly definable in terms of the n-function. If it is the case that, for *all* domains D, the t-function required to make $\langle D, n, t \rangle$ a model for S is uniquely determined by n, then there is good reason to think that the t-function will prove to be explicitly definable in terms of the n-function. For the analogous situation in axiomatized deductive theories in the first order predicate calculus, Beth's definability theorem [3], shows this to be the case. But if it is only the case that for *some* domains D, the t-function is uniquely determined by the n-function, then there is no reason to expect that the t-function will prove to be definable in terms of the n-function. Indeed, it is easy to see that, in this situation, an application of Padoa's principle will always suffice to prove that such definability is not possible.

A failure to appreciate this point has led to much confusion in discussions of the mass function in classical particle mechanics. The source of the confusion is roughly this. Mach ([23], pp. 180–185) noted that, for physical systems consisting of two particles, the mass-ratio required to make the system a model for classical particle mechanics with third law, central forces is uniquely determined by the paths of the particles. Mach claimed that this showed that 'mass' could be defined in terms of 'position'. Although Mach may indeed have pointed out an important fact about the mass function and its role in classical particle mechanics, it is misleading to say that he showed that 'mass' or 'the mass-ratio', is explicitly definable in terms of 'position'. He merely showed that in a two element domain the mass ratio is uniquely determined by the position function. He did not show this to be the case for *all* domains. A more detailed discussion of these issues will be provided in Chapter VI.

Having considered these two formal questions connected with sentences of form (2), we can now return to our consideration of the adequacy of sentences of this form to the task of providing a logical reconstruction of theories of mathematical physics. Consider again the situation described in Chapter I. We are attempting to use 'is an S' to elucidate the logical structure of theory θ, where t is θ-theoretical and n is θ-non-theoretical. For each particular application i, we propose now to render the claim that θ applies to i as a sentence of form (2) where 'Q' denotes a possible partial model for S. We have already noted that this appears to offer a solution to the problem of theoretical terms. But this is not sufficient to assure us that a logical reconstruction of this sort will always be adequate. Still other difficulties may be encountered.

To see one difficulty, assume that we have been successful in rendering every claim that theory θ applies to i as a sentence of form (2). In this case, every empirical law of θ is a consequence of some statement made with a sentence of form (2). But remember now that the t-function was recognized to be θ-theoretical because every means of measuring t-function values, appearing in the existing exposition of the theory, presupposed the truth of some empirical law of θ. How might we expect this to be reflected in our logical reconstruction? Intuitively, the relation between the laws of θ and means of measuring t-function values is an important feature of the logical structure of the theory. Any logical reconstruction which did not provide a way of understanding this relation would clearly be inadequate. Thus we should naturally expect to be able to find claims of form (2) in our logical reconstruction which can be identified as justifying certain ways of measuring t-function values appearing in the existing exposition of θ. What we must seek first then is some way of characterizing those claims of form (2) which correspond to (entail) the laws in the existing exposition which justify ways of measuring t-function values. Second, we must explain, in some sense, how these claims of form (2) justify practitioners of the theory θ using these measured t-function values in the way they do.

The obvious line to take here is this. Some claims of form (2) *may* justify calculating certain t-function values from "observed" n-function values. This is to say that the t-function values required to satisfy some claims of form (2) might be uniquely determined by the n-function values. Thus, one who believed this claim to be true could infer what certain t-function values must be, from certain n-function values. One might plausibly expect that an existing account of a means of measuring t-function values could be construed this way – roughly, using a measuring apparatus believed to be a model for S, one calculates the t-function values needed to make it so from "observed" n-function values. We need not even insist that the t-function required to satisfy this claim of form (2) be unique. In a case in which the t-functions appearing in the existing exposition are recognized to be unique only up to a constant multiple, i.e. where the "t-property" is measured by a ratio scale, we should not expect to find means of measuring values of the t-function which determine more than ratios of t-function values. In this case, the situation we have already noted (example E1–b) where the ratio of certain t-function values is uniquely determined, might very well serve to justify a way of measuring t-function values which appear in the existing exposition of θ.

An example of the sort of measurement procedure one might hope to account for in this fashion is the measurement of mass-ratios by means of a "reaction-car" or "Fletcher trolley" experiment [31]. In this experiment, two "trolleys" rolling on the same track are connected by a spring. Their motion is assumed to be approximately frictionless and thus the entire system is assumed to be a model for classical particle mechanics with third law central forces. The mass ratio of the two trolleys is then calculated from their accelerations. Though this is not a "practical" means of determining mass ratios, it does serve as a clear illustration of line suggested in the preceding paragraph. One might also hope that more practically useful methods of measuring mass ratios, like the analytical balance, would yield to a similar analysis.

This seems to be a fairly plausible way to *identify* the claims of form (2) which justify ways of measuring t-function values. They are, roughly, those where the t-function values, or perhaps t-function ratios, are uniquely determined by n-function values. But can we further *explain* how those who believe some, or all, the claims of theory θ to be true are justified in using the measured t-function values in the ways they do? It is reasonably clear that there are theories of mathematical physics in which the measured values of theoretical functions are used in non-trivial ways to make predictions and test hypotheses. In classical particle mechanics, the measured values of masses and forces obtained from one application of the theory are commonly thought to be relevant to at least some other applications of the theory. For example, the mass-ratio of two bodies determined by an analytical balance are thought to be relevant to their behavior as members of a compound pendulum system. Force functions, like gravitational and electromagnetic forces determined by a Cavendish balance, are thought to be relevant to specifying the functions appearing in other applications of the theory. We use the measured values of forces and masses to predict the behavior of physical systems quite distinct from those systems employed in making the measurements. We use them to test hypotheses that certain force laws hold in certain situations. Can we account for these sorts of uses of measured values of theoretical functions within the framework of a logical reconstruction employing only claims of form (2)?

It appears evident that we cannot. To see this one has only to note that the t-function values calculated in the way we have sketched, have no significance beyond the single application of the theory used in their calculation. Suppose we use a claim of form (2) for $\langle D_i, n_i \rangle$ to justify calculating t-function values from the "observed" values of n_i. What we get is the t-function ratio needed to make $\langle D_i, n_i, t \rangle$ a model for S. Clearly, these t-function values tell us nothing about the n-function values on D_i that we did not know already. We can make no further prediction about other n-function values on D_i. This is not surprising. But more than this, these t-function values tell us nothing new about the n-function values on any other domain D_j. If we render *all* the claims that θ applies to i as claims of form (2), there is no reason to expect that the value of the t-function for individual x in D_i will be related in any way to the value of the t-function for this individual if it occurs in D_j. Thus, even if we have reason to believe some claim of form (2) to be true and can use it to calculate a t-function ratio, we could never use this ratio to make predictions about *other* situations in which we have reason to believe a claim of form (2), nor could we use it to test the hypothesis that θ applies in another situation. It is difficult to see any way that claims of form (2) alone could provide justification for using measured t-function values in applications beyond those employed in calculating them. Indeed, the calculation of t-function values must be regarded as a rather trivial enterprise on this account of the logical structure of θ.

One might suggest here that the difficulty we encounter in accounting for non-trivial uses of measured values for theoretical functions is occasioned not by the employment of sentences of form (2), but rather by the view that the empirical content of theory consists of more than one application claim. It might be suggested that there is really only *one* claim of form (2) needed to render the empirical content of the theory. This would appear to avoid the somewhat arbitrary and *ad hoc* aspect of the theoretical functions being so-to-speak "tailor made" for each application. This suggestion seems particularly appealing for theories like classical particle mechanics where there is a strong intuitive feeling that they are (or were) claimed to apply to "everything". One might think that the empirical content of classical particle mechanics could be rendered as a claim of roughly the following sort. Given the paths of all material particles in the universe, throughout all eternity, there exist mass and force functions which make these paths into a model for some axiomatization of classical particle mechanics, plus appropriate force laws. The appeal of this line is considerably diminished when one tries to see how to formulate it precisely. Even if this could be done, it is by no means obvious how one would employ a cosmic claim like this to account for rather pedestrian uses of the theory like predicting the results of billiard ball collisions. Further, there is no obvious way in which this claim would fare better than a variety of application claims in explaining how empirical laws of the theory function to justify ways of measuring values of theoretical functions and the uses to which these values are put.

Another suggestion is that we might modify the Ramsey sentence slightly by claiming that all possible partial models *of a certain kind* are extendible to models for S. This is, in an obvious way, more like the claims found in empirical theories. After all, an analytical balance is a *kind* of physical system, not one particular physical system. Indeed, much of our intuitive discussion has been carried on as if we were considering a generalized Ramsey sentence of this sort. However, it should be clear that the situation for the generalized sentence is no different than for the singular. If we employ a physical system of a certain kind to measure a t-function value, we would surely require that these measured values be applicable to some physical systems of other kinds. The generalized Ramsey sentence will not provide for this.

The difficulty with the Ramsey solution to the problem of theoretical terms is that roughly speaking, it solves only half of the problem. It shows us how the theoretical terms figure in the empirical claims of the theory even though they are not "interpreted". It shows us how they can be employed to rule out certain observable states of affairs as being inconsistent with the truth of the theory's claims. Indeed, there is some reason to think that there are cases where this work could not be done without theoretical terms, i.e. when they are not Ramsey eliminable. However, the Ramsey solution fails to account for the fact that statements made with theoretical terms (values of theoretical functions) are relevant to prediction and hypothesis testing. It fails to account for the fact that we are frequently concerned with determining, by "empirical" means, the truth-value of statements containing only theoretical terms, and that we employ such statements, in non-trivial ways. We are not merely interested in whether it is *possible* to "fill in" theoretical relations or functions behind the observed facts. We are frequently interested in exactly what theoretical relations are required to do this. We make use of this information in contexts other than that in which it was obtained. The Ramsey solution offers us no account of why we should have this sort of interest in theoretical relations or functions, nor of the use we make of this information.

In the next chapter we will examine another sentence constructed with 'is an S'. This sentence will retain all the virtues of the Ramsey sentence (2) in avoiding the problem of theoretical terms. But it will appear to be more adequate than the Ramsey sentence to the task of logical reconstruction in that it provides a possibility of accounting for non-trivial measurements of values of theoretical functions.

NOTES

[1] Ramsey ([35], p. 219) considers a notion of eliminability very close to what I call 'Ramsey eliminability'. However, I do not wish to imply, by using his name, that this is precisely what Ramsey had in mind. A brief discussion of Ramsey's view appears below.
[2] Capital letters with bars above denote predicates. The same letter without the bar denotes a relation assigned to this predicate.
[3] I owe these examples to Professor Dana Scott. For pointing out the errors of an earlier treatment, I am indebted to Professor Herbert Simon and Dr. Raimo Tuomela.
[4] The proof of this is provided by mathematical induction on the "length" of sentences containing $\bar{O}$ and identity.

CHAPTER IV

THE RAMSEY VIEW EMENDED

A third proposal for using set-theoretic predicates to make empirical claims is considered in this chapter. This proposal is an emendation of the Ramsey view considered in the preceding chapter. It succeeds in avoiding the problem of theoretical terms in the same way as the Ramsey solution. The question of eliminability of theoretical terms arises in essentially the same way. However, this proposal goes beyond the Ramsey solution in providing an account for non-trivial measurements of theoretical function values. In addition, this proposal provides the key to a precise understanding of some other features of theories of mathematical physics – for example, the role of crucial experiments in the rejection of such theories.

We have noted that the Ramsey solution to the problem of theoretical terms is defective in that it fails to provide an adequate means of accounting for the use that is made of measured, or calculated, values of theoretical terms. There appears to be no way of providing a rationale for the practice of exploiting values of theoretical functions obtained in one application of the theory to draw conclusions about other applications of the theory. The source of this defect is readily seen to be the independence of the theoretical functions (in the claims of form (2)) associated with different applications of the theory. This failure to account for the non-trivial use of calculated values of theoretical functions is perhaps ultimately the source of the strong intuitive feeling that the theoretical functions introduced to produce a model for the mathematical formalism can not be so completely *ad-hoc* as the Ramsey solution makes them. We feel that there must be some relation among the theoretical functions used in different applications of the theory to satisfy the claim of form (2).

Recognizing these facts, an obvious suggestion is this. We should simply agree that a collection of claims of form (2), one for each application, is not an adequate rendering of the empirical content of the theory in question. In addition to these claims, there is the further claim that some relations will hold among the various different theoretical functions produced to satisfy these claims of form (2). This is to say, we introduce, as an additional part of the empirical content of the theory, the claim that the entire array of theoretical functions which satisfy the Ramsey claims of form (2) also satisfy certain other *constraints*. These constraints will require that certain relations hold among the values of theoretical functions employed in different applications of the theory. The particular nature of the constraints might naturally be expected to differ for different theories. A part of clarifying the logical structure of a particular theory would be ferreting out and making precise the constraints relevant to this theory. This task would be on a par with the task of clarifying the mathematical structure of the theory. Both would be required to provide a precise statement of the empirical content of the theory.

We want now to examine the possibility that this suggestion, when carried out in detail for a particular theory, will yield an adequate logical reconstruction of that theory. The strategy will be the same as in the previous two chapters. We will first attempt to formulate precisely, using our example of a set-theoretic predicate 'is an S', a claim of the form suggested. Then we will try to see what could be expected from this claim, viewed as *the* central empirical claim of some theory of mathematical physics.

Our first task is to find some general way of formulating a claim that the array of t-functions used to satisfy a number of claims of form (2) also satisfy certain constraints. 'General' in the sense that it will work for anything we might conceivably want to regard as a constraint. To make clear what it is we are seeking, and to give some intuitive content to the notion of a constraint, it seems appropriate to begin by giving some examples of ways in which the array of t-functions employed in different applications might be constrained. These examples will then serve to suggest an appropriate general format for constraint claims.

Let D_i be the set of individuals associated with application i of the theory in question, and let

$$\bar{D} = \{D_1, D_2, \ldots, D_i, \ldots\}$$

be the (perhaps infinite) set of domains of individuals, each associated with some application of the theory. (Subsequently, we shall occasionally refer to members of $\bar{D}$ as 'intended applications of the theory'.) Note that there is no reason to expect that the intersections of members of $\bar{D}$ will be empty. The same individual *may* appear in several different applications of the theory. Indeed, some members of $\bar{D}$ may be sub-sets of others. Let

$$\varDelta = \bigcup D_i, D_i \in \bar{D}.$$

Intuitively, $\varDelta$ is the set of all individuals that the theory "talks about". It is important to realize that $\varDelta$ need not be a member of $\bar{D}$, though it *may* be. This is to say, there need be no *single* application of the theory which has $\varDelta$ as its domain of individuals. It is at least possible to claim that a theory of the sort we are considering applies to D_i and D_j without claiming that it applies to the union of D_i and D_j. This is not to say that the algebraic structure of $\bar{D}$ is irrelevant to the empirical content of the theory in question. It may be that, for some theories, properties that $\bar{D}$ has, e.g. closure under union and intersection, are useful in specifying the membership of $\bar{D}$ and thus play a role in making the empirical claims of the theory. But there are no reasons for assuming, in general, that $\bar{D}$ has some particular algebraic structure.

Now, for all $D_i \in \bar{D}$, let t_i be a function from D_i into the real numbers and let

$$\bar{t} = \{t_1, t_2, \ldots, t_i, \ldots\}.$$

Keeping in mind that each member of $\bar{t}$ is alleged to satisfy a claim of form (2) for its domain, let us consider what we might count as constraints on $\bar{t}$.

First, suppose that we simply require that

$$t = \bigcup t_i, t_i \in \bar{t}$$

be a *function* from $\varDelta$ into the real numbers. This amounts roughly to this. If the same individual appears in more than one application of the theory, then we are constrained to assign the same t-function value to this individual in every application, i.e. in every domain, in which it appears. This constraint reflects the intuitive idea that the t-function measures an "intrinsic" property of the individuals. That is, it measures a property of the individuals which remains unchanged as they are (perhaps) moved about in space and time to become participants in various applications of the theory.

In addition to requiring that t be a function, we might also place certain restrictions on its values. One way of doing this is the following. Suppose that there is a binary operation $\circ$, defined on $\varDelta$. We might constrain $\bar{t}$ by requiring that, for all $x, y \in \varDelta$

$$t(x \circ y) = t(x) + t(y).$$

This amounts to requiring that the t-function be an "extensive quantity" with respect to the operation ∘. Intuitively, ∘ might be regarded as some kind of concatenation operation on Δ. However, in stating a restriction of this sort, we need not specify any of the formal properties of the ∘-operation. We need only presuppose that truth conditions for statements involving this operation are known. Of course, it will be obvious, for certain operations, that this restriction could not possibly be satisfied. For example, if the operation were known to be idempotent it would be evident that the restriction could not be satisfied since t-function values must be positive (D1–4). In such cases, the claim that $\bar{t}$ satisfies this constraint is simply false and this may be regarded as an empirical fact about the ∘-operation.

Roughly, the claim that we want to make with these constraints is this. Given the set of domains $\bar{D}$ and the n-functions on each domain, there is a function t defined on Δ and extensive with respect to the ∘-operation which determines a set of functions $\bar{t}$ on the members of $\bar{D}$ such that each $\langle D_i, t_i, n_i \rangle$ is a model for S. That is, we want to claim that we can extend each $\langle D_i, n_i \rangle$, $D_i \in \bar{D}$, to produce a model for S, and moreover that we can do it in such a way that the t-function can be regarded as measuring an intrinsic property of the individuals and is extensive with respect to the ∘-operation.

This is an example of the sort of claim to be examined in this chapter. The particular constraints in the example have not been chosen arbitrarily. They are intended to be such as to make the t-function play a role roughly analogous to that of the mass-function in mechanics. By describing briefly, what I take this analogy to be, I hope to make it at least plausible that such a claim might be the central empirical claim of some physical theory. A detailed discussion of the role of the mass function, from this point of view, will be found in Chapter VI.

I want to suggest that the role of the mass function in the empirical claims of classical mechanics is very roughly this. We "postulate" that there is a property of all physical objects – mass – which is extensive with respect to concatenation, and that the function measuring this property allows us to make the motions of all physical systems into models for classical particle mechanics. This view of the mass function is quite similar to that of Pendse and Simon mentioned in the preceding chapter. There is however, an important difference. On their view, the fact that mass is extensive with respect to concatenation would be a straightforward empirical fact, presumably discovered by calculating values of the mass function in particular cases.[1]

On the view suggested here, we do not simply discover that the mass function happens to be extensive. Rather, we postulate that it is so. That is, we claim there is an extensive mass function that allows us to regard the motions of all bodies as models for classical particle mechanics. We seek to verify this claim by actually exhibiting such a mass function. In doing this we may calculate particular values for the mass function. But, if such a calculation yields values that are not extensive, it appears, at least in some cases, possible to question whether we are really measuring "mass". Indeed, I would go so far as to suggest that there is no single, crucial experiment which would prove conclusively that mass was not extensive. Rather, a number (perhaps a large number) of unsuccessful attempts to find an extensive mass function might convince us that our claim that such a function exists is false.

It must be emphasized that this view of the role of the mass function in classical particle mechanics is in no way incompatible with the fact that different systems of units are employed to measure mass. Typically there will be more than one array of t-functions that satisfy a claim of the sort we are suggesting. This is surely the case with the mass function. Each one of these different mass functions that satisfy this claim corresponds to a different system of units for measuring mass. The claim we are suggesting is, roughly, that there is *at least* one scale for measuring mass that "works" in all applications of the theory and in which mass is extensive with respect to particle concatenation.

It is easy to see that the t-function is playing a role in our example somewhat like that I have attributed to the mass function in classical mechanics. That the t-function is extensive with respect to the ∘-operation is not a straight-forward empirical fact. We do not discover this by "directly" measuring t-function values, nor even by calculating t-function values required to satisfy some Ramsey claim of form (2). Rather, we claim that we can find some extensive (with respect to ∘) t-function which makes every domain in $\bar{D}$ a model for our theory. Very shortly, we will see that viewing our claim as the central empirical claim of the theory provides a way of accounting for non-trivial calculations of t-function values. Further, it will be seen that on this account, it might happen that no one experiment could conclusively prove that the t-function was not extensive. This is again analogous to what I claim to be the case for the mass function, and indeed this account will serve to clarify those situations where we allegedly might throw out the calculations, rather than give up to claim that mass is extensive. Thus, if my sketch of the role of the mass function is even remotely plausible, there is some reason to think that claims like our example are worth investigating.

One feature of the view of the logical structure of theories of mathematical physics now being suggested needs to be made explicit. On this view, the empirical content of the theory is essentially only *one* claim. The claim is roughly this. Values of theoretical functions can be found to make the observed facts in a variety of situations models for the mathematical structure associated with the theory, and moreover, these values satisfy certain constraints in addition to those imposed by this mathematical structure. In contrast to the present view, one might take the empirical content of a theory to be a conjunction of Ramsey claims of form (2). In this latter case, one could envision *parts* of the theory – different Ramsey claims – being accepted or rejected independently of one another. On the present view, such piecemeal acceptance or rejection of the theory is apparently not possible. One must accept or reject the claim of the theory *in-toto*.

This feature of the view suggested here appears, at least superficially, to be in accord with certain "holistic" views of scientific theories – views like that propounded by Duhem [10] for physical theories and, more recently by Kuhn [21] for scientific theories in general. One tenet, among others, of these views is that acceptance or rejection of a theory is an all-or-nothing decision. It will be argued later that this resemblance is more than superficial. For theories of mathematical physics, at least, the view suggested here seems to provide us with a way of clarifying some central tenets of the holistic view. To the extent that these tenets are based on sound insights into the nature of such theories, this will provide some reason to believe that the view of logical structure suggested here is correct.

It is reasonably clear that we should allow for the possibility of constraints different from those in this example. In particular, there are functions appearing in theories of mathematical physics which do not purport to measure "intrinsic" properties of the individuals. That is, the values of these functions for a particular individual might be expected to differ as the individual appears in different applications of the theory. Some of these functions might be theoretical with respect to the theory in question. The force function in classical mechanics is an example of such a function.

We do not expect that the forces acting on a particle will remain the same in all applications of the theory in which this particle appears. Entropy in classical thermodynamics is another example. Even though we do not expect an individual to have the same value for these functions in all applications in which it appears, it may still not be possible to assign values to these functions in one application with complete disregard for the values assigned in some other application. For example, we usually assume that the elastic properties of bodies are roughly constant and this puts certain constraints on the values that we may assign to the force function in various applications of classical mechanics. This suggests that it might be possible to account for the role of the force function along the line being pursued here. But to do so, it is clear that our notion of a constraint must allow for the possibility of constraints on functions which do not measure "intrinsic" properties.

The explicit nature of the constraints that one might plausibly take to be put on the force function appears to be somewhat more complicated than for the mass function. I know of no way to construct analogous constraints on the t-function in our example. However, one can easily think of constraints which do not require that individuals have the same t-function value in all domains in $\bar{D}$. For example, one might require this. If $D_i \subset D_j$ then, for all $x \in D_i$, $t_i(x) = t_j(x)$. This would allow that D_i and D_j intersect and yet individuals in their intersection have different values for t_i and t_j. Another example is this. Require that the product of the t-function values for individual x in every domain in which x appears be equal to some constant C. The first example may be roughly similar to what one would require who wished to claim that the theory in question applies to all "sub-systems" of any physical system to which it applies. The second example has no apparent intuitive similarities with any feature of a familiar physical theory.

With these examples in mind, we can make the notion of a constraint more precise with the following definition.

(D4) If $\bar{D} = \{D_1, D_2, \ldots, D_i, \ldots\}$ is an ordered set of non-empty sets, $\Delta = \bigcup D_i$, $D_i \in \bar{D}$ and $\bar{t} = \{t_1, t_2, \ldots, t_i, \ldots\}$ is an ordered set of functions such that $t_i \in \bar{t}$ is a function from $D_i \in \bar{D}$ into the real numbers then $\bar{t}$ *is constrained by* $\langle R, \rho \rangle$ if and only if

(1) R is an n-ary relation whose field is Δ,
(2) ρ is an n-ary relation whose field is the real numbers; and

(3) If $R(x_1, x_2, \ldots, x_n)$ and $x_j \in D_{t_j}$; $D_{t_j} \in \bar{D}$, $1 \leq j \leq n$
$\rho(t_{i_1}(x_1), t_{i_2}(x_2), \ldots, t_{i_n}(x_n))$.

It is not difficult to see how our previous examples of constraints can be cast into the form characterized by (D4). We assure that t is a function on Δ by requiring that $\bar{t}$ be constrained by $\langle =, = \rangle$. We assure that t is extensive with respect to the $\circ$-operation by requiring in addition that $\bar{t}$ be constrained by $\langle R_0, \rho_+ \rangle$ where $R_0(x, y, z)$ if and only if $x \circ y = z$ and $\rho_+(x, y, z)$ if and only if $x + y = z$. The constraint that requires t-function values on a domain that is a sub-set of D_j to be the same as the t_j values of its members can be expressed as follows. For all integers i and j, let $R_{ij}(x, y)$ if and only if $x = y$, $D_i \subset D_j$ and $x \in D_i$. Then we get the desired constraint by requiring that $\bar{t}$ be constrained by $\langle R_{i_j}, = \rangle$ for all values of i and j.

Subsequently we shall call the ordered pair $\langle R, \rho \rangle$ 'a constraint on $\bar{t}$', keeping in mind that more than one such ordered pair may be required to characterize what we have intuitively recognized as a constraint. Instead of saying that $\bar{t}$ is constrained by $\langle R, \rho \rangle$, we shall sometimes say equivalently that $\bar{t}$ satisfies the constraint $\langle R, \rho \rangle$.

Later (Chapter VII, (D27)) we will provide a somewhat more general characterization of the notion of a constraint. It is useful to note now that the notion of a constraint defined by (D4) has the following two properties. First, the sets of models for 'is an S_0' the t-functions of which satisfy some constraint $\langle R, \rho \rangle$ contain all possible configurations of n-functions. That is, until we require that some relation between t and n-functions hold, constraints on t-functions do not restrict the possible n-function values. This is to say, the constraint is *on* the t-function and not *on* the n-function. Second, given any constraint $\langle R, \rho \rangle$, and any t-function, say t_i, there is *some* set of t-functions containing t_i which satisfies $\langle R, \rho \rangle$. This is just to say that constraints are on *sets* of t-functions, rather than on the functions themselves. It is these two features of constraints that we will subsequently exploit to provide a more general characterization.

To describe precisely the general form of the claims we have been considering let us introduce the following notation. For any set $\bar{Q}$ of possible partial models for S, we may define the predicate 'is an $E_{\bar{Q}}^-$' in the following way.

(D5)   If $\bar{Q} = \{\langle D_1, n_1 \rangle, \langle D_2, n_2 \rangle, \ldots, \langle D_i, n_i \rangle, \ldots\}$ and $\langle D_i, n_i \rangle$ is a P_0 then $\bar{x}$ is an $E_{\bar{Q}}^-$ if and only if there exists a $\bar{t}$ such that

(1) $\bar{t} = \{t_1, t_2, \ldots, t_i, \ldots\}$ and t_i is a function from D_i into the real numbers;

(2) $\bar{x} = \{\langle D_1, t_1, n_1 \rangle, \langle D_2, t_2, n_2 \rangle, \ldots, \langle D_i, t_i, n_i \rangle, \ldots\}$

That is, if $\bar{x}$ is an $E_{\bar{Q}}$ then $\bar{x}$ is a set of extensions of the possible partial models in $\bar{Q}$.

Let us also define the two predicates 'is an $\bar{S}$' and 'is an $\bar{S}_0$' in the following way.

(D6) $\bar{x}$ is an $\bar{S}(\bar{S}_0)$ if and only if there exists a

$\bar{D} = \{D_1, D_2, \ldots, D_i, \ldots\}$
$\bar{n} = \{n_1, n_2, \ldots, n_i, \ldots\}$
$\bar{t} = \{t_1, t_2, \ldots, t_i, \ldots\}$

such that:

(1) $\bar{x} = \{\langle D_1, n_1, t_1 \rangle, \langle D_2, n_2, t_2 \rangle, \ldots, \langle D_i, n_i, t_i \rangle, \ldots\}$;

(2) for all $x_i = \langle D_i, t_i, n_i \rangle \in \bar{x}$, x_i is an $S(S_0)$.

Thus $\bar{x}$ is an $\bar{S}$ says that $\bar{x}$ is a set of models for S and $\bar{x}$ is an $\bar{S}_0$ says that $\bar{x}$ is a set of possible models for S.

If $\bar{x}$ is an $E_{\bar{Q}}$, let us agree that '$C(\bar{x}, R, \rho)$' means that the set of t-functions $\bar{t}$ – the set of second members of members of $\bar{x}$ – is constrained by $\langle R, \rho \rangle$.

Consider now a sentence of the form:

(3) $(\exists \bar{x})[\bar{x}$ is an $E_{\bar{Q}} \wedge \bar{x}$ is an $\bar{S} \wedge C(\bar{x}, R, \rho)]$.

where $\bar{Q}$ is a set of possible partial models for S and $\langle R, \rho \rangle$ is a constraint on functions defined on the domains of the members of $\bar{Q}$. Intuitively, (3) says that there is some set of extensions of members of $\bar{Q}$ all of which are models for S and, moreover, the set of t-functions used to produce this set of extensions satisfies the constraint $\langle R, \rho \rangle$. If we regard '$\bar{Q}$' as a description of the observed facts about a number of situations, then (3) says that these observed facts may be "filled out" – in some way which satisfies $\langle R, \rho \rangle$ – to make each situation a model for S.

As before, we want to ask this question. Is it reasonable to expect that sentences of form (3) might be used to make empirical claims? Our examples have suggested that they might. Indeed, it was suggested that *one* sentence of form (3) might be used to make *the* central empirical claim of a theory of mathematical physics. We want now to follow up these suggestions by giving a detailed account of how such a claim might be expected to function in a scientific theory. This seems most expediently done by beginning with the simplest possible example of a claim of form (3). Consider a case where $\bar{D}$ has a finite number of members – r, i.e. there are only r of intended applications of the theory. Let the only constraint be $I = \langle =, = \rangle$, i.e. we demand only that the t-function measure an intrinsic property of the individuals. In this case, call the claim of form (3), for some $\bar{Q}$, '(3I)'.

What does claim (3I) entail about the values of the n_i functions on members of $\bar{D}$? First, note that any claim of form (3) entails a claim of form (2) for each $Q_i = \langle D_i, n_i \rangle \in \bar{Q}$. In addition, the set of functions $\bar{t} = \{t_1, t_2, \ldots, t_r\}$, members of which satisfy these claims of form (2), must be such that $t = \bigcup t_i$ is a function on $\Delta = \bigcup D_i$. Let t be any function from Δ into the real numbers. What properties must t have in order that its values on members of $\bar{D}$ are t-function values satisfying (3I)? Clearly, it is necessary and sufficient that its values satisfy (D1–4) and (D1–5) for each member of $\bar{D}$.

For (D1–5), this means that it is necessary that t-values satisfy a set of r homogeneous linear equations in N unknowns:

(δ) $$\sum_{[x \in D_i]} n_i(x) t(x) = 0, \ 1 \leq i \leq r$$

where N is the number of distinct individuals in Δ. (Remember, that the n-function values are "given" and that $n_i(x)$ need not equal $n_j(x)$.) But (D1–4) entails that the t-function values be strictly greater than zero. This means that the t-values must be non-trivial solution to (δ). A necessary and sufficient condition for the existence of a non-trivial solution to (δ) is the following:

(A) The rank of the matrix;
$M = (v_i(x_j)), \ 1 \leq i \leq r; \ 1 \leq j \leq N$
where $v_i(x_j) = n_i(x_j)$ if $x_j \in D_i$,
and $= 0$ otherwise;
is less than r.

Thus (A) is a *necessary condition* for the existence of a t-function on Δ satisfying (D1–4) and (5) for all D_i, and thus a necessary condition for the truth of (3I).

But (D1–4) also entails that:

(B) For all $\langle D_i, n_i \rangle \in \bar{Q}$, not all $n_i(x)$, $x \in D_i$ have the same sign.

Thus (B) is a further necessary condition for the truth of (3I). It is also clear that (A) and (B) together are sufficient for the truth of (3I). That is '(A) and (B)' is logically equivalent to (3I).

It is interesting to note that (A) and (B) are stated entirely in terms of conditions on n-function values. This suggests that a concept of eliminability of theoretical functions from claims of form (3), analogous to Ramsey eliminability, might be fruitfully introduced. Let us consider this possibility briefly.

Let us continue to assume that $\bar{D}$ has r members. As in the previous chapter, let 'is an S^i' be defined by some restriction of (D0), and let 'is an $\bar{S}^i$' be defined as in (D6). Let $V(S^i, \langle R, \rho \rangle)$ be the set of all r-tuples of the form $\bar{Q} = \langle \langle D_1, n_1 \rangle, \langle D_2, n_2 \rangle, \ldots, \langle D_r, n_r \rangle \rangle$ where $D_j \in \bar{D}$ and $\langle D_j, n_j \rangle$ is a possible partial model for S^i and such that

(3^t) $(\exists \bar{x})[\bar{x} \text{ is and } E_{\bar{Q}} \wedge \bar{x} \text{ is an } \bar{S}^i \wedge C(\bar{x}, R, \rho)]$

is true. Then raise this question. Under what circumstances, i.e. for what S^i's, does there exist a restriction of (D2), defining 'is a P^i', and a constraint $\langle \hat{R}, \hat{\rho} \rangle$ on $\bar{n} = \{n_i, n_j, \ldots, n_r\}$ such that $V(S^i, \langle R, \rho \rangle) = U(P^i, \langle \hat{R}, \hat{\rho} \rangle)$ where $U(P^i, \langle \hat{R}, \hat{\rho} \rangle)$ is the set of all r-tuples of the form $\bar{Q}$ such that

(4^t) $\bar{Q}$ is a $\bar{P}^i$ and $C(\bar{Q}, \hat{R}, \hat{\rho})$

is true ('is a $\bar{P}^i$' being defined in a way analogous to (D6) and '$C(\bar{x}, \hat{R}, \hat{\rho})$' meaning that $\bar{n}$ satisfies $\langle \hat{R}, \hat{\rho} \rangle$)? That is, for what S^i's can we find a predicate which involves only the n-function and a constraint on n-function values such that (4^t) determines the same set of r-tuples of form $\bar{Q}$ as does (3^t)? For S^i's where this could be done we might naturally say that the t-function was *Ramsey eliminable* from sentences of form (3^t).

It seems clear that t is Ramsey eliminable from our example (3I). To show this explicitly, one would have to cast (A) in the form of a constraint on $\bar{n}$. This is tedious, but not difficult. It should be noted that t's being Ramsey eliminable from sentences of form (2) does not entail that it is likewise Ramsey eliminable from all sentences of form (3) containing the same predicate. Intuitively, the constraints on the n-function may not be describable without reference to the t-function. The notion of Ramsey eliminability here is obviously subject to the same limitations on precise formulation as previously.

It should also be noted here that we have only Ramsey eliminated the t-function from (3^t) by employing a claim (4^t) which places constraints on the n-functions. We could have demanded that 'is a P^i' be such that $V(S^i,$

⟨R, ρ⟩) = the set of all sub-sets of the set of models for 'is a P^i'. That is we could have demanded that we find some restriction of (D2) which "picked out" the same sets of models as (3^i) without using *either* theoretical functions *or* constraints on the non-theoretical functions. It should be clear that we can not meet this demand. Any restriction we place on the *n*-function is going to rule out *some* set of models that (4^i) allows. This follows from the second feature of constraints we noted – that constraints are on sets of functions rather than functions themselves. Some question might be raised as to whether it is intuitively acceptable to formulate the central empirical claim of a theory by placing constraints on the non-theoretical functions. Were it the case that this was intuitively unacceptable, then (4^i) would not count as a sentence which Ramsey eliminates the *t*-function from (3^i) and indeed no such sentence could be found. We shall consider the question of the intuitive acceptability of formulations involving constraints on non-theoretical functions further in Chapter VII. For the time being, it suffices to note that the effect of rather simple constraints on theoretical functions has only been reproduced by employing quite complex constraints on the non-theoretical functions. This is surely *a* good reason for resorting to theoretical functions, even if one *could* get on without them.

A second point to note about (A) and (B) is that they say something about every member of $\bar{Q}$. In particular, there is an equation in (δ) for each member of $\bar{Q}$. However, both (A) and (B) entail a similar condition for every sub-set of $\bar{Q}$. This fact is important to the understanding of subsequent examples. Further, it suggests how we might extend this treatment to a $\bar{Q}$ with an infinite number of members. We just require that conditions like (A) and (B) hold for every finite sub-set of $\bar{Q}$.

To see more clearly what (A) and (B) amount to, consider three cases:

(i) $N > r$;   (ii) $N = r$;   (iii) $N < r$.

In case (i), the rank of M is always less than r, thus there is always a non-trivial solution to (δ) and (B) is the only significant requirement on the values of the *n*-functions. In case (ii), the rank of M is less than r if and only if the determinant

$$|v_i(x_j)| = 0$$

While in case (iii), the rank of M is less than r if and only if every $r \times r$ determinant formed by deleting rows and columns from M is equal to zero. Thus in cases (ii) and (iii), (A) entails that some arithmetic relation hold among the values of the n_i-functions.

What this means is roughly the following. In case (i), where we have more individuals than intended applications, (3I) is simply equivalent to a conjunction of claims of form (2) – one for each intended application. The additional constraint places no additional restrictions on n-function values and is thus irrelevant, from the point of view of excluding possible observed states of affairs from the class of those consistent with the empirical content of the theory. In case (ii) and (iii), the situation is different. There might be some set of possible practical models for S, $\bar{Q}$, – i.e. some set of observed states of affairs – which satisfied (B) and yet the n-function values failed to satisfy the conditions on M. Thus, in these cases, (3I) is saying more about n-function values on members of $\bar{D}$ than can be said by a conjunction of claims of form (2). Roughly, it is saying that the n-function values in different intended applications of the theory must be related in certain ways.

It should be noted that in (ii) and (iii) there must be *some* sharing of individuals among applications of the theory (in (ii)) because no one-element domain can satisfy a claim of form (2). Intuitively, it is this overlapping of intended applications that allows our constraint to put restrictions on n-function values. In case (i), the domains of various intended applications *may be* mutually disjoint. If so, it is obvious that our constraint on t-function values is totally vacuous. There may also be sharing of individuals in this case too. But the overlapping here is not sufficient to place any restriction on n-function values. Our constraint, though not totally vacuous, is vacuous in so far as n-function values are concerned. These facts suggest that the particular structure of the set of intended applications $\bar{D}$ may be quite significant in determining the "empirical import" of any particular constraint on the t-function.

It is also worth noting here that the predicate 'is an S' might be such that a claim of form (2) was "vacuously" true, in the sense that *any* Q could be extended to a model for S. Even in this case a claim of form (3) might not be "vacuous". It might be that the necessity to satisfy the constraints rules out the possibility of finding t-functions to satisfy (3) for some $\bar{Q}$, even though it is possible to find some t-function to satisfy (2) for each member of $\bar{Q}$. This possibility will prove to be relevant to a consideration of classical particle mechanics.

Thus far we have seen that (3I) can place restrictions on sets of possible

78   THE LOGICAL STRUCTURE OF MATHEMATICAL PHYSICS

partial models beyond those that can be placed on them by conjunctions of claims of form (2). This is certainly enough to make (3I) a candidate for an empirical claim of some theory. But we have yet to see explicitly how taking (3I) as *the* central empirical claim of a theory might fill the gap we noted in the Ramsey solution. How can (3I) serve to justify non-trivial calculations of t-function values? This can be best understood by looking at some simple examples where $\bar{D}$ contains sets with small numbers of members. To understand these examples, one must remember that a *necessary* condition for the truth of (A) and (B) is that analogous conditions hold for any *sub-set* of $\bar{Q}$.

(E3–a) $\bar{D}$ contains $D_1=\{x, y\}$, $D_2=\{x, z\}$ and $D_3=\{y, z\}$. (It may contain more).
(3I) entails that

(i) $\dfrac{n_1(x)}{n_1(y)} \cdot \dfrac{n_3(y)}{n_3(z)} = -\dfrac{n_2(x)}{n_2(z)},$

since $r=N=3$ and (i) follows immediately from (A). Alternatively, (i) may be arrived at in this way.
Remember ((E1–b)) that claim (2) for D_1 entails

(ii) $\dfrac{n_1(x)}{n_1(y)} = -\dfrac{t_1(y)}{t_1(x)}$

and corresponding results for $i=2, 3$. If we assume these t-functions satisfy $\langle =, = \rangle$, we have

$t_1(y) = t_3(y)$
$t_1(x) = t_2(x)$
$t_2(z) = t_3(z),$

and consequently that

(iii) $\dfrac{t_1(y)}{t_1(x)} \cdot \dfrac{t_3(z)}{t_3(y)} = \dfrac{t_2(z)}{t_2(x)},$

which leads directly to (i).

The second way of arriving at (E3–a–i) is, of course, just a retracing, for a particular case, of the general arguments which lead to (A). But this way of arriving at it illustrated one way claim (3I) might function in a theory. Take any two-element domain in $\bar{D}$, D_i, together with n_i, the given n-function on that domain, and discover whether or not a t-function, t_i, can be found to make $\langle D_i, t_i, n_i \rangle$ a model for S. This simply amounts to checking to see if the n_i values are either both zero or of opposite sign. If the n_i values are not both zero, we can go further and calculate what the ratio of the values of the function t_i must be in terms of the values of the function n_i. Having done this for two two-element domains (D_1 and D_3 above) which contain one member (y) in common, we can then calculate the ratio of the t-function values for a third two-element domain (D_2) which contains one member in common with each of the two initial domains. This calculation is justified by the fact that all three domains belong to a $\bar{D}$ for which claim (3I) is assumed to be true. We can then use the t-function ratio for D_2 to calculate the perhaps unknown, but observable, n-function ratio for D_2.

(E3–b) $\bar{D}$ contains $D_1=\{x, z\}$, $D_2=\{y, z\}$, $D_3=\{x, y, z\}$. (3I) entails that the n-functions on D_1, D_2, and D_3 are such that

(i) $n_3(z) = \dfrac{n_2(z)}{n_2(y)} n_3(y) + \dfrac{n_1(z)}{n_1(x)} n_3(x),$

for $r=N=3$ and (i) follows from (A). We may also arrive at (i) in this way. If there is a function t_3 such that $\langle D_3, t_3, n_3 \rangle$ is a model for S then

(ii) $n_3(z) = \left(-\dfrac{t_3(y)}{t_3(z)}\right) n_3(y) + \left(-\dfrac{t_3(x)}{t_3(z)}\right) n_3(x).$

If (3I) is true, we may calculate the values of $t_3(y)/t_3(z)$ and $t_3(x)/t_3(z)$ in terms of the values of the n-functions on D_1 and D_2 and, by substituting these into (ii), obtain (i).

One might plausibly say of this situation that the domains D_1 and D_2 provide us with an opportunity for "observing" the ratio of the t-function values, for y and z and x and z respectively. The claim (3I) "justifies" these methods of observing t-function values in the following sense. If (3I) is true then the claim of form (2) is true of both $\langle D_1, n_1 \rangle$ and $\langle D_2, n_2 \rangle$. This alone justifies the use of the "laws of the theory" – in this case (D1–5) – to calculate the values of the t-function ratios. The fact that we can make such a calculation does not alone warrant saying we have observed t-function values. So far as D_1 and D_2 are concerned, these calculations are trivial. The calculated values tell us nothing we do not already know about n_1 and n_2. But (3I) makes these values significant in a way they could never be if we only made a claim like (2) for each member of $\bar{D}$. The truth of (3I), justifies our employing these "observed values" of the t-function ratios to make predictions about the n-function values for D_3. For example, if we knew $n_3(x)$ and $n_3(y)$ these ratios would allow us to predict $n_3(z)$. This relevance of the "observed values" to domains distinct from those in which they were calculated appears to be precisely what is required to warrant calling these calculations 'observations'.

We are not limited to the use of two-element domains to calculate t-function ratios. Let $\bar{D}_1 = \{D_1, D_2, \ldots, D_r\}$ to be a sub-set of $\bar{D}$ and let $\Delta' = \bigcup D_i$, $D_i \in \bar{D}'$. Then the ratios $t(x_i)/t(x_1)$, $1 \le i \le n$, where n is the number of members of Δ', are determined uniquely by the n-function values for members $\bar{D}'$ if and only if

(C) The rank of the matrix

$(v_i(x_j))$, $1 \le i \le r'$; $1 \le j \le n$

is equal to the rank of

$(v_i(x_j))$, $1 \le i \le r'$; $2 \le j \le n$.

An analogous condition can be formulated for the ratios $t(x_i)/t(x_j)$ for any value of j, $1 \le j \le n$, i.e. for any $x_j \in \Delta'$. In general, these conditions can only be satisfied – the ranks of the appropriate matrices can only be equal – if $n-1 \le r'$, i.e. only if the domains in $\bar{D}'$ overlap. But they will not *necessarily* be satisfied when such overlapping occurs. Whether they are still depends on the particular values of the n-function.

From this, it is clear that, for some $\bar{D}$'s, there may be several possibilities for calculating the same t-function ratio. There may be more than one sub-set of $\bar{D}$ like $\bar{D}'$ in which a condition like C is satisfied which guarantees that some of $t(x_i)/t(x_j)$ are uniquely determined. For example, if $D_1 = \{x, y\}$, $D_2 = \{x, y, z\}$ and $D_3 = \{y, z\}$ are in $\bar{D}$ then both D_1 and $\bar{D}' = \{D_2, D_3\}$ could, provided the n-function values were suitable, yield a method of calculating $t(x)/t(y)$. In general, whether there is *any* method of calculating a particular t-function ratio and how many different methods there are, will depend upon the way the members of $\bar{D}$ share individuals.

With these examples in mind, we can begin to understand the role that a claim like (3I) might have in a theory. It is at least plausible to think that (3I), for some appropriate set $\bar{Q}$ of alleged applications, might be *the* central empirical claim of the theory. 'Central' in the sense that all "scientific activity" connected with the theory could be represented either as attempts to verify or falsify this claim or as attempts to make predictions based on it. On one widely accepted understanding of 'confirmation' and 'prediction', this would suggest that all empirical statements of the theory might be logical consequences of (3I). This suggestion provides a concise way of describing the logical structure of the theory and for some theories it might be accurate. But, for reasons I will detail shortly, even if accurate, it may fail to bring out an essential feature of the empirical content of the theory. Nevertheless, it is important to understand how (3I) might be verified or falsified and how, in turn, it might be used to warrant predictions.

First consider how we might falsify (3I). One straight-forward way would be to find some member of $\bar{Q}$ for which the Ramsey claim (2) is false. The constraint $\langle =, =\rangle$ plays no role here. Another way to falsify (3I) would be to find two members of $\bar{D}$, D_i, and D_j, such that, if the claim (2) is satisfied, then the corresponding claim for D_j and the constraint $\langle =, =\rangle$ can not both be satisfied. This just amounts to showing that the observed values of n_i and n_j are such that both (A) and (B) can not be true. This is a special case of a more general method of falsifying (3I), by showing essentially that some sub-set of $\bar{Q}$ is such that (A) and (B) can not be true of the n-function values in $\bar{Q}$. In these latter cases it is possible to view the process leading to falsification of (3I) in the following way. We provisionally assume that (3I) is true and use it to "justify" the calculation of t-function ratios by finding members of $\bar{D}$ which overlap in the requisite ways. Then we use these calculated t-function values to "predict" n-function values on some *other* member of $\bar{D}$. If we find these predictions incorrect, then (3I) is falsified.

Thus far we have said nothing about the way that $\bar{Q}$ is described. It could be described by simply listing its members. It could also be described by naming a property possessed by all and only its members. That is, we could simply list the physical systems to which we claim the theory to apply, or we could name their defining characteristic. If we did the latter then the question of whether a particular individual was actually in $\bar{Q}$ might be one to be answered by empirical investigation. In this case one might view the procedure of falsification in still another way. One might say that we "seek ways to observe crucial t-function ratios". That is, we seek, by empirical investigation, to discover members of $\bar{Q}$ whose domains overlap in just the way required to allow us to calculate the t-function ratios we need for a "crucial" test of the theory in some other, already discovered, member of $\bar{Q}$. Whether we are actually able to come up with a crucial test of (3I) depends, of course, on how skillful we are in discovering members of $\bar{Q}$ which allow us to calculate the necessary t-function ratios. Claims like (3I) may not come with procedures for testing "written on their forehead".

As to the question of verifying (3I), it is clear that conclusive verification is only possible when $\bar{Q}$ is described in such a way that it is possible to know when all its members have been examined. Otherwise, there is always the possibility that, yet to be discovered, members of $\bar{Q}$ will serve to falsify the claim. In such cases it would be natural to regard (3I) as more and more strongly *confirmed* in proportion to the number of situations discovered in which the claim might have been falsified but, in fact, was not. That is, every time the claim passes a crucial test, we become a bit more confident that it will pass *all* such tests that we can devise.

The notion of prediction used in the preceding discussion can be elaborated by referring again to example (E3–b). Assume that (3I) is true and that there is good reason to believe that $D_3 = \{x, y, z\}$ is in $\bar{D}$. If we knew the value of n_3 for x and y then we could calculate $n_3(z)$ from (ii) provided we knew $t(y)/t(z)$ and $t(x)/t(z)$. But we have no way of "directly observing" these ratios. We can, however, discover their values if we can find other domains in $\bar{D}$ which overlap with D_3 in an appropriate way. For example, $D_1 = \{x, y\}$, $D_2 = \{x, z\}$ might do, as might $D_4 = \{y, z\}$, D_1 and also $D_5 = \{x, z, w\}$, $D_6 = \{x, y, w\}$, $D_7 = \{w, y, z\}$, provided the n-function values on each of these sets of domains satisfy a condition like C. The important thing to note here is that (ii), unlike (i), is not simply a statement about the relations among n-function values for some specified set of domains. It says nothing about any *specific* member of $\bar{D}$ except D_3. But, together with (3I), (ii) tells us that *any* domains in $\bar{D}$ which are discovered to overlap with D_3 in an appropriate way will have their n-function values related in a particular way to the values of n_3. If we are interested in calculating $n_3(z)$ from $n_3(x)$ and $n_3(y)$, (ii) helps us only if we can find such domains in $\bar{D}$. In this sense, (ii) might be said to determine a *program* for discovering $n_3(z)$. It tells us that we can calculate it if we can determine certain n-function ratios, but it does not tell us how to determine them. Possibly there are many different ways to determine these ratios, and possibly there is no way to determine them – this depends upon what $\bar{D}$ is like. If $\bar{Q}$ is described by naming a defining characteristic of its members, then whether (and if so, exactly how) we can use (ii) to calculate $n(z)$ will typically be discovered by empirical investigation of $\bar{Q}$.

Thus far we have considered a claim of form (3) containing only the simplest sort of constraint. Even in this case, we have seen that such a claim affords possibilities for depicting features of theories which can not be reached by Ramsey claims. It is perhaps doubtful whether such a simple constraint as $\langle =, = \rangle$ would be sufficient to account for the role of a theoretical function in any real theory. Nevertheless, it appears that no essentially new conceptual features are introduced by dealing with more complex constraints. It is, however, fruitful to consider briefly how a claim like (3) would work in which the t-function was constrained to be extensive. This is a somewhat more realistic example since it is at least plausible to think that just such constraints operate with respect to the mass function in classical mechanics.

Consider a claim of form (3), for some given $\bar{Q}$, in which the t-function is constrained both by $\langle =, = \rangle$ and by $\langle R_0, \rho_+ \rangle$: That is, t must be a function on Δ and also extensive with respect to $\circ$. Call this claim '(3IE)'. To see the force of this additional constraint let us first look at an example.

(E3–c) $\bar{D}$ contains $D_1=\{x, y\}$, $D_2=\{x, z\}$ and $D_3=\{x, y\circ z\}$. As in (E3–a), we can calculate t-function ratios for each of these domains. In particular:

$$\frac{n_3(x)}{n_3(y\circ z)} = -\frac{t_3(y\circ z)}{t_3(x)}$$

$$= -\frac{t(y\circ z)}{t(x)} \quad \text{by } \langle =, = \rangle$$

$$= -\frac{t(y)}{t(x)} - \frac{t(z)}{t(x)} \quad \text{by } \langle R_0, \rho_+ \rangle$$

$$= \frac{n_1(x)}{n_1(y)} + \frac{n_2(x)}{n_2(z)}$$

In example (E3–c) it is clear that the effect of $\langle R_0, \rho_+ \rangle$ is to impose additional relations between the values of n-functions on different domains. Without this constraint, the values of n_3 would be completely unrelated to the values of n_1 and n_2.

In general, the effect of adding $\langle R_0, \rho_+ \rangle$ is to add further equations of the form:

$$\sum_{i=1}^{k} t(x_i) - t[(x_1 \circ x_2) \circ x_3) \cdots \circ x_k)] = 0$$

to the system of equations (δ) to produce a new system (δ'). Just how many such equations are added and their explicit nature will depend upon the "structure" of $\bar{D}$ with respect to the $\circ$-operation. Condition (A) is, of course, not a sufficient condition for the existence of a non-trivial solution to the new system of equations (δ), though it is still necessary. Thus, the net effect of $\langle R_0, \rho_+ \rangle$ is to require that there be relations among the n-function values in different domains in addition to those required by $\langle =, = \rangle$.

Given any particular $\bar{Q}$ for which the structure of $\bar{D}$ with respect to the $\circ$-operation is known, it will be possible to give a necessary and sufficient condition (A'), analogous to (A), for the existence of a non-trivial solution to (δ). Conditions (A') and (B) together will be necessary and sufficient for the truth of (3IE). This will be stated entirely in terms of n-function values. Thus (3IE) is essentially no different from (3I) with regard to the eliminability of the t-function. Also (A') and (B) will entail analogous conditions for any sub-set of $\bar{Q}$.

Corresponding to the new relations among n-function values introduced by $\langle R_0, \rho_+ \rangle$, there are introduced new possibilities for non-trivial observation of n-function values. Obviously, if we have calculated or observed already the t-values of x and y we can, without further information about n-function values in any domain, calculate the t-value for $x \circ y$. We need not determine the mass of a locomotive "directly" by suspending it from one arm of a balance. We can determine the mass of its parts and add.

Something should perhaps be said about the way this account of the measurement of calculation of t-function values fits in with the traditional accounts of measurement like those mentioned in Chapter II ([4], [17], [44]). In such accounts it is customary to distinguish between fundamental and derived measurements. Fundamental measurements are those which rest directly on some empirical relational system like the extensive system described in Chapter II. Derived measurements are obtained by calculating the values of numerical functions whose arguments are values obtained from fundamental measurement. Into which of these catagories does the measurement of t-function values (according to our account) fall?

At first glance, it might appear to be an example of derived measurement since we do calculate t-function ratios from n-function values and these are (presumably) obtained via fundamental measurement. However, there is no numerical function which carries the n-function value of a particular individual into the t-function value for that individual. It is easy to see

that the best we will ever be able to do with our method is to calculate
t-function ratios. This is simply because any constant multiple of a t-function which satisfies a claim of form (2) will also satisfy this claim. Moreover, there is no *single* numerical function which allows us to calculate
t-function ratios from n-function values in all circumstances. Whether
these ratios are uniquely determined by n-function values and, if so,
exactly how, depends on the domains we are considering. This is just
another way of saying that neither the t-function nor any t-function ratio can
be *defined* in terms of the n-function in the way that density may be defined
in terms of weight and volume. For such a definition to be possible the
t-function (or t-function ratios) would have to be uniquely determined by
n-function values in *every* domain. This seems to rule out regarding measurement of t-values as derived measurement. However, this conclusion
depends crucially on an account of derived measurement which requires
that there be a *functional* relation between the values of the derived
quantity and those of the quantities from which it is derived. That is, the
values of the derived quantity must be uniquely determined by the values
of the quantity from which it is derived. There are accounts of derived
measurement which weaken this requirement ([44], pp. 17–20). In these
accounts it is only required that there be *some* relation holding between
the values of these quantities. The values of the derived quantity need not
be uniquely determined. It is possible to regard the measurement of t-values as an instance of derived measurement in this weaker sense. For
domains in which claim (2) is true, t-values are derived from n-values. The
relevant relation in this case is just the relation that must hold if the t and
n-values are to provide a model for 'is an S'.

What about fundamental measurement? There is no obvious way in
which a relational property of the individuals is involved in our determination of t-function ratios. However, in an attempt to cast our procedure into
the framework of fundamental measurement, one might introduce a
relation T, defined in the following way:

$T(x, y)$ if and only if $t(x)/t(y) \leq 1$.

Then all our methods of calculating t-function ratios *could be* regarded as
means of discovering truth-values of T-statements. It would be a simple
matter to show that $\langle \Delta, T \rangle$ is a relational system sufficient to determine a
class of numerical functions on Δ. If, in addition, we constrained t-function
values to be extensive with respect to the $\circ$-operation then $\langle \Delta, T, \circ \rangle$ would

turn out to be an extensive relational system. We could, in this way, at least for claims involving the $\langle =, = \rangle$ constraint, regard the calculation of t-function values as an instance of fundamental measurement. However, it is not obvious that anything in the way of clarity about the role of the t-function in the theory is to be gained by this. Indeed, we do intuitively feel that when we determine mass ratios we are determining the truth value of some statement involving the relation 'is at least as massive as'. But, if the account of the role of the mass function in classical mechanics suggested here is correct, this intuitive feeling is misleading. It is misleading in that it tends to obscure the fact that the existence of an extensive mass function is a kind of working hypothesis, not an obvious fact about material bodies.

At this point it is also worthwhile to note that the notion of a constraint that we have been considering also appears to be applicable to the theories of measurement that "underlie" theories of mathematical physics.[2] To see how this is so, consider the following definition.

(ES) x is an *extensive scale system* if and only if there exists an A, R, ∘, and h such that:

(1) $x = \langle A, R, \circ, h \rangle$;
(2) A is a non-empty set;
(3) R is a binary relation on A;
(4) ∘ is a function from $A \times A$ into A;
(5) h is a function from A into the positive real numbers such that, for all x, y, A:
 (i) xRy if and only if $h(x) \leq h(y)$;
 (ii) $h(x \circ y) = h(x) + h(y)$.

Intuitively, an extensive scale system is just a set of individuals together with a binary relation R and binary operation ∘ on this set and some positive valued numerical function that "agrees with" R and is extensive with respect to ∘. The relation between an extensive scale system and an extensive system (defined by (DE) in Chapter II), is this. If $\langle A, R, \circ \rangle$ is an extensive system then there exists at least one h such that $\langle A, R, \circ, h \rangle$ is an extensive scale system and, moreover, all $\bar{h}$'s such that $\langle A, R, \circ, \bar{h} \rangle$ is an extensive scale system are positive multiples of h.

Now consider all the models for the three axioms (ES1)–(ES3) in which it is plausible to identify R as the "is-shorter-than-or-just-as-long-as" relation on the individuals in (A). In our terminology, these models are possible models for an extensive system and partial possible models for an extensive scale system. Intuitively, they comprise the range of intended applications for the theory of length. In different models in this collection, the means actually employed to determine the truth value of R-statements may differ. In some models we may be able to simply lay the individuals end-to-end; in others we may employ complex "indirect" methods to compare two individuals. The same may be true of statements involving the ∘-operation. Even though we do intuitively regard all these as methods for determining facts about essentially the same property of the individuals – length – one might insist that the R-relation (and perhaps also the ∘-operation) is "really" interpreted in different ways in models in which different means are used to determine the truth values of R-statements. If one is convinced of this, then a rough way to understand the empirical content of the theory of length is the following. This theory simply claims that all the different empirical procedures used to determine the truth value of R-statements do, in fact, determine the same property. That is, the theory claims that it is possible to regard all the different possible "operational definitions" of length as definitions of the same quantity. How can we say more precisely what this claim is?

To begin, consider the structure of the domains in this class of models. It is clear that some of these domains are going to overlap. Some will overlap in trivial ways. For example, it appears that any sub-set of the domain of a model will also be the domain of a model. But they may also be expected to overlap in non-trivial ways. For example, consider two domains of models (A_1) and (A_2) whose union is not the domain of a model. Intuitively, this might be because there is no way of "comparing" every individual in (A_1) with every individual in (A_2). It might be, however, that (A_1) and (A_2) share some individuals. Some individual, like a movable measuring rod, might be comparable with all the individuals in both (A_1) and (A_2). In this example, it might be the case that the methods of comparison in (A_1) and (A_2) were, intuitively, the same – say laying end-to-end. It is just that physical limitations prevent its *general* application across the domains. But this need not be the case. Different means might be employed to determine the truth values of R-statements in (A_1) and (A_2). Yet, it might be that there are some individuals – those shared by (A_1) and (A_2) – to which *both* these means are applicable. An object on the earth's surface may be compared "directly" with some other objects on the earth's surface and indirectly, via optical means, to an object like the moon. At this point also note that it is very unlikely that *all* these models will turn out to be extensive systems. Some of them will, for example, have finite domains. Indeed, it may be that the only ones that are extensive systems are ones whose domains contain "non-physical" objects (see Chapter II, p. 22).

What empirical claim does the theory of length make for the class of models just described? It appears to be this. There is some way to fill out all these models, by adding an h-function, to produce a model for an extensive scale system. Moreover, there is some way to do this so that individuals may be assigned the same h-value in all the domains in which they appear. Roughly, there is some way of assigning lengths to individuals which allows us to regard length as an intrinsic property of the individual – a property that is independent of the various different means that may be employed to determine it. Intuitively, the reason we regard various different empirical processes as means of determining the same thing – length – is that we believe that they are consistent in the way just described. When two different processes can be applied to the same individual they yield the same result. It is just this intuition that is captured in claiming that it is possible to find *one* length function which will serve for the entire range of intended applications.

It should be evident now how the notion of a constraint applies to this situation. The claim just mentioned can be rendered as a claim of form (3) in which the partial possible models are characterized by (ES1)–(ES3), the predicate analogous to 'is an S' is 'is an extensive scale system' and the constraint is $\langle =, = \rangle$. In this situation the h-function plays the role of a theoretical function with respect to the non-theoretical relation R and operation ∘. The set of partial possible models, $\bar{Q}$, in this claim consists of all those models for (ES1)–(ES3) in which R and ∘ are interpreted respectively as "is-shorter-than-or-just-as-long-as" and "laid-end-to-end-with". It is useful to note that this range of intended applications can be described in two ways. First, it is a set of partial possible models for which we have some intuitive reason to claim that R and ∘ are interpreted in the same way in each member. Second, it is a set of partial possible models throughout which it is intuitively plausible to claim that the same constraints are imposed on the theoretical function h. Thus, it appears that there is some relation between our inclination to regard non-theoretical properties as essentially the same in different situations and our inclination to impose constraints on theoretical functions that operate "across" these different situations.

The usual accounts of the measurement of length fit into this account of the theory of length in the following way. Presumably, some of the intended applications for the theory of length will be extensive systems. In these applications the theory of measurement of extensive quantities assures us that suitable h-function exists and that it is unique up to a positive multiple. Moreover, we can actually determine the values of this function by examining the R-relation and ∘-operation in this application. In effect, we can determine h-function ratios in this application and the $\langle =, = \rangle$ constraint allows us to infer that these same ratios must hold when the same individuals appear in different applications. In this sense, the intended applications for the theory that are extensive systems provide means of measuring length. Two intended applications with overlapping domains and a different "interpretation" for the R-relation, both extensive systems, provide a crucial test for the empirical claim of the theory of length. The possibility, mentioned before (Chapter II, p. 22), that there are no extensive systems whose domains consist exclusively of physical objects should be recalled at this point.

Thus far we have seen that a claim of form (3) allows us to account for non-trivial calculations of values for theoretical functions. For this reason alone, it might be expected that this claim would fare better than a Ramsey claim of form (2), or a collection of such claims, in exhibiting the logical structure of a theory of mathematical physics. It has also been suggested that there are other reasons to expect this. It was suggested that regarding the empirical content of a theory to be given by a single claim of form (3) illuminates certain features of a "holistic" view of scientific theories, and further that these features might be based on sound insight into the nature of theories of mathematical physics. The time has come to deal with this suggestion in more detail. I want to emphasize at the outset that I am not concerned to defend, in its entirety, any "holistic" view of scientific theories. I am interested only in certain tenets of these views. Moreover, I am interested in these tenets *only* as they apply to theories of mathematical physics. Whether they have wider application is of no concern to me here. The primary purpose of the following discussion is to add plausibility to the view of logical structure now being considered. Many of the points touched upon in the discussion will be considered in more detail in Chapter VIII when we come to talk of the dynamic aspect of physical theorizing.

The features of "holistic" views that I want to consider may be summarized as follows, where 'theory' should be read to mean 'theory of mathematical physics'.

(I) Theories are accepted or rejected as a whole, not piece by piece.
(II) Theories are not rejected as a result of single crucial experiments.
(III) No clear distinction can be made between what a theory claims and what is evidence for these claims.

These formulations are rough and imprecise. I doubt that any philosopher has maintained such theses without amplification and qualification. However, I think that no injustice is done in saying that (I)–(III) express, albeit very roughly, central tenets amplified and defended in the work of both Duhem [10] and Kuhn [21]. The roughness of *these* formulations will not hinder the task at hand – that of understanding how our view of logical structure yields a more precise formulation.

Thesis (I) might be simply understood to mean that theories are not *sets* of statements, at least some of which are logically independent of others, but rather only *one* statement. This understanding is certainly in accord with our view that a statement of form (3) may be *the one* empirical claim of the theory. However, this way of understanding (I) appears to run counter to some common ways of regarding theories. We often speak of the range of application of theories being extended, of new ways to apply a theory being discovered, and of a theory being applied to newly discovered phenomena. For example, we speak of discovering how to apply quantum mechanics to crystalline slides. Are we not, in these situations, speaking of *adding* new statements (new pieces) to an already existing body of theory? Understood in this way, thesis (I) appears to conflict with our intuitive ideas about the way the *same* theory grows and develops.

There is, however, another way to understand (I). Suppose that the central claim of the theory θ is some claim of form (3), containing a particular set-theoretic predicate, some particular constraints, and a particular $\bar{Q}$. One way we would understand the growth of θ is this. As we discover more applications for the theory, we simply add them to $\bar{Q}$ thereby obtaining a *new* claim with the same logical form as the old one. The theory does grow in a piecemeal fashion as new applications are discovered, but it still remains the case that all the empirical content of the theory is captured, at any given time, by *one* claim of form (3). Indeed, this claim always retains the same set-theoretic predicate and the same constraints; only the set $\bar{Q}$ changes. However, this piecemeal growth does not take place independently of the initial content of the theory's claims. Adding a new member to $\bar{Q}$ requires us to examine existing members to see that the constraints on the values of theoretical functions may still be satisfied. One might then say that 'accepting the theory θ' does not mean that one accepts some particular claim of form (3) with a fixed range of applications $\bar{Q}$. Rather, it means that we commit ourselves to try to deal with some, perhaps not definitely specified, class of phenomena with *some* claim of form (3) with fixed set-theoretic predicate and constraints. That is we commit ourselves to using a certain mathematical structure with certain constraints on its interpretations, but we do not commit ourselves, in advance, as to precisely what physical systems this formalism applies. We can then understand (I) to mean that accepting a theory, in this sense, is an all-or-nothing decision.

Understood in this way, thesis (I) is very plausible. We might be convinced, by actually doing the requisite calculations, that a claim of form (3) is true for some small, finite $\bar{Q}$. This leads us to conjecture that it might remain true when we add other physical systems to $\bar{Q}$ that are, in some sense, "like" those already in $\bar{Q}$. But we need not be able to specify, at the outset, without further research, just what this sense of "likeness" is. That is, we may have good reason to suspect that we have a mathematical formalism at hand which will prove to be a powerful tool in dealing with the world. Yet, we have no very precise conception of how the details of its applications will be worked out. It is also true that further applications of the theory can not be added without taking due account of those to which we are already committed. In adding a new application we must always check to see if this can be done in such a way as to maintain the constraints on values of theoretical functions. In this sense our initial choice of a claim of form (3) determines, at least partially, the course of the theories' further development. This too might be looked upon as a kind of all-or-nothing acceptance. But it is perhaps more accurate to say simply that questions about whether particular physical systems are in $\bar{Q}$ may not be independent of one another.

We have seen that taking the empirical content of a theory θ to be given by a claim of form (3) allows us to distinguish two senses of 'accepting θ'. One sense is that one comes to believe some particular claim of form (3). The other sense is that one becomes committed to finding evidence for *some* claim of form (3) with a specified set-theoretic predicate and constraints. Making this distinction also allows us to understand thesis (II) more clearly. We have devoted some discussion already to showing how *particular* claims of form (3) might be subject to decisive falsification by crucial experiments. If this is so, then clearly (II) is false if we understand 'accepting θ' in the first way. On the other hand, it appears that (II) might well be true if we understand 'accepting θ' in the second way. Suppose we are in a situation where the set of intended applications $\bar{Q}$ is only "provisionally" specified and find a sub-set of $\bar{Q}$ which falsifies our claim of form (3). It is always open to us to modify the membership of $\bar{Q}$ rather than give up completely trying to find some claim of form (3) that is tenable. This is to say, roughly, that we are committed to carving up some portion of the world in *some* way – producing some $\bar{Q}$ – for which a claim of form (3) is true. Failure at *one* attempt to do this need not convince us that further attempts will be similarly unsuccessful. Single crucial experiments may cause us to reject particular claims of form (3), but, typically, only repeated failure to make *some* claim of form (3) work will convince that we should quit trying. We may have firm views about some of the things which must be in $\bar{Q}$ but less firm views about others. This would explain why, for example, we might throw out a certain calculation of t-function values rather than admit that the t-function was not extensive. We are committed to trying to use a claim of form (3) with the $\langle R_0, \rho_+ \rangle$ constraint, but we are not firmly committed to including in $\bar{Q}$ all the systems required to make this calculation.

Let us now consider thesis (III). At one point Kuhn expresses (III) in the following way:

> .. theories do not evolve piecemeal to fit facts that were there all the time. Rather, they emerge together with the facts they fit from a revolutionary reformulation of the preceeding scientific tradition ... ([21], p. 140).

Suppose the empirical content of a theory θ to be given by a claim of form (3). Then, so far as "facts" described in terms of θ-theoretical functions are concerned, something like what Kuhn says appears to be true. It would certainly be impossible to have any facts about t-function values prior to the formulation and, at least tenative, acceptance of something like (3I) or (3IE). If talk of t-functions appeared *only* in the theory θ and if a single claim of form (3) were the only claim of θ then we simply wouldn't know what to make of talk about t-function values unless we accepted θ. But this alone surely does not indicate that there is something peculiar, "revolutionary" and perhaps even irrational, as Kuhn's formulation might suggest, about our coming to accept θ. In no sense do we "manufacture" the facts to fit our theory. Rather, once we understand that the claim of θ has the logical form of (3), we can give a quite ordinary account of how we come to accept it. This account makes clear exactly how we are justified in describing "brute facts" which may be evidence for the theory θ in terms of concepts which are only "understandable" if θ is true. The intuitive idea of the account is not very subtle. It suggests that we adopt a claim of form (3) provisionally and see how we fare with it – see if we can make sense out of the data with the theoretical concepts it provides.

This understanding of (III) suggests a natural way of understanding another claim, closely related to (III). It is sometimes held that the meaning of theoretical terms (at least) does not remain fixed, but rather changes as the content of the theory changes. On the present account, there is at least one clear sense in which the meaning of theoretical terms changes when the "range" of the theory changes. Consider a consequence of a claim of form (3) for some particular $\bar{Q}$, say a claim of form (2) for $Q_i \in \bar{Q}$. The truth conditions for this claim of form (2) will typically change when elements are added to or deleted from $\bar{Q}$. This is because the particular way in which the constraints in (3) impinge on the t-function on D_i is generally dependent on how the members of D_i overlap with other members of $\bar{D}$. In this sense then, the truth conditions of statements about t-function values may change as the "range" of the theory changes, and thus one might say that the meaning of the corresponding theoretical term changes.

One aspect of this discussion requires some further attention. In what has just been said, there is a strong suggestion that it may be misleading to regard theories of mathematical physics as simply sets of statements. 'Misleading' in that, by doing so, we may be led to overlook some important facts about how the enterprise of "doing mathematical physics" is carried on. Physicists may be, in some sense, committed to working with a certain mathematical formalism, have some idea of constraints on the application of this formalism, and yet not be willing to make more than "tentative" empirical claims using this formalism. In one sense 'having a theory' may mean only that one has this mathematical formalism and a conviction that it *can* be used to say what certain chunks of the world are like. It would perhaps be unlikely that one would have a theory, in this sense, without *some* empirical claims with the formalism. The point is that these particular empirical claims, or at least some of them, might be given up and yet the commitment to use this same mathematical formalism retained. This suggests that being clear about the formalism and the constraints on its application is more central to understanding what it is like to practice a certain branch of mathematical physics than being clear about any particular empirical claims that might be made. It would be misleading to *identify* the theory with some particular empirical claims (say a particular claim of form (3) and its consequences) because these claims might change and yet what is characteristic of this branch of scientific activity remain the same. All this may be admitted, nevertheless it is important to be clear about what sort of empirical claims *could* be made with the formalism of a particular theory – what their logical form is, what might be evidence for them, and how they might be used to make predictions. The end-product of physical theorizing is empirical claims. We can not understand fully what it is like to *do* mathematical physics until we see how it leads to empirical claims. It is to this end that logical reconstruction is directed.

Taking account of this, one might say that the aim of a logical reconstruction of a particular theory of mathematical physics is threefold. First, it aims at clarifying the mathematical structure of the theory and the constraints on its application. Second, it aims at clarifying the logical form of the empirical claims made with this mathematical formalism. And third, it aims at exhibiting particular instances of empirical claims of this form that have been or are presently being made by practitioners of the theory. There is some reason to believe that the second enterprise can be carried on with some degree of generality, i.e. independently of any particular theory. It is just this that we have been pursuing in this and the preceding two chapters, and will continue to pursue in the next chapter. In Chapter VI we shall consider the first and third enterprises with reference to classical particle mechanics.

The discussion in this chapter suggests very strongly that the empirical content of a theory of mathematical physics may be construed as a claim of form (3). By so construing it, we can give a complete account of the role of theoretical functions in the claim of the theory. We can understand the meaning of statements involving these functions and we can understand what we are doing when we speak of measuring their values. Moreover, construing the empirical content of the theory in this way, allows us to describe the relation between the empirical content of the theory and the mathematical formalism of the theory in a way which illuminates some, otherwise imprecise, insights. In the next chapter this account will be extended to allow the possibility that a "family" of set-theoretic predicates might be employed to express the empirical content of one theory.

NOTES

[1] This was also apparently the view of Mach and his view is endorsed by Nagel ([29], p. 192 ff).
[2] I am indebted to Mr. Gary Bower for calling my attention to this fact.

CHAPTER V

THEORETICAL FUNCTIONS WITH SPECIAL FORMS

In this chapter one final proposal for using set-theoretic predicates to make empirical claims is considered. The motivation for considering this proposal is provided by examples of situations in which we apparently "postulate" or "hypothesize" that theoretical functions have some special form in certain applications of a theory. It appears that a logical reconstruction of theories with such claims will require something more than claims of form (3). The proposal for dealing with such theories is roughly this. A number of predicates are employed, all of them defined by restrictions of the definition of the same basic predicate. These predicates are used to construct a sentence which says that there are theoretical functions which make all intended applications models for the basic predicate, make some designated sub-sets of intended applications models for the "restrictions" of this basic predicate, *and* satisfy certain constraints. The "restrictions" of the basic predicate are to characterize various special forms that the theoretical function is hypothesized to have. This claim is regarded as the central empirical claim of the theory.

The motivation for pursuing further the question of how one might use a set-theoretic predicate to make empirical claims is provided by examples of certain claims made about the values of theoretical functions. Consider the force function in classical particle mechanics. In the usual expositions of this theory, claims of roughly the following sort are made. (a) In applications of classical particle mechanics to physical systems where *only* gravitational forces are present, the forces will be inverse-square, central forces. (b) A wide class of applications of classical particle mechanics is characterized by the forces in those applications satisfying Newton's third law. Indeed, the applications mentioned in (a) are a sub-class of these applications. (c) In still another class of applications – charged particles subject only to an electromagnetic field – the forces do not satisfy the third law, but rather a law of the form:

$$\bar{f} = q[\bar{E} + 1/c(\dot{\bar{r}} \times \bar{B})].$$

THEORETICAL FUNCTIONS WITH SPECIAL FORMS   97

Common to all these claims appears to be roughly this. The force function is alleged to have some special properties in certain applications – properties that we do not claim that it has in *all* applications. In (a) and (c) the force function in certain independently identified classes of applications is claimed to have a special form. That is to say, the force function, in applications of this kind, is claimed to have some properties in addition to those that force functions always have. In (b) a class of applications of the theory is singled out by the special form of the force function in such applications. This is to say, the class of applications is "defined" by the special form of the force function.

This example suggests that it may be fruitful to consider situations of the following sort. We have a theory θ whose mathematical structure is given by the predicate 'is an S' where t and n seem to be respectively θ-theoretical and θ-non-theoretical functions. Further, suppose that existing expositions of θ contain claims about t-function values analogous to those about the force function in our examples. That is, they contain claims that the t-function has a certain form in some independently described set of applications of θ, or claims that some set of applications is singled out by a particular form of the t-function. For example, there might be claims that t-function values, in each application of a certain kind, all sum to a constant K, or that they form a geometric progression. Intuitively, these are claims that t-function values in some application have a special form not shared by t-functions in all applications. How are we to deal with such claims in our logical reconstruction of θ?

To see more clearly what the alternatives might be for dealing with such claims, suppose we attempt to regard a claim of form (3), for a set of intended applications $\bar{Q}$, as *the* central empirical claim of θ. How might we then regard claims that the t-function has a special form in some sub-set $\bar{Q}' \subset \bar{Q}$? One way is this. Suppose that the n-function values on members of $\bar{Q}'$ were such that only t-functions having certain special properties could satisfy a claim of form (2) for these applications. For example, suppose the n-function values in every member $\langle D_i, n_i \rangle$ of $\bar{Q}'$ were either $+1$ or -1. Then the t-function values satisfying (2) for $\langle D_i, n_i \rangle$ would have to be such that D_i could be partitioned into two parts, the sum of t-values on each part being equal. In this case, the claim that t-function values in $\bar{Q}'$ had this particular form would simply be a consequence of (3) and the fact that n-function values in $\bar{Q}'$ had a particular form. The constraints in (3), play no role here. We simply make use of the fact that (3) entails a claim of form (2) for every member of $\bar{Q}$. Indeed, $\bar{Q}'$ might consist of a single application. It is not essential to this account that we have n-function values in more than one application.

Another way claims that the t-function values have a special form in some sub-class of applications might be accounted for in a logical reconstruction of this sort is the following. The claim of form (3), together with observed n-function values on members of $\bar{Q}' \subset \bar{Q}$, might allow us to calculate explicitly the t-function values in some $Q \in \bar{Q}'$. This would just amount to "observing" the values of the t-function in Q in the way described in detail in the preceding chapter. Having done this, one might discover that these t-function values had some special form. Here again this claim would be a consequence of (3) and some facts about n-function values. In this case, however, the constraints in (3) are relevant and an essential feature in the situation is that n-function values in more than one application are employed.

Both of the ways we have just seen of accounting for claims that the t-function has a special form in certain applications might naturally be described in the following way. Assuming that the claim of form (3) is true, there is "direct" empirical evidence that the t-function values have a special form in certain applications. 'Direct' in the sense that observed n-function values entail that t-functions satisfying (3) have this form. There might, however, be less direct evidence for claims of this sort about t-function values. For example, we might have this sort of direct evidence for claims about the form of the t-function in some few applications and generalize these claims to other applications which appear to be similar in some relevant way. This is to say, we might discover by actual calculation that the t-function values in some applications have a certain form and generalize this claim to similar applications without being in a position to make calculations of t-function values in these applications. But even in this case, our account of claims that the t-function has a special form presupposes that, in *some* applications, this claim is substantiated by direct calculations of t-function values.

These suggestions about how we might account for claims that the t-function has a special form in certain applications, when a claim of form (3) is the central claim of the theory, are straight-forward and obvious. For some claims of this sort about the values of theoretical functions in real theories, an account along these lines seems possible. For example, Newton showed that any particle, whose path is a conic section and whose motion along that path obeys Kepler's second law, must be acted upon by a resultant force directed toward one focus of the conic whose magnitude is inversely proportional to the square of the particle's distance from this focus. This is to say roughly this. If the motion of such a particle is to provide a model for classical particle mechanics whose crucial axiom is roughly, $\vec{f} = m\vec{a}$, then the force function $\vec{f}$ must have a particular form. This appears to be an example of the first suggested account where values of the non-theoretical function in perhaps one application, together with the claim of form (3) lead to a claim about the form of the theoretical function which satisfies (3). A situation in which the force law of a spring is determined in some relatively simple situation and then claims are made, on this basis, about the force function in more complicated situations involving the same spring, or relevantly similar springs, appear to provide examples respectively of the second and third suggestions.

These straight-forward possibilities we have suggested do not, however, appear to be sufficient to deal with all claims of this sort about the values of theoretical functions. Frequently we appear to *assume, conjecture*, or *postulate* that values of theoretical functions have a special form in certain applications. This is to say, we adopt, as a kind of working hypothesis, a claim of this sort and see if it is *consistent* with what we can observe about these applications. In some cases, at least, we do not base this hypothesis on *any* evidence gleaned from calculating what the values of the theoretical function *must* be in some application(s). We make no claim that this hypothesis is the *only* one that will serve to explain the observed facts. The only claim is that the hypothesis is consistent with the observed facts and that it will continue to be when new observations are made. Again Newton's work provides an example. Roughly speaking, Newton *postulated* that every "particle" in the solar system (at least) exerts a force on every other "particle" which is attractive, directed along the line connecting them, proportional to the square of the distance between them. There is no readily apparent way to view this assumption as somehow justified by calculations of values of the force functions required to make simple applications models for the theory. Intuitively, its only justification appears to be that it is a reasonably simple hypothesis and in reasonably good agreement with the facts about the paths of bodies in the solar system.

It does not appear likely that all claims of the sort just described where it is *postulated* that the theoretical function has a special form in certain applications can be satisfactorily accounted for in a logical reconstruction by any of the methods that have been suggested. That is, they cannot be regarded as consequences of a single claim of form (3) plus some facts about the values of non-theoretical functions, nor as inductive generalizations from such consequences. Clearly they can not be regarded simply as consequences of a claim like (3) and observed facts because an essential feature of such claims is that the form the theoretical function is alleged to have is not *uniquely* determined by the observed values of non-theoretical functions. Nor does it appear that *all* such claims can be regarded as inductive generalizations of such consequences (though perhaps some can). Newton's claim about the gravitational force does not appear to be grounded on *any* actual calculations of force function values. If we can not regard these claims as consequences of a single claim of form (3) plus some observed facts, then how are we to account for them in a logical reconstruction?

One obvious suggestion for dealing with such claims is this. We might attempt to render the empirical content of the theory as several claims of form (3) constructed with different set-theoretic predicates and perhaps different constraints. Each predicate would be a different restriction of the same basic predicate, determining a different special form of the theoretical function. Indeed, claims of form (2) constructed with various restrictions of the basic predicate might also be allowed. The basic predicate would characterize the properties that the theoretical function was alleged to have in *all* applications of the theory, while each restriction would characterize a special form that it was alleged to have in certain applications, or kinds of applications.

There is no apparent reason to suspect that such an approach could not possibly work for *some* theories. There is, however, reason to expect that, in some cases at least, it might encounter the same difficulty as the proposal that the empirical content of the theory be rendered by a collection of Ramsey claims of form (2). The difficulty is roughly that there is no connection between the theoretical functions which satisfy the claims of form (3) associated with different sub-classes of applications of the theory. If we use two different claims of form (3) to render the content of the theory, we might be able to account for non-trivial calculations of theoretical function values. But, intuitively speaking, we would have no reason to suspect that the theoretical functions in applications associated with different claims of form (3) were measures of the *same* "physical quantity". The reason for this is that the theoretical function values whose calculation is justified by one claim of form (3) are in no way relevant to whether any other claim of form (3) is true. We could not, for example, use them to make predictions about the values of non-theoretical functions in applications associated with some other claim of form (3). One might even be led to say that each claim of form (3) is (a claim of) a different physical theory, in the sense that the truth of one is completely independent of the truth of any other.

These considerations suggest that it is at least possible that a logical reconstruction of some theories of mathematical physics will require something other than claims like (2) and (3). If we are dealing with a theory in which the theoretical function(s) are apparently postulated to have special forms in different applications, *and* if there are relations among the theoretical function values in these different applications, then it appears that we must use something more than simply a collection (conjunction) of claims of form (2) and (3). What we need is, roughly, a claim which brings together different claims of form (2) and (3), asserting the existence of theoretical functions with different special forms in various applications. More precisely, we want to impose something like *constraints* on the theoretical functions which are claimed to satisfy different claims of form (2) and (3) made with different set-theoretic predicates. Each of these claims will be constructed with a predicate which is a restriction of the same basic predicate so that each can be regarded as simply imposing a special form on the theoretical function. But, in addition, we want to claim that the theoretical functions that satisfy these claims are not completely independent of one another. They must satisfy constraints analogous to those which operate in a single claim of form (3) to "bring together" a number of claims of form (2) into a mutually dependent network of claims. To see how this might work in detail, it is helpful to look at an example.

Consider the predicate 'is an S' defined by (D1). Let 'is an S^i' $i = 1, 2, \ldots$ be a predicate defined by a restriction of (D1), (D1i). Here 'restriction' is used in the same sense as in Chapter III (p. 41) to mean roughly a definition obtained by adding additional axioms, involving the same primitive symbols, to the original definition. It is subject to the same impreciseness noted there. Likewise, let 'is an S^{ij}' $j = 1, 2, \ldots$ be a predicate defined by a restriction of (D1i), (D1ij), and so on for 'is an $S^{ijk\cdots}$'. We shall also say that the *predicate* 'is an S^i' is a restriction of the *predicate* 'is an S'.

Further, let us suppose that the axioms added to produce the definitions of the restrictions 'is an $S^{ijk\cdots}$' all "involve" the t-function, that is, that they can not be formulated in terms of the n-function alone. Intuitively this makes it possible to regard all the predicates 'is an $S^{ijk\cdots}$' as specifying that the t-function has properties in addition to those that 'is an S' requires it to have, or as specifying a special form for the t-function. As noted in Chapter III, it will not always be clear whether or not an axiom contains a reference to a given primitive symbol. It is reasonably clear that axioms

(D1i.6) $\sum_{x \in D} t(x) = 1$

(D1j.6)   For all $x, y \in D$

$$|t(x) - t(y)| = |n(x) - n(y)|$$

can be regarded as specifying a special form of the t-function when added to (D1). It is also reasonably clear that axiom

(D1k.6) $\sum_{x \in D} n(x) = 1$

can not be so regarded. So far as the present discussion is concerned, there is no need to agonize about trying to make this requirement more precise. If a predicate did not obviously satisfy the requirement, there would be little reason to believe that it could be used in a logical reconstruction of theory where claims that the t-function has a special form appear. We simply assume that we are dealing with predicates that reflect in obvious ways the intuitive content of the theory.

As before, let $\bar{Q} = \{\langle D_1, n_1 \rangle, \langle D_2, n_2 \rangle, \ldots, \langle D_i, n_i \rangle, \ldots\}$ be a set of possible partial models for S, i.e. each $\langle D_i, n_i \rangle$ is a P_0. For all $i = 1, 2, \ldots$ let $\bar{Q}^i \subset \bar{Q}$, and for all $j = 1, 2, \ldots$ let $\bar{Q}^{ij} \subset \bar{Q}^i$. In general, let $\bar{Q}^{i_1 i_2 \cdots i_n + 1} \subset \bar{Q}^{i_1 i_2 \cdots i_n}$. Let $\bar{D} = \{D_1, D_2, \ldots, D_i, \ldots\}$ and $\bar{D}^{ijk\cdots} = \{D \mid (\exists n)(\langle D, n \rangle \in Q^{ijk\cdots})\}$. Further, if $\bar{x}$ is an $E\bar{Q}$ (see D5), then let $\bar{x}^i$ be the sub-set of $\bar{x}$ such that $\bar{x}^i$ is an $E_{\bar{Q}^i}$ (obviously, such a sub-set exists, for any $\bar{Q}^i \subset \bar{Q}$). Likewise, let $\bar{x}^{ij} \subset \bar{x}^i$ be such that $\bar{x}^{ij}$ is an $E_{\bar{Q}^{ij}}$ and so on for $\bar{x}^{ijk\cdots}$. As before, if $\bar{x}$ is an $E_{\bar{Q}^{ijk}}$, '$C(\bar{x}, R, \rho)$' means that the set of t-functions $\bar{t}^{ijk\cdots}$ on $\bar{D}^{ijk\cdots}$, added to make the members of $\bar{Q}^{ijk\cdots}$ possible models for S, satisfy the constraint $\langle R, \rho \rangle$.

Consider now a sentence of the following form:

(4) $(\exists \bar{x})[\bar{x}$ is an $E_{\bar{Q}} \wedge \bar{x}$ is an $S \wedge \bar{x}^i$ is an $S^i \wedge C(\bar{x}, R, \rho)]$,

where S^i is some restriction of S specifying a special form for the t-function, $\bar{Q}$ is some set of possible partial models for S, $\bar{Q}^i \subset \bar{Q}$, and $\langle R, \rho \rangle$ is a constraint on functions defined on the domains of the members of $\bar{Q}$. Intuitively, (4) says that there is some set of extensions of members of $\bar{Q}$ all of which are models for S; the extensions in $\bar{x}$ of members of $\bar{Q}^i$ are not only models for S, but also models for S^i, and moreover the set of t-functions used to produce this set of extensions satisfies the constraint $\langle R, \rho \rangle$.

To understand how a sentence like (4) might be used to make an empirical claim, note first that, whatever the constraint $\langle R, \rho \rangle$ is like, (4) entails both

(3) $\qquad (\exists \bar{x})[\bar{x}$ is an $E_{\bar{Q}} \wedge \bar{x}$ is an $S \wedge C(\bar{x}, R, \rho)]$

and

(3i) $\qquad (\exists \bar{x})[\bar{x}$ is an $E_{\bar{Q}^i} \wedge \bar{x}$ is an $S^i \wedge C(\bar{x}, R, \rho)]$

where '$C(\bar{x}, R, \rho)$' in (3i) means that the set of t-functions $\bar{t}^i$ on members of $\bar{D}^i$, used to extend the members of $\bar{Q}^i$ to possible models for S (and thus for S^i) satisfy $\langle R, \rho \rangle$. But, the constraints *may be* such that the conjunction of (3) and (3i) does not entail (4). There may exist a set of t-functions $\bar{t}^i$ on members of $\bar{D}^i$ which make all members of $\bar{Q}^i$ models for S^i (and *a fortiori* models for S) and also satisfy $\langle R, \rho \rangle$; there may exist $\bar{t}$ on members of $\bar{D}$ which make all members of $\bar{Q}$ models for $\bar{S}$ and satisfy $\langle R, \rho \rangle$, but $\bar{t}^i$ may not be a sub-set of $\bar{t}$, as (4) requires. We can not, for example, start with a set of t-functions $\bar{t}^i$ extending the members of $\bar{Q}^i$ to models for S^i and satisfying $\langle R, \rho \rangle$, add t-functions on the members of $\bar{D} - \bar{D}^i$ extending the members of $\bar{Q} - \bar{Q}^i$ to models for S, *and* be assured that the resulting set of t-functions $\bar{t}$ on the members of $\bar{D}$ satisfy $\langle R, \rho \rangle$. If, for example, the constraint were $\langle =, => \rangle$ and $D_1 \in \bar{D}^i$ shared an element x with $D_2 \in \bar{D} - \bar{D}^i$, then it might not be possible to assign the same value to $t_1(x)$ and $t_2(x)$, as required to satisfy $\langle =, => \rangle$, and also make both $\langle D_1, t_1, n_1 \rangle$ and $\langle D_2, t_2, n_2 \rangle$ models for S^i and S respectively.

Thus (4) entails the claim (3) that we intuitively recognize as simply a claim that there exist t-functions *all* of which have the properties demanded by the basic mathematical structure of the theory, and also entails the claim (3i) that we intuitively recognize as a claim that the t-function has a special form in certain applications of the theory. Yet, these two claims together do not exhaust the content of (4) – they are not equivalent to (4). In addition, (4) requires that the t-functions satisfying (3i) be a sub-set of those that satisfy (3). That is, it requires that there be *one* set of t-functions

$\bar{t}$, satisfying constraints that are a part of "the concept of t", which makes all members of $\bar{Q}$ models for S *and* makes all members of $\bar{Q}^i$ models for S^i. It is the requirement that $\bar{t}$ satisfy the constraints $\langle R, \rho \rangle$ which, intuitively speaking, makes the claims of (4) about members of $\bar{Q}-\bar{Q}^i$ and $\bar{Q}^i$ "hang together". More precisely, it says that t-function values we choose to make members of $\bar{Q}-\bar{Q}^i$ models for S may not be independent of t-function values we choose to make members of $\bar{Q}^i$ models for S^i, and vice versa. It is this "additional" content of (4) that allows us intuitively to regard the t-functions alleged to exist by (3) and (3i) to be theoretical functions of the *same* theory. It is this that provides for the possibility the t-function values on members of $\bar{Q}^i$ might be useful in making predictions about n-function values on members of $\bar{Q}-\bar{Q}^i$.

It is important to note that claim (4) may provide possibilities of observing t-function values beyond those provided by claim (3). Suppose that 'is an S^i' is defined by (D1i), obtained from (D1) by adding axiom (D1i.6) and let the constraint in (4) be $\langle =, = \rangle$. Consider the following example.

(E4)   Suppose $D_1 = \{x, y\} \in \bar{D}'$. (4) entails (3i) and (3i) entails

$$n_1(x)t(x) + n_1(y)t(y) = 0$$

and $t(x) + t(y) = 1$,

so that

$$t(x) = \frac{n(y)}{n(y) - n(x)}, \qquad t(y) = \frac{n(x)}{n(x) - n(y)}.$$

Thus (4) justifies a calculation of t-function *values*, in some cases, rather than just t-function *ratios*. Moreover, if either x or y occur in any other member of $\bar{D}$, this calculation is non-trivial. It might, for example, provide us a means for predicting unknown n-function values on some member of $\bar{D}$.

This rather simple-minded example serves to bring to light a very important possibility. It might be the case that the predicate 'is an S' and the constraints $\langle R, \rho \rangle$ are such that the claim (3) justifies *no* non-trivial calculations to t-function values. Yet, it may still be that the claim (3i) does so. This is to say that the claim that the t-function in all applications has the properties demanded by the basic mathematical structure of the theory may not be sufficient to justify non-trivial calculations of t-function values (or even of t-function ratios). It may be that, as a matter of fact, the domains in $\bar{D}$ just do not happen to overlap in the appropriate way. Or it may be that there simply is no possibility that n-function values on *any* domain could uniquely determine what the t-function values or ratios would have to be to produce a model for S. Intuitively, S might not put very stringent conditions on the t-function. Nevertheless, by postulating that the t-function has special forms in certain applications *and* treating this claim in the way suggested here, we may still account for non-trivial calculations of t-function values.

This situation can be described in a way which has an air of paradox about it. We can not make meaningful calculations of t-function values unless we assume that certain "t-laws" are true in some applications. But, how are such assumptions ever to be warranted? How can we have any reason to believe that these "t-laws" hold in these applications if we have not observed at least some t-function values? A bit of reflection, however, should suffice to convince us that this situation is no different, in this respect, than the situation in which a single claim of form (3) suffices to justify non-trivial calculations of t-function values. In either case we can not speak significantly of t-function values until we have, at least provisionally, accepted the overall claim of the theory. In the situation just described, this overall claim entails that there exist t-functions having a special form in certain applications. This is not to be understood as a claim which is supported by observations of t-function values, just as the claim that there exist t-function values which make all applications models for S is not to be understood in this way. It is rather to be understood as a kind of "working hypothesis" – something we adopt provisionally and see how far we can get. The significant point to be understood is that the hypotheses we make about t-function values in certain applications may be relevant in determining what is to count as an observation of t-function values. There is nothing paradoxical about this when one understands precisely how these "observations" function in relation to the empirical claim of the theory.

Thus far we have seen that sentence (4) might plausibly be used to express the empirical content of a theory in which the t-function is postulated to have a special form in *one* set of intended applications. Clearly we can generalize (4) to deal with theories where the t-function is postulated to have different forms in different sub-sets of intended applications, and where its form is possibly restricted still further in some sub-sets of these sub-sets. The force function in classical particle mechanics is apparently an example of this situation.

Sentence (4) is obviously a special case of a sentence of the form:

(5) $\quad (\exists \bar{x})[\bar{x}$ is an $E_{\bar{Q}} \wedge \bar{x}$ is an $S \wedge$
$\bar{x}^{i_1}$ is an $S^{i_1} \wedge \bar{x}^{i_1 j_1}$ is an $S^{i_1 j_1} \wedge \cdots \wedge$
$\bar{x}^{i_2}$ is an $S^{i_2} \wedge \bar{x}^{i_2 j_1}$ is an $S^{i_2 j_1} \wedge \cdots \wedge$
$\vdots$
$\bar{x}^{i_n}$ is an $S^{i_n} \wedge \bar{x}^{i_n j_1}$ is an $S^{i_n j_1} \wedge \cdots \wedge$
$C(\bar{x}, R, \rho)]$.

This sentence says, roughly, that there is one set of t-functions, satisfying the constraint $\langle R, \rho \rangle$ which seems to extend all members of $\bar{Q}$ to models for S and, in addition, extends members of various sub-sets of $\bar{Q}$ to models for various restrictions of S.

The suggestion being made here is that *the* central empirical claim of theories in which the theoretical function is postulated to have various special forms in different applications may be rendered as a claim of form (5). This claim of form (5) will entail a number of claims of form (3) – one for the entire range of intended applications $\bar{Q}$, and one for each sub-set of $\bar{Q}$ in which the t-function is claimed to have a special form. But, just as in the case of (4), the conjunction of these claims of form (3) will not entail (5). Intuitively, the main force of the present suggestion is that the various hypotheses about special forms of the theoretical function are not independent of one another. The theory can not be viewed as simply a conjunction of independent hypotheses about the form of the t-function in different applications. An addition of such a hypothesis to the theory requires that we check to see that the present content of the theory remains tenable when we add the new hypothesis. It is the interdependence of these hypotheses that allows us to exploit them in making non-trivial calculations of t-function values. In turn, the possibility of doing this supports our intuitive feeling that the theoretical function, in all applications of the theory, is a measure of the same "physical quantity" rather than some *ad hoc* construction which just happens to fit the data.

One way to understand (5) is to compare it with (3) in the following way. A claim of form (3) brings together a variety of claims of form (2) by requiring that the t-functions which satisfy them also satisfy certain constraints. A claim for form (5) brings together, in the same way, a collection of claims of form (3) made with the same constraint but different set-theoretic predicates characterizing different special forms for the t-function.

Of course, some of these claims of form (3) could degenerate into claims of form (2) if there was only one application in which the t-function was alleged to have the special form in question. It might also be that there were some intended applications in which the t-function was not claimed to have any special form. This is why the second conjunct of (5) is included. In general, using a claim like (5) allows us to bring together both claims of form (2) made with S and restrictions of S and claims of form (3) made with restrictions of S.

Two features about claims of form (5) are worth noting. It was pointed out in the preceding chapter that one might be committed to using a claim of form (3), with a given set-theoretic predicate and constraint, to make an empirical claim, and yet be a bit vague about just exactly what the range of applications $\bar{Q}$ in this claim should be. With claims of form (5) an additional possibility for impreciseness is available. One could be reasonably confident that a given mathematical structure and certain special forms of the t-function were adequate for dealing with some phenomena – one could be committed to using a claim of form (5) to make a claim about these phenomena. Yet, one might not be as confident about exactly where to draw the lines between the applications in which different "t-laws" were postulated to hold. This feature of claims of form (5) allows us to see how hypotheses that the t-function has special forms in certain applications might be maintained despite putative empirical evidence to the contrary. One might envision making repeated attempts at "carving up" the range of intended applications, trying to find one partition that made the claim of form (5) true. In this way, one might hang on to the claim that the t-function had a special form in some set of applications by adding or deleting members of this set and perhaps redrawing the boundaries of other sets of applications mentioned in (5).

The second feature of (5) to be noted is this. Suppose that the predicate 'is an S' and the constraint $\langle R, \rho \rangle$ were such that (3) was trivially true. That is, for any possible n-function values in members of $\bar{Q}$, t-functions satisfying $\langle R, \rho \rangle$ could be found to make each member of $\bar{Q}$ a model for S. In this case it might still be that (5) was *not trivially* true. There might be possible n-function values in members of $\bar{Q}$ which would make it impossible to find t-functions satisfying (5). In this case, one might naturally be led to say that the predicate 'is an S' can only be used to make empirical claims if we postulate specific t-function laws, or specific forms for the t-function. It is, however, important to understand that this does not mean that, in order to make a significant empirical claim, we must postulate specific forms for the t-function in *all* applications. It might be that the $\bar{Q}^{i_1}, \bar{Q}^{i_2}, \ldots, \bar{Q}^{i_n}$ in (5) do not exhaust $\bar{Q}$, i.e. there are some members of $\bar{Q}$ for which we make no claim that the t-function has a special form. Further, it might be that we could not delete these members from $\bar{Q}$ without weakening (5). That is, there could be n-function values in members of $\bar{Q} - \bigcup_\alpha \bar{Q}^{i_\alpha}$ which would make (5) false. The reason is this: Although there might be, for all possible n-function values, t-functions making all members of $\bar{Q} - \bigcup_\alpha \bar{Q}^{i_\alpha}$ models for S, there might not be t-functions which, together with the t-functions needed to make the members of $\bar{Q}^{ijk\cdots}$ models for S^{ijk}, satisfied $\langle R, \rho \rangle$. Roughly speaking, the simple claim that some applications can be made models for S may not be trivial when there are constraints on the t-functions that can be used "connecting" these applications with others in which special forms for the t-function are postulated.

There is at least one possibility of generalizing (5) that should be mentioned briefly. Thus far we have supposed that there was *one* constraint to be satisfied by all the t-functions used to make intended applications models for S and its restrictions. We could, however, consider claims in which the t-functions for some applications were required to satisfy *additional* constraints. For example, we might require that all t-functions satisfy $\langle =, =\rangle$ and further that all t-functions for some $\bar{Q}' \subset \bar{Q}$ satisfy $\langle R_0, \rho_+ \rangle$ – that they be extensive with respect to the ∘-operation. This generalization provides additional possibilities for dealing with an existing theory's claims about observing values of theoretical functions.

This discussion of claims of form (5) completes our examination of various possibilities for using set-theoretic predicates to make empirical statements. No claim is made that this examination has covered all possibilities, nor even all plausible possibilities. It has covered enough to let us see one way these predicates may be used that avoids the problem of theoretical terms and permits an adequate account of observations of theoretical function values. In addition, we have seen a way to formulate sentences with these predicates which appear to be adequate renderings of claims that postulate special forms for theoretical functions. It is, of course, not obvious that these possibilities are sufficient to deal with *all* theories of mathematical physics. Indeed, it is not yet clear that they are completely adequate to provide a logical reconstruction of *any* significant theory. So far only fragmentary and sketchy examples have been cited aimed at making it at least plausible that claims like those we have considered are adequate for some theories at least. In the next chapter we will consider classical particle mechanics in some detail and try to show how the possibilities we have examined fare when we attempt to use them to provide a logical reconstruction of this theory.

CHAPTER VI

CLASSICAL PARTICLE MECHANICS

Thus far we have considered some simple examples of set-theoretic predicates to illustrate various alternative ways that such a predicate might be used to elucidate the logical structure of a scientific theory. We want now to see how these alternatives fare when we attempt to use them in providing a logical reconstruction of a real theory of mathematical physics. The theory to be considered is classical, or Newtonian, particle mechanics. Our procedure will be roughly this. We will settle upon a likely candidate for a set-theoretic predicate to characterize the mathematical structure of this theory. Then we will try to use this predicate to render the empirical content of this theory in each of the ways discussed in the preceding four chapters. This will serve to illustrate, in a more concrete way, the difficulties with some of these methods, and ultimately to provide at least a sketch of an adequate logical reconstruction of this theory. This sketch, together with the notion of Ramsey eliminability, will provide a means of treating in a systematic and perspicuous way some frequently raised questions about the status of the concepts of mass and force. It will also serve to illuminate questions about the measurability of masses and forces and the status of specific force laws in measuring forces.

Just exactly what is classical particle mechanics? What is its mathematical structure like? What are some paradigm examples of the claims this theory makes about the way the world is? In short, just what theory is it that is to serve as our example of a theory of mathematical physics?

Standard treatments of classical mechanics (e.g. [20]) sub-divide the discipline into four parts: particle mechanics, rigid body mechanics, the mechanics of deformable bodies, and the mechanics of liquids and gases. Each of these sub-divisions is supposed to deal with the motion of a different kind of thing – particle mechanics with the motion of particles, rigid body mechanics with the motion of systems of particles whose position relative to one another is fixed, etc.... In addition, there are apparently significant differences in the mathematical apparatus used in

CLASSICAL PARTICLE MECHANICS 111

each of the sub-divisions. Functions like torque and moment of inertia appear in the exposition of rigid body mechanics and not in particle mechanics. The basic partial differential equations of hydrodynamics contain different functions and have a different mathematical form from those of either rigid body mechanics or particle mechanics. These facts, together with the view of theories of mathematical physics that has been sketched, suggest that it might be fruitful to regard each of these sub-divisions of classical mechanics as a separate physical theory – each with its own mathematical formalism and range of applications.

Of course, the sub-divisions of classical mechanics are related in some way. There is something about them that makes it intuitively reasonable to say that they are all "parts" of the same theory. If we take the view that each of these sub-divisions is a distinct theory of mathematical physics, in our sense of 'theory', then we are obliged to explain the relation between these theories. The intuitive nature of this relation is apparent in most expositions of classical mechanics. Particle mechanics is usually expounded first, and then the concepts of the other sub-divisions explained in terms of the concepts of particle mechanics. This suggests that particle mechanics is, in some sense, more basic than the other sub-divisions, or that the other sub-divisions can be "reduced" to particle mechanics. Just how this intuitive notion of one theory's being reduced to another can be made precise is one of the topics of the next chapter. For the present, it is sufficient to note that we shall restrict our attention to classical particle mechanics, laying aside the question of precisely in what sense, if any, the rest of the classical mechanics is reducible to it.

In this connection, it should be noted that the set-theoretic predicates we propose to depict of the mathematical structure of classical particle mechanics all commit us to taking the mass of a particle to be constant in time. There are obviously applications of this theory where mass is not taken to be constant, e.g. to rockets with a diminishing fuel supply. One could simply say that our set-theoretic predicates are too simple to account for all applications of classical particle mechanics. We need one which allows that mass might be a function of time. This may well be true. But there is some intuitive reason to believe that any problem we could treat using a time-dependent mass function, we could also treat using time-independent masses, provided we took the appropriate sorts of things to be "particles". That is there is some reason to think that a theory with time-dependent masses is "reducible" to a theory with time-independent masses. At any rate, this would seem to be a promising line to try in attempting to provide a logical reconstruction to the entire theory of classical mechanics.

Suppose we agree to regard classical particle mechanics as a distinct physical theory characterized by a particular mathematical formalism and range of applications. Our first task is to become clear about exactly what this formalism is. Here we immediately encounter a difficulty. There is apparently not just one mathematical formalism associated with classical particle mechanics, but several. This fact is commonly expressed by saying that there are several "equivalent" formulations of classical particle mechanics. Among these are: the familiar Newtonian formulation in which mass, force and position are the primitive concepts; the Langrangian formulation in which generalized positions, forces and the kinetic energy function are primitives and Langrange's equation is the central relation; the Hamiltonian formulation in which generalized position and momenta and the Hamiltonian function are primitives and Hamilton's equations the central relation. On the view of theories of mathematical physics we have sketched, the appropriate thing to say about these and other different formulations of classical particle mechanics appears to be this. They are distinct theories – characterized by distinct mathematical structures – having the same range of applications – systems of particles in motion. Moreover, these distinct theories are, in some appropriate sense, "equivalent". The task of making precise this relation of equivalence between theories of mathematical physics will be deferred until the next chapter. For the present, it suffices to point out that we shall restrict our attention to the Newtonian formulation of classical particle mechanics. This is to say, we will attempt to make precise the mathematical apparatus associated with *this* formulation and make clear the way this apparatus is used to make empirical claims.

Most expositions of classical particle mechanics do not explicitly discuss the range of application of the theory they expound. They simply say that it deals with the motion of "particles", without being explicit as to what is to count as a particle. Joos ([20], p. 81) is a bit more explicit. He says that classical particle mechanics deals ". . . with the motion of bodies whose extension in space may be neglected". Though more explicit, this is still not much help unless one is given some idea of the circumstances in which a body's extension in space may be neglected. A better understanding of the sorts of physical systems to which classical particle mechanics is applied can be obtained by looking at the exemplary applications presented in expositions of the theory, and perhaps in the exercises accompanying some expositions. Clearly, it is claimed to apply to the motion of bodies in the solar system – both to the solar system as a whole and to various of its sub-systems like the system of Jupiter and its moons and the earth–moon system. It is claimed to apply to projectiles and freely falling bodies. It is claimed to apply to systems of "blocks" in various configurations: sliding on planes connected by cords, springs, etc.; suspended from "frictionless" pulleys; wired on the end of a cord. This brief list is obviously not exhaustive. However, it should serve to give some intuitive idea of what classical particle mechanics is "about".

On the view of logical reconstruction outlined in Chapter I, anyone committed to giving a complete logical reconstruction of classical particle mechanics would be faced with the task of giving some systematic account of the range of applications. Only a casual look at the applications of this theory, like the one above, suggests that this would be a formidable undertaking. This is a task I will attempt to avoid, in so far as possible. Only as much will be said about the explicit nature of the range of applications as is required to understand how constraints on the theoretical functions operate. In this way, among others, what is provided in this chapter falls short of being a complete logical reconstruction of classical particle mechanics. It might properly be termed 'a sketch of a logical reconstruction'. My reason for providing only a sketch is simple. I am unable to produce more. It may be that the task of providing a complete and systematic account of the range of applications is simply very difficult. However for reasons suggested near the end of Chapter IV, it may be that there is an inherent vagueness about the range of applications. This possibility will be examined in more detail in Chapter VIII.

With this intuitive understanding of what classical particle mechanics in the Newtonian formulation is like, we can begin to see how one might attempt to produce a logical reconstruction of this theory. In doing this we will have occasion to use several definitions of set-theoretic predicates. These are all essentially due to McKinsey *et al.* [26]. That these predicates characterize mathematical structures relevant to the Newtonian formulation of classical particle mechanics will be almost obvious. Very little will be offered in the way of explicit argument to this effect. The particular set-theoretic predicates used here are not the only plausible candidates for use in a logical reconstruction. They were chosen over others largely because they most closely resemble informal expositions of the theory. For a discussion of some alternatives see the article by McKinsey *et al.* cited above.

First, consider the predicate 'is a PM' (is a particle mechanics) defined by

(D7) x is a PM if and only if there exists a $P, T, \bar{s}, m, \bar{f}$ such that:

(1) $x = \langle P, T, \bar{s}, m, \bar{f} \rangle$;
(2) P is a non-empty, finite set;
(3) T is an interval of real numbers;
(4) $\bar{s}$ is a function from $P \times T$ into the set of ordered triples of real numbers such that, for all $p \in P, t \in T, D^2\bar{s}(p, t)$ exists;
(5) m is a function from P into the real numbers such that, for all $p \in P, m(p) > 0$;
(6) $\bar{f}$ is a function from $P \times T \times I$ into the set of ordered triples of real numbers such that, for $p \in P, t \in T$,

$$\sum_{i \in I} \bar{f}(p, t, i)$$

is absolutely convergent.

In (D7) we employ the following notation. If f is a real or vector valued function of one real variable, we denote the derivative of f at a point x in its domain by '$Df(x)$'. We denote the function whose value at every point where $Df(x)$ exists is $Df(x)$ by 'Df'. We denote $DDf(x)$ and DDf respectively by '$D^2f(x)$' and 'D^2f'. For any $p \in P$ let s_p be the function whose value at t is $\bar{s}(p, t)$. We denote $D\bar{s}_p(t)$ by '$D\bar{s}(p, t)$'. Similar notation will be used subsequently for other functions which have other than real numbers as arguments. Here we denote the set of all positive integers by 'I' and subsequently we shall denote the set consisting of the first n integers by 'I_n'.

The predicate 'is a PM' characterizes the basic set-theoretic structure common to all the various structures we will consider. They will all be defined by restrictions of (D7). In our previous terminology, anything that is a PM is a possible model for the set-theoretic structures associated with classical particle mechanics and the various restrictions of this structure associated with different force laws. (D7) is analogous to (D0) in the preceding discussion.

The first restriction of (D7) we consider simply adds Newton's second law. We define 'is a CPM' (is a classical particle mechanics) by:

(D8)　　x is a CPM if and only if:
(1) x is a PM;
(2) For all $p \in P$ and $t \in T$,
$$m(p)D^2\bar{s}(p, t) = \sum_{i \in I} \bar{f}(p, t, i).$$

Intuitively, 'is a CPM' characterizes the set-theoretic structure of systems of particles which move in accordance with Newton's second law. A natural question to ask about 'is a CPM' is this. Can any empirical claim of classical particle mechanics be made with a sentence of the form:

(1-CPM) Q is a CPM,

where 'Q' is a definite description of something that is a PM?

It is reasonably clear that 'Q' being a definite description of a PM is a necessary condition for (1-CPM) to plausibly be taken as a claim of *any* theory. (1-CPM) hardly makes sense otherwise. But it is not sufficient to make (1-CPM) plausible as a claim of classical particle mechanics. Many entities that are PM's are obviously not the sort of things that classical particle mechanics 'talks about'. But is it possible to find *some* things that are PM's which we can recognize as intended applications of classical particle mechanics?

The answer seems pretty clearly to be that we can. Take P to be the set of 'particles' associated with some intended application of classical particle mechanics – the set of things whose motion this application is alleged to explain. Let T be some interval of real numbers which is isomorphic to the time interval during which this application treats the motion of the members of P, the <-relation on T corresponding to the earlier-than-relation on this time interval. Roughly, T is a time scale for this application. Let $\bar{s}(p, t)$ be the position vector of particle p at time t, with respect to some spatial coordinate system. Let $m(p)$ be the mass of the particle p on some mass scale, and let $\bar{f}(p, t, i)$ be the i-th force acting on the particle p at time t, on some scale for measuring forces. It is certainly plausible to expect that these entities, for any application of classical particle mechanics, will constitute a PM. Further, it is plausible to think that it is just this sort of model for PM that we want to claim also satisfies Newton's second law.

It is important to understand that the entities that are T's and $\bar{s}$'s in models for PM are simply intervals of real numbers and functions from P into the set of ordered triples of real numbers. If this model for PM is to be an intended application of classical particle mechanics it is necessary that these entities be related to physical theories about time and geometry in roughly the way sketched above. That is, the numbers in T must correspond to instants of time and the members of $\bar{s}(p, t)$ must correspond to distances between p and some fixed point at time t. To explain exactly how one is assured that any given candidate for T or $\bar{s}$ has these properties requires a logical reconstruction of these theories of time and geometry which "underlie" classical particle mechanics. Providing such logical reconstructions is a non-trivial task. It is by no means obvious that intuitively acceptable ones can be given which yield numbers having the properties (D7) requires of T and $\bar{s}$. For example, it is not obvious that a *physical* theory of the earlier-than-relation would guarantee an uncountably infinite number of time instants. A similar question might be raised about the requirement that $\bar{s}$ be twice differentiable with respect to time. Failure to provide an explicit account of these underlying theories is another way in which what is offered in this chapter falls short of a complete logical reconstruction of classical particle mechanics. Moreover, even the part that we do reconstruct remains open to the objection that it takes as "raw materials" mathematical entities that can not actually be "produced" by these underlying theories.

Though it seems plausible to think that *some* empirical claims of classical particle mechanics might be made with sentences of form (1-CPM), it does not appear likely that *all* could be. The reason for this has already been mentioned in Chapter II. It appears very likely that both mass and force, the functions m and $\bar{f}$, are theoretical with respect to classical particle mechanics. This is to say that in every application of classical particle mechanics there are some arguments of these functions with the following property. There is no acceptable method of determining the values of these functions for such arguments which does not presuppose that classical particle mechanics does, in fact, apply to some physical systems. If we attempt to express the empirical content of this theory as a collection of claims of form (1-CPM), one for each intended application of the theory, it is clear that there is a problem in understanding how we obtain evidence for these claims. We can apparently never have good reason to believe that some particular numbers are values of the m_1 and $\bar{f}_1$ functions described by 'Q' unless we have good reason to believe that some claim of the form 'Q_j is a CPM' is true. This difficulty was described, in general, in Chapter II and labeled 'the problem of theoretical terms'. Thus, if we are correct in saying that mass and force are theoretical with respect to classical particle mechanics, it appears that claims of form (1-CPM) alone are not sufficient to provide a logical reconstruction of this theory.

Of course one might dispute the claim that mass and force are theoretical with respect to classical particle mechanics. As was emphasized in Chapter II, this is a substantive claim about the existing expositions of this theory. It might be refuted by producing an example of an acceptable, generally applicable method of determining values of these functions which did not presuppose that classical particle mechanics applied to any physical system. In the case of force, I know of no, even remotely plausible, candidate for such a method. All means of measuring forces, known to me, appear to rest, in a quite straight-forward way, on the assumption that Newton's second law is true in some physical system, and indeed also on the assumption that some particular force law holds.

In the case of mass, the absence of such a method may be less obvious. It is often suggested for example, that 'mass' might simply be "defined" as that quantity determined by using an analytical balance in the customary way. This way of determining mass values would not then presuppose the truth of *any* physical laws. A moment's reflection, however, shows that this proposal could not account for the fact that we can, and quite frequently do, question whether or not any particular analytical balance is determining mass-ratios accurately. Moreover, one way of understanding what we are doing when we raise such questions is this. We are asking whether this particular analytical balance is such that, when the bodies on its pans are regarded as particles, we obtain a model for CPM in which the force function has a special form – a form which allows us to calculate the mass-ratio of the "particles" from their observed motion. To discover whether the force function has this form we are typically forced to consider rather subtle questions about whether, and in what way, the balance and its parts may be regarded as models for other branches of mechanics such as rigid body mechanics. To give a full account of this matter would take us far afield. But this much is clear. When we claim that an analytical balance determines mass-ratios accurately, we are claiming, at least, that there is a way of regarding this physical system as a model for CPM.

Suppose we agree that claims of the form (1-CPM) will not be completely adequate for producing a logical reconstruction of classical particle mechanics. The obvious next step is to try to exploit the Ramsey solution for the problem of theoretical terms – that is, try to use a sentence like (2), treating both m and $\vec{f}$ like t was treated in the previous examples. To this end, let us define the predicate 'is a PK' (is a particle kinematics) in the following way:

(D9) x is a PK if and only if, there exists a P, T and $\vec{s}$ such that:
   (1) $x = \langle P, T, \vec{s} \rangle$;
   (2) P is a non-empty finite set;
   (3) T is an interval of real numbers
   (4) $\vec{s}$ is a function from $P \times T$ into the set of ordered triples of real numbers such that, for all $p \in P$, $t \in T$, $D^2\vec{s}(p, t)$ exists.

(D9) is analogous to (D2) in our previous discussion of the Ramsey solution. Anything that is a PK is a possible partial model for CPM. Likewise, analogous to (D3) we may define, for any PK, $\langle P_0, T_0, \vec{s}_0 \rangle$, the predicate 'is an $E_{\langle P_0, T_0, \vec{s}_0 \rangle}$' in the following way:

(D10)   If $\langle P_0, T_0, \vec{s}_0 \rangle$ is a PK, then x is an $E_{\langle P_0, T_0, \vec{s}_0 \rangle}$ if and only if there exists an m and $\vec{f}$ such that:
   (1) $x = \langle P_0, T_0, \vec{s}_0, m, \vec{f} \rangle$;
   (2) m is a function from P into the real numbers such that, for all $p \in P$, $m(p) > 0$;
   (3) $\vec{f}$ is a function from $P \times T \times I$ into the set of ordered triples of real numbers such that, for all $p \in P$, $t \in T$,

   $$\sum_{i \in I} \vec{f}(p, t, i)$$

   is absolutely convergent.

Consider now a sentence of the form

(2-CPM) $(\exists x)(x$ is an $E_Q \wedge x$ is a CPM$)$

where Q is a PK. If $Q = \langle P_0, T_0, \vec{s}_0 \rangle$ is such that $\vec{s}(p, t)$ may be regarded as the position vector of p at time t, then (2-CPM) says roughly this. Given the paths of all the particles in P_0, during the time interval T_0, there is some way to fill in mass and force functions to satisfy Newton's second law.

Is it plausible to think that (2-CPM) might be used to make an empirical claim? Clearly it is not, for (2-CPM) is trivially true. It is easy to see that, no matter what model for PK we choose, it is always possible to produce m and $\vec{f}$ functions which extend it to a model for (CPM). It is also clear that, in a quite trivial sense, both m and $\vec{f}$ are Ramsey eliminable from the claim (2-CPM). So long as we restrict our attention to the predicate 'is a CPM', this fact appears to count decisively against using a Ramsey claim like (2) in a logical reconstruction of classical particle mechanics. It does not, of course, show that a claim like (2) made with some other predicate would not fare better.

It must be understood that the inadequacy of the Ramsey solution here is quite distinct from the general inability of this solution to account for non-trivial calculations of theoretical function values. Nevertheless, it is possible that the present difficulty *too* might be avoided by using a sentence like (3) with appropriate constraints on the m and $\bar{f}$ functions.

To follow up this suggestion, let us consider how we might describe the range of intended applications of classical particle mechanics. Let

$$\bar{P} = \{P_1, P_2, \ldots, P_i, \ldots\}$$

be a set of non-empty, finite sets. Suppose that each $P_i \in \bar{P}$ is a set of particles whose motion, during some time intervals, classical particle mechanics is alleged to explain. Let T_i^j be an interval of real numbers which "represents" the j-th time interval during which classical particle mechanics is alleged to explain the motion of the particles in P_i ('j' may range over all real numbers). Then, if

$$\bar{D} = \{\langle P_i, T_i^j \rangle\}, \quad P_i \in \bar{P},$$

the set of intended applications of classical particle mechanics is the set of models for PK:

$$\bar{Q} = \{\langle P_i, T_i^j, \bar{s}_i^j \rangle\}, \quad \langle P_i, T_i^j \rangle \in \bar{D},$$

where $\bar{s}_i^j$ "represents" the position function of the particles in P_i during T_i^j.

It seems clear intuitively that there is some interesting structure on the set $\bar{D}$ which might be described. For example, it appears that if we claim that classical particle mechanics applies to $\langle P_i, T_i^j, \bar{s}_i^j \rangle$ and T_i^k is a sub-interval of T_i^j, then we claim also that it applies to $\langle P_i, T_i^k, \bar{s}_i^k \rangle$. Likewise, if P is a sub-set of P_i, we would apparently claim that classical particle mechanics also applies to $\langle P, T_i^j, \bar{s} \rangle$. A complete logical reconstruction of classical particle mechanics would contain a specification of the membership of $\bar{D}$ and a systematic account of such structural features as these.

An example of a possible account of this sort is the following. One might maintain that "the universe" is composed of a large, but finite number of indivisible atomic particles. Let π be the set of all such particles. Classical particle mechanics is alleged to explain the motion of the members of π during a time interval τ which extends to infinity in both directions. Moreover, it is alleged to explain the motion of any sub-set of π during any sub-interval of τ. Thus $\bar{P} = 2^\pi$ and has the structure of a Boolean algebra, and $\bar{D} = \bar{P} \times$ (the set of all sub-intervals of the real numbers).

This account of the range of application of classical particle mechanics appears to be implicit in many expositions of the theory, including that of Newton. However, a hard-headed empiricist might object to it on at least two grounds. First, we are given no explicit account of the nature of the particles in π. No ordinary physical objects are known to be indivisible. Second, even if we understood what these "ultimate particles" were, it is clear that we sometimes apply classical particle mechanics to explain the motion of sets of ordinary physical objects that can be broken into parts to which we would also claim the theory applies. We do not *just* claim that the theory applies to the motion of sets of ultimate particles.

The second objection might be countered in this way. It can be shown that when the force function has a certain form – when the "internal" forces that particles exert on each other are third law, central forces – a set of particles may be regarded as a single particle located at the center of mass of the set, having the total mass of the set and acted upon by the vector sum of the external forces acting on all the particles in the set. (For a precise statement of this result, see below (T1), p. 130. If we now assume that ordinary physical objects are sets of the ultimate particles mentioned in this account, then we may say this. For each application of classical particle mechanics where ordinary physical objects are taken as particles, there is some application, in the range of applications this account specifies, which "underlies" it in the way sketched here.

But, the empiricist might still stick by his first objection and claim that we *only* apply classical particle mechanics to ordinary, non-ultimate physical objects. He might argue for this claim on some epistemological grounds, or simply support it by citing as evidence actual expositions of the theory. He might say that the talk of "ultimate particles" in these expositions serves, at best, as a device to aid us in picking out those sets of ordinary physical objects whose motion we claim to be explained by this theory. In particular, it suggests a principle for constructing new applications from previously given applications. It does this in two ways. First, it says that if $\langle P, T \rangle$ is in $\bar{D}$, then so are all the members of $2^P \times$ (the set of all sub-intervals of T). Second, together with the theorem cited in the previous paragraph, it says that if $\langle P, T \rangle$ is in $\bar{D}$, and the forces have the appropriate form, then by *concatenating* elements of sub-sets of P one may generate new members of $\bar{D}$.

We shall not pursue this question any further now. It is sufficient, for present purposes, to note that this discussion strongly suggests that *any* adequate account of the range of intended applications and its structure will have the following features. It is possible that the same particle appear in different applications of the theory, and that the same particle at the same time appear in different applications. Moreover, there is a kind of concatenation operation defined on the union of the sets of particles associated with every application. These facts will be useful in understanding the kind of constraints that might be imposed on the m and f functions in a claim like (3).

The general notion of a constraint on the m-functions in various applications can be defined in a way precisely analogous to the way we defined a constraint on the t-function in the preceding example. In (D4), replace $\bar{D}$ by $\bar{P}$, $\varDelta$ by $\pi = \bigcup P_i$, $P_i \in \bar{P}$ and $\bar{t}$ by $\bar{m} = \{m_1, m_2, \ldots, m_i, \ldots\}$. This yields a definition of '$\bar{m}$ is constrained by $\langle R, \rho \rangle$'.

The situation with respect to constraints on the $\bar{f}$-function is a bit more complicated since the arguments of this function are ordered triples $\langle p, t, i \rangle$, $p \in P$, $t \in T$, $i \in I$.

Consider the following definition:

(D11) If $\bar{P} = \{P_1, P_2, \ldots, P_k, \ldots\}$, $k \in I_n \subseteq I$, is a set of non-empty sets; for each $k \in I_n$, and j in some (perhaps non-denumerably infinite) set of real numbers J_k, T_k^j is an interval of real numbers; $\bar{D} = \{\langle P_k, T_k^j \rangle\}$, $1 \leq k \leq n$, $j \in J_k$; $\varDelta = \bigcup_{k,j} [P_k \times T_k^j \times I]$, $\langle P_k, T_k^j \rangle \in \bar{D}$; and $\bar{f} = \{\bar{f}_k^j\}$ is a set of functions such that $\bar{f}_k^j$ is a function from $P_k \times T_k^j \times I$ into the set of ordered triples of real numbers then $\bar{f}$ *is constrained by* $\langle R, \bar{\rho} \rangle$ if and only if:

(1) R is an n-ary relation whose field is $\varDelta$.

(2) ρ is an n-ary relation whose field is the set of ordered triples of real numbers, and

(3) If $R(x_1, x_2, \ldots, x_n)$ and
$x_i = \langle p_i, t_i, i_i \rangle \in P_{k_i} \times T_{k_i}^{j_i} \times I$
then
$\rho(\bar{f}_{k_1}^{j_1}(p_1, t_1, i_1), \bar{f}_{k_2}^{j_2}(p_2, t_2, i_2), \ldots, \bar{f}_{k_n}^{j_n}(p_n, t_n, i_n))$.

In the case that $\langle R, \bar{\rho} \rangle = \langle =, = \rangle$, the constraint on $\bar{f}$ amounts to this. If the same particle p at the same instant of time t appears in several

122   THE LOGICAL STRUCTURE OF MATHEMATICAL PHYSICS

applications, then for any $i \in I$, the ordered triple $\bar{f}(p, t, i)$ will be the same in all these applications. That is, the same forces (numbered in the same order) are required to act on p at t, in every application in which p at time t appears. This is equivalent to requiring that $\bigcup_{k,j} \bar{f}_k^j$ be a function on Δ.

We may formulate a claim like (3) for CPM precisely by introducing the following definitions analogous to (D5) and (D6).

(D12)   If $\bar{Q} = \{\langle P_i, T_i^j, \bar{s}_i^j \rangle\}, i \in I_n \subseteq I; j \in J_i$ is a set of real numbers and $\langle P_i, T_i^j, \bar{s}_i^j \rangle$ is a PK; then $\bar{x}$ is an $E_{\bar{Q}}$ if and only if there exists an $\bar{m}$ and $\bar{f}$ such that:
(1) $\bar{m} = \{m_1, m_2, \ldots, m_i, \ldots\}$
(2) $f = \{\bar{f}_i^j\}$
(3) $\bar{x} = \{\langle P_i, T_i^j, \bar{s}_i^j, m_i, \bar{f}_i^j \rangle\}$
(4) For all $i \in I, j \in J_i$
   $\langle P_i, T_i^j, \bar{s}_i^j, m_i, \bar{f}_i^j \rangle$ is an $E_{\langle P_i, T_i^j, \bar{s}_i^j \rangle}$

(D12) simply says that, if $\bar{Q}$ is a set of possible partial models for CPM, then '$\bar{x}$ is an $E_{\bar{Q}}$' means that $\bar{x}$ is a set of possible models for CPM constructed by "filling out" the members of $\bar{Q}$.

(D15) $\bar{x}$ is a $\overline{\text{CPM}}$ if and only if there exists a

$\bar{P} = \{P_1, P_2, \ldots, P_i, \ldots\}$ and a
$\bar{m} = \{m_1, m_2, \ldots, m_i, \ldots\}, i \in I_n \subseteq I$, and for all $i \in I_n$, there exist
$\bar{T}_i = \{T_i^j\}$
$\bar{s}_i = \{\bar{s}_i^j\}$
$\bar{f}_i = \{\bar{f}_i^j\}$
where j is in some set of real numbers J_i such that
(1) $\bar{x} = \{\langle P_i, T_i^j, \bar{s}_i^j, m_i, \bar{f}_i^j \rangle\}$
(2) for all $i \in I_n, j \in J_i$
   $\langle P_i, T_i^j, \bar{s}_i^j, m_i, \bar{f}_i^j \rangle$ is a CPM

(D13) says roughly that '$\bar{x}$ is a $\overline{\text{CPM}}$' means that $\bar{x}$ is a set of models for CPM.

Let us also agree that, when $\bar{Q}$ is a set of possible partial models for CPM and $\bar{x}$ is an $E_{\bar{Q}}$, then '$C(\bar{x}, m, R, \rho)$' means that the m-functions in the various members of $\bar{x}$ satisfy the constraint $\langle R, \rho \rangle$ and '$C(\bar{x}, \bar{f}, R', \bar{\rho})$' means that the $\bar{f}$-functions in members of $\bar{x}$ satisfy the constraint $\langle R', \bar{\rho} \rangle$.

Consider now a sentence of the form

(3-CPM) $(\exists \bar{x})[\bar{x}$ is an $E_{\bar{Q}} \wedge \bar{x}$ is a $\overline{\text{CPM}} \wedge$
$C(\bar{x}, m, R, \rho) \wedge C(\bar{x}, \bar{f} \ R', \bar{\rho})]$,

where $\bar{Q}$ is some set of possible partial models for CPM (models for PK), $\langle R, \rho \rangle$ is a constraint (in the sense of (D4)) on the set of functions $\bar{m}$, members of which appear as the fourth element in members of $\bar{x}$, and $\langle R', \bar{\rho} \rangle$ is a constraint (in the sense of (D11)) on the set of functions f, members of which appear as the fifth element in members of $\bar{x}$. Can we find some such $\bar{Q}, \langle R, \rho \rangle$ and $\langle R', \bar{\rho} \rangle$ which would make it possible to regard (3-CPM) as *the* central empirical claim of classical particle mechanics?

Roughly what we want to do is this. Find a set of sets of particles $\{P_i\}$ and a set of time intervals $\{T_i^j\}$ such that it is plausible to think that the empirical content of classical particle mechanics is roughly this. The motion of any set of particles P_i during any time interval T_i^j can be accounted for by the laws of classical particle mechanics. This is to say, each triple $\langle P_i, T_i^j, \bar{s}_i^j \rangle$ is a kinematical description of a physical system whose motion is accounted for by classical particle mechanics. Classical particle mechanics is *applied* to explain this motion by assigning masses and forces to the particles in P_i so that, taken together with the observed position functions $\bar{s}_i^j$, these give us a model for CPM. But the masses and forces we assign can not be entirely arbitrary and *ad-hoc*. It is an essential feature of these concepts, as they are employed in classical particle mechanics, that the assignments of masses and forces in different applications of this theory satisfy certain constraints. These constraints serve to rule out, for particular applications, assignments that we would intuitively regard as arbitrary or *ad-hoc*. By doing this they *may* make the claim that a particular kinematical description of a physical system can be extended to a model for CPM a non-trivial one. We shall say no more here specifying the sets $\{P_i\}$ and $\{T_i^j\}$ but we will examine some possible constraints on the mass and force functions to see if any plausible ones can be found for which (3-CPM) at least does not turn out to be trivially true.

It seems fairly clear that one constraint we shall want to place on the mass function is $\langle =, = \rangle$. This is just to say that we ordinarily regard mass as an intrinsic property of a particle that remains the same at whatever time and in whatever environment we happen to find the particle. It should be clear that whether or not this constraint can be satisfied may depend on when we are willing to say two "things" are one and the same particle.

That is, it may depend on how we decide to carve the world up into particles in specifying the range of intended applications. If we had unlimited freedom in deciding what to call one and the same particle, then clearly this constraint could always be satisfied. But presumably we do not have such unlimited freedom. There are surely some strong intuitive *limits* on the identity conditions for the things that we regard as particles in applications of classical particle mechanics. It might still be the case, however, that these limits do not completely determine these identity conditions so that one could envision accommodations being made in the range of intended applications designed to preserve mass as an intrinsic property of particles.

Earlier (Chapter IV) it was suggested that it might be fruitful to regard the mass function assignments to be constrained to be extensive with respect to a concatenation operation on the set of all things we count as particles. It was noted then that this view conflicts with the commonly accepted view that it is a straight-forward empirical fact – discovered by measuring mass ratios – that mass is extensive with respect to concatenation. I know of no "knock down argument" which shows the accepted view to be incorrect. However, the immediately preceding discussion of the range of intended applications of classical particle mechanics suggests an account of the concatenation operation on particles which supports the view that the extensiveness of mass should be regarded as a constraint.

Roughly, the account is this. Suppose we have a model for CPM whose set of particles is P. Construct a new set of "particles" by partitioning P into $P_1, P_2, \ldots, P_n$ and regarding P_i as a particle whose position is the position of the center of mass of its members. If the motion of P during T is a model for CPM where the internal forces are third-law, central forces then we know the following. These new "particles" *can be* made into a model for CPM by taking the mass of P_i to be the sum of the masses (in the initial model) and the force on P_i to be the vector sum of the external forces (in the initial model) on its members. But note, this may not be the *only* way of making the motion of the new "particles" a model for CPM. We might find *other* mass and force functions which would serve equally well. A trivial example is a two-body central force system whose center of mass is at rest. Any forces whose vector sum is zero, together with any arbitrary mass will make the motion of this composite "particle" a model for CPM. However, it seems intuitively clear, that even though these other possibilities exist, we would decline to consider them when we set about to find masses and forces to make all our intended applications models for CPM. This is to say roughly this. Whenever new applications are built up from old by concatenation of particles, we constrain ourselves to assign masses and forces in the way described above.

This is probably not a completely adequate account of why we assign masses to the concatenations of particles in the way we do. For example, it does not account for situations in which the motion of concatenations of particles is considered at different times the motion of their components. Also, it appears that we concatenate particles and assign masses to the concatenations in this way even when the assumption of third law, central forces does not hold. This account does not illuminate these situations. But, even though incomplete, the account does strongly suggest that extensiveness with respect to concatenation is a property we *demand* of assignments of mass functions, rather that one we find they just happen to have.

In the case of the force function, it also appears that the $\langle =, = \rangle$ constraint is a natural one to impose. It requires, roughly, that, at any given time, the same forces are acting on a particle, in whatever situation we consider the particle to be, for purposes of applying classical particle mechanics to explain its motion. For example, we might consider the motion of a two-particle system $P = \{p_1, p_2\}$ and a three particle system $P' = \{p_1, p_2, p_3\}$ during the same time interval T. This constraint requires us to use force functions whose values for p_1 and p_2 are the same when we seek to make both these motions models for CPM. Except in cases where we explicitly neglect the forces due to some particles – as when we neglect the sun in treating the earth–moon system – it appears that we do impose such a constraint.

Though we do appear to constrain our force function assignments in this way, this constraint appears to be less significant than the similar constraint on mass assignments. Intuitively, the reason is this. The constraint on the mass functions allows us to make inferences about the masses of particles in applications to motions taking place at different times. In contrast, the constraint on the force functions allows us to make inferences essentially only about applications to motions of a set of particles and its sub-sets during the same time interval. We can infer nothing about what happens to the forces on these particles when we put them in another environment at a different time.

One consequence of constraining the force function by $\langle =, =\rangle$ is that it places a kind of continuity requirement on these assignments. If we apply classical particle mechanics to overlapping intervals in the history of the same set of particles, we must use the same forces during the time the applications overlap.

Let us first consider placing only the constraint $\langle =, =\rangle$ on both the mass and force functions. Are these constraints alone enough to make (3-CPM) a plausible candidate for an empirical claim of classical particle mechanics? First note that (3-CPM) with these constraints is not trivially true, in the sense of being true for *any* position functions on the members of $P \times T$. Clearly, (3-CPM) with these constraints entails that the position functions satisfy a condition precisely analogous to the constraint on the force functions. Roughly, the same particle at the same time cannot have different values of the position function in different applications of classical particle mechanics. However, this is the *only* condition on the position functions that follows from (3-CPM) with these constraints. *Any* position functions meeting this condition will satisfy (3-CPM) with these constraints. Whether one takes this condition on the position functions as an empirical condition or as a part of the meaning of 'position', it is obvious that this condition can not plausibly be taken as something that classical particle mechanics tells us about "the way the world is".

We could, of course, add the constraint that the mass function be extensive with respect to concatenation. But this would not significantly change the situation. By suitably choosing the force functions, we could still make (3-CPM) true for any position functions meeting the condition of the previous paragraph. Indeed, it is fairly obvious that we can never change this situation by merely adding more constraints to the mass function. The only way to rule out more position functions is to add more constraints to the force functions.

This fact suggests that a natural course, at this point, would be to embark on an exhaustive discussion of various possibilities for constraining force functions which might lead to a claim of form (3-CPM) that could plausibly be taken as an empirical claim of classical particle mechanics. There are two reasons why this course does not appear attractive. The first, which we shall only mention here and return to later, is that constraints on the force functions appear to be intimately connected with postulating specific force laws. The second is that, even if we succeeded in coming up with some constraints on the force functions which made (3-CPM) a significant empirical claim, it is doubtful that we would take this claim to be *the* central empirical claim of classical particle mechanics. The reason is this.

If a claim of form (3-CPM) were *the* central empirical claim of classical particle mechanics, we should expect this claim to justify observations of values of mass and force functions in the way described in Chapter IV. In the case of the mass function this means that in some intended application it must be the case that the mass-ratios (for at least some of the particles) required to make this application a model for CPM be uniquely determined by the position function. It is clear that, so long as we are allowed to fill in the force function in any way we please, this can never be the case. For any observed acceleration function and any arbitrarily chosen mass, we can adjust the magnitude of the force vector to satisfy the second law (D8-2). More precisely, the m and $\bar{f}$-functions required to satisfy a claim of form (2-CPM) are never uniquely determined – not even up to a constant multiple.

Might it not be, however, that constraints on force function assignments would prevent such arbitrary choice of the magnitude of the force vector in some applications? This could be, but *only* if some values of the force function in some application were known. Constraints alone will not rule out the choice of any force function in any one particular application. But, knowing some values of the force function in some application presupposes that there is some (perhaps, some other) application in which some values of the force function are uniquely determined by the position function. But, this cannot be the case so long as we can assign values to the mass function arbitrarily. The makings of a vicious circle are clear. A claim of form (3-CPM), no matter what the constraints are, could not justify observations of values of mass and force functions.

Realizing this inadequacy of (3-CPM) – taken as *the* central empirical claim of classical particle mechanics – one might suggest the following line. The predicate 'is a CPM' is simply an inadequate or incomplete account of the mathematical structure of classical particle mechanics (in the Newtonian formulation). We should seek another predicate to use in a claim of form (3). This predicate must be such that it: (a) differs enough from 'is a CPM' to provide an adequate account of measuring values of mass and force functions; (b) can plausibly be regarded as characterizing *some* mathematical structure associated with classical particle mechanics; and (c) characterizes a mathematical structure possessed by *all* applications of classical particle mechanics. We can find predicates which satisfy (a) and (b) easily enough. They are restrictions of 'is a CPM' which, intuitively speaking, require that the force function have a special form, for example, that the forces be inverse-square, central. The difficulty with these predicates is that they do not satisfy (c). They all characterize special force laws which we might claim to hold in some applications of classical particle mechanics, but certainly not in all. It is not obvious that there are no predicates, other than these, which satisfy (a) and (b). I know of no argument demonstrating conclusively that there are none. However, it is clear that any predicate which satisfies (a) must, intuitively speaking, require more than the second law. Further, it appears that the only additional requirements that would satisfy (b) are force laws. That is, the only additional requirements we actually do place on applications of classical particle mechanics are certain force laws. Intuitively, this means that the only way we provide justification for measuring values of mass and force functions is by "postulating" that specific force laws hold in certain applications. The difficulty with using a claim of form (3) is that no *one* of these force laws are we willing to claim holds in *all* applications of the theory.

This discussion suggests that the appropriate way to depict the logical structure of classical particle mechanics is with a claim of form (5) made with 'is a CPM' and various restrictions of it characterizing specific force laws. In carrying this out in detail, there are two central issues to be met. First, just exactly what restrictions of 'is a CPM' are employed and how do they function to provide for the possibility of measuring values of mass and force functions? In particular, how do they introduce the possibility that there are applications in which the values of these functions or their ratios are uniquely determined by the value of the position function? Second, just exactly what constraints on the assignment of mass and force functions are employed and how do they function to make these determinations of mass and force values non-trivial? That is, what is the nature of the constraints that make these values applicable outside the application in which they are determined.

Insofar as possible, I shall attempt to deal with these two questions separately. We shall begin with the first question by considering claims of form (2) made with various restrictions of 'is a CPM'. Our object will be roughly to investigate the conditions under which information about the values of the position function is sufficient to uniquely determine values of the mass and/or force function required to satisfy these claims. Then we shall consider the possibility of introducing constraints which "bring together" these various claims of form (2) into a claim of form (4).

First, let us consider the predicate 'is an NCPM' (Newtonian classical particle mechanics) which characterizes the mathematical structure of an application in which all forces are directed along lines connecting particles and may be "paired-off" in such a way that members of the pairs are equal in magnitude and opposite in direction (satisfy Newton's third law).

It is expedient to define this predicate in a piecemeal fashion. Consider first:

(D13) If $\langle P, T, \bar{s}, m, \bar{f} \rangle$ is a CPM and $A \subseteq P \times I$ then

(1) A is balanced by $\langle A_1, A_2, C \rangle$ if and only if:
 (i) A_1 and A_2 are disjoint sets whose union is A;
 (ii) C is a one–one mapping from A_1 onto A_2 such that if $\langle q, j \rangle = C(\langle p, i \rangle)$ then $q \neq p$.

(2) A is balanceable if and only if there exist A_1, A_2, and C such that A is balanced by $\langle A_1, A_2, C \rangle$.

Intuitively, a balanceable sub-set of $P \times I$ is one whose members can be "paired-off" into "mutual interaction" or "internal" pairs. If $\langle q, i \rangle = C(\langle p, i \rangle)$ we may think of $\bar{f}(p, t, i)$ as a force that q "exerts" on p at t (D13-1-ii) rules out the possibility that a particle exerts a force on itself by requiring that $q \neq p$ when $\langle q, i \rangle = C(\langle p, j \rangle)$. Note though that one particle may exert several forces on another particle.

We can now say that a sub-set A of $P \times I$ is balanceable with Newtonian forces if there is some way to pair-off the forces in A so that the mutual interaction pairs are third-law central forces.

(D14) If $\langle P, T, \bar{s}, m, \bar{f} \rangle$ is a CPM and $A \subseteq P \times I$ then A is *balanceable with Newtonian forces* if and only if there exist A_1, A_2, C such that:

(1) A is balanced by $\langle A_1, A_2, C \rangle$;
(2) If $\langle q, j \rangle = C(\langle p, i \rangle)$ then, for all $t \in T$,
 (i) $\bar{f}(p, t, i) = -\bar{f}(q, t, j)$
 (ii) $\bar{s}(p, t) \times \bar{f}(p, t, i) = -\bar{s}(q, t) \times \bar{f}(q, t, i)$.

We then say that an NCPM is a CPM in which $P \times I$ is balanceable with Newtonian forces.

(D15) $x = \langle P, T, \bar{s}, m, \bar{f} \rangle$ *is an* NCPM if and only if:
   (1) x is a CPM;
   (2) $P \times I$ is balanceable with Newtonian forces.

At this point we can give a precise statement of the theorem mentioned earlier in the discussion of "particle concatenation".

(T1)   If $\langle P, T, \bar{s}, m, \bar{f} \rangle$ is a CPM and $A \subseteq P \times I$ is balanceable with Newtonian forces, then for all p', if

   (1) $P' = \{p'\}$

   (2) $m'(p') = \sum_{p \in P} m(p)$

   (3) $\bar{s}'(p', t) = \dfrac{\sum_{p \in P} m(p) \bar{s}(p, t)}{\sum_{p \in P} m(p)}$

   (4) $\bar{f}'(p', t, 1) = \sum_{\langle p, i \rangle \notin A} \bar{f}(p, t, i)$

   (5) $\bar{f}'(p', t, i) = 0, i > 1$,

   then $\langle P', T, \bar{s}', m', \bar{f}' \rangle$ is a CPM.

Intuitively, (T1) says that, given a model of CPM in which the "internal" forces – the forces that the particles exert on each other – are third-law central forces, a new model for CPM can be constructed by considering the initial *set* of particles P to be a particle p' whose mass is the total mass of particles in P; whose position is the center of mass of P; and acted upon by the sum of the "external" forces on the members of P.

Now, consider a claim of form (2) made with 'is an NCPM'.

(2-NCPM) $(\exists x)(x$ is an $E_Q \wedge x$ is an NCPM)

where Q is a PK and a member of the range of intended applications of classical particle mechanics. 'Is an E_Q' is defined by (D10).

Clearly (2-NCPM) is not trivially true. For example, it is false in any two-particle system where the accelerations of the particles are not directed along the straight line connecting the particles.

We are interested in discovering if there are *any* PK's such that the m and $\bar{f}$-functions satisfying (2-NCPM) are determined up to a constant multiple (or perhaps even uniquely) by the values of the $\bar{s}$-function. To the end of answering this question for the m-function, it is useful to note that the $\bar{f}$-function is Ramsey eliminable from (2-NCPM).

To see this consider the following definition of the predicate 'is an MPM' (is a momentum particle mechanics).

(D16) x is an MPM if and only if there exists a $P, T, \bar{s}$, and m such that

      (1) $x = \langle P, T, \bar{s}, m \rangle$
      (2) $\langle P, T, \bar{s} \rangle$ is a PK
      (3) m is a function from P into the real numbers such that, for all $p \in P, m(p) > 0$
      (4) For all $t \in T$:

         (a) $\sum_{p \in P} m(p) D \times \bar{s}(p, t) = \bar{O};$

         (b) $\sum_{p \in P} m(p) D \times \bar{s}(p, t) \times \bar{s}(p, t) = \bar{O}.$

A momentum particle mechanics (MPM) is essentially a particle kinematics (PK) with a mass function added to it in which linear (D16-4-a) and angular (D16-4-b) momentum are conserved.

Analogous to (D10), we can define for any PK, Q, an extension of Q to a possible model for MPM.

(D17)   If $\langle P_0, T_0, \bar{s}_0 \rangle$ is a PK, then x is an $\hat{E}_{\langle P_0, T_0, \bar{s}_0 \rangle}$ if and only if there exists an m such that:

      (1) $x = \langle P_0, T_0, \bar{s}_0, m \rangle$
      (2) m is a function from P into the real numbers such that, for all $p \in P m(p) > 0$.

Consider now the sentence

(2-MPM) $(\exists x)(x$ is an $\hat{E}_Q \wedge x$ is an MPM$)$,

where Q is a PK. (2-MPM) says that there is some way of adding a mass function to the observed paths of particles in Q so that linear and angular momentum are conserved.

It can easily be shown that (2-MPM) and (2-NCPM) are logically equivalent, in the following sense. Any MPM is extendable to a model for NCPM. Further, if $\langle P, T, \bar{s}, m, \bar{f} \rangle$ is a model for NCPM, then $\langle P, T, \bar{s}, m \rangle$ is a model for MPM. Thus the sub-sets of the set of all possible PK's, determined by each of these sentences, are co-extensive. That is, the set of all PK's extendible to a model for NCPM is identical with the set of all PK's extendible to a model for MPM. By an obvious extension of our notion of Ramsey eliminability (Chapter III, pp. 79 ff.) to situations in which there is more than one theoretical function, it seems natural to say that the $\bar{f}$-function is Ramsey eliminable from (2-NCPM).

Further, if the mass functions which will serve to extend Q_0 to a model for MPM are determined, up to a constant multiple, by $\bar{s}_0$, then so are those which will serve (in part) to extend Q_0 to a model for NCPM. Thus we have only to ask whether there are any PK's such that the m-functions satisfying (2-MPM) are determined up to a constant multiple by the $\bar{s}$ function.

Before considering this question, it is perhaps in order to pursue the question of Ramsey eliminability a bit further. It is natural to ask, at this point, whether the m-function is Ramsey eliminable from (2-MPM).

Claim (2-MPM) is equivalent to the claim that the infinite set of linear vector equations in unknowns $m(p)$, $p \in p$, of forms (D16-4-a) and (b), for all $t \in T$, has a solution for which $m(p) > 0$, for all $p \in P$. Call this system of equations Σ_T^P. This claim entails the claim that, for every finite set $T' \subset T$, the corresponding set of linear vector equations has a solution with $m(p) > 0$, for all $p \in P$. A necessary condition for the existence of such a solution, for a finite set T', is easy to state entirely in terms of the values of $D\bar{s}$ and $\bar{s}$. Each vector equation of the forms (D16-4-a) and (b) is equivalent respectively to three scalar equations of forms:

$$\sum_{p \in P} m(p)[D\bar{s}(p, t)]_i = 0$$
$$\sum_{p \in P} m(p)[D\bar{s}(p, t) \times \bar{s}(p, t)]_i = 0. \qquad i = 1, 2, 3$$

Thus we have a system of $6\eta(T')$ linear equations in $\eta(P)$ unknowns. A necessary condition for the system's having a solution with $m(p) > 0$, for all $p \in P$, is that the rank of the coefficient matrix be less than $6\eta(T')$.

This assures a solution for which not all $m(p) = 0$. Sufficient conditions which assure that all $m(p) > 0$ are harder to come by. It is not obvious that such sufficient conditions could not be stated entirely in terms of values of $\bar{s}$ and $D\bar{s}$. But some consideration of means of actually doing this should serve to convince one that it is going to be difficult to avoid talking about "something like" an m-function. Here again, though, the vagueness of the notion of conditions "involving reference to the m-function" plagues us. When we leave the finite sets of equations and try to specify sufficient conditions for a solution of the infinite set of equations with $m(p) > 0$, it is even less clear that we could avoid slipping in something that was functioning in the same way as the m-function. These remarks are certainly not offered as conclusive proof that the m-function is not Ramsey eliminable from (2-MPM). At best, they are an admission that I do not know how to Ramsey eliminate it and a conjecture that this cannot be done in an intuitively acceptable way.

Let us now return to the question of the uniqueness of the mass functions satisfying (2-MPM), for a given PK. If there exists a solution to $\sum_T^P$ with $m(p) > 0$ for all $p \in P$, it is clearly never unique, nor is it, in general, unique up to a constant multiple. However, Simon ([37], p. 892) has shown that the following condition is both necessary and sufficient for the uniqueness of the solution up to a constant multiple. This means, of course, that mass ratios are uniquely determined.

(CS)   If $\sum_T^{P'}$ has a solution with $m(p) > 0$, for all $p \in P$, it is unique up to a constant multiple, if and only if there exists no $P' \subset P$ such that $\sum_T^{P'}$ has a solution with $m(p) \neq 0$, for all $p \in P'$.

Simon calls PK's for which (2-MPM) is true and (CS) is satisfied '*holomorphic*'. Note that *necessary* conditions for $\sum_T^{P'}$'s having a solution with $m(p) \neq 0$ can be given in terms of a *finite* coefficient matrix whose elements are values of $\bar{s}$ and $D\bar{s}$. This means that necessary conditions, at least, for the uniqueness of mass ratios can be expressed entirely in terms of paths and their derivatives.

A natural question to ask, at this point, is whether any PK's that we actually claim to be extendible to models for NCPM (or equivalently MPM) are in fact holomorphic. Simon ([40]) opines that, among systems (PK's) which we truly claim to be extendible to models for MPM, holomorphicity is the rule rather than the exception, i.e. (CS) is usually satisfied in these systems. Why might one think this?

One way that (CS) might be false is this. Suppose that P_0 could be partitioned into two disjoint sub-sets, $P_1 \cup P_2 = P_0$, such that both $Q_1 = \langle P_1, T_0, \bar{s}_1 \rangle$ and $Q_2 = \langle P_2, T_0, \bar{s}_2 \rangle$ were extendible to models for MPM, as well as $\langle P_0, T_0, \bar{s}_0 \rangle$. Here there is clearly a $P' \subset P_0$ such that $\sum_T^{P'}$ has a solution with $m(p) > 0$, for both P_1 and P_2 are such, and (CS) tells us that the mass ratios which extend $\langle P_0, T_0, \bar{s}_0 \rangle$ to a model for MPM are not uniquely determined. Intuitively, this means that Q_1 and Q_2 are "isolated from each other". Masses extending Q_1 to a model for MPM can be chosen independently of the values of $\bar{s}_2$ and conversely. By an appropriate choice of m_1 and m_2 *arbitrary* values of mass ratios for particles respectively in P_1 and P_2 can be produced. Thus (CS) surely fails to hold for Q_0 if there are "isolated" sub-systems of Q_0 for which claims of form (2-MPM) are true. It *may* fail to hold, even if there are none of these. Simon's intuition that (CS) usually holds appears to be grounded in the belief that there are usually no "isolated" sub-systems of those we claim to be extendible to models for MPM. Intuitively, the reason for this is that, in the systems we actually choose, *all* the particles interact with one another. If we could partition the system into two sub-sets of particles whose members only "interacted" with each other we would say that this was *two* intended applications of MPM, rather than one.

There are other possibilities for observing mass ratios which do not require that the entire path of the particles be observed. Let T' be a finite sub-set of T, and $\sum_{T'}^P$ the finite set of linear vector equations of forms (D16-4-a) and (b), for all $t \in T'$. If $\sum_{T'}^P$ has a solution with $m(p) > 0$, for all $p \in P$, a necessary and sufficient condition for this solution's being unique up to a constant multiple can be given.

(CT) If $\sum_{T'}^P$ has a solution with $m(p) > 0$, for all $p \in P$, then the ratios

$$\frac{m(p_l)}{m(p_1)}, \quad p_l \in P$$

are uniquely determined, if and only if the rank of the matrix

$$\begin{pmatrix} [D\bar{s}(p_l, t_j)]_i \\ [D\bar{s}(p_l, t_j) \times \bar{s}(p_l, t_j)]_i \end{pmatrix} \quad \begin{matrix} l=\eta(P) & j=\eta(T') & i=3 \\ l=1 & j=1 & i=1 \end{matrix}$$

with $\eta(P)$ columns and $6\eta(T')$ rows is equal to the rank of the matrix

$$\begin{pmatrix} [D\bar{s}(p_l, t_j)]_i \\ [D\bar{s}(p_l, t_j) \times \bar{s}(p_l, t_j)]_i \end{pmatrix} \quad \begin{matrix} l=\eta(P) & j=\eta(T') & i=3 \\ l=2 & j=1 & i=1. \end{matrix}$$

with $\eta(P)-1$ columns and $6\eta(T')$ rows.

An analogous condition can be formulated for the ratios

$$\frac{m(p_l)}{m(p_k)}, \quad k=2, 3, \ldots.$$

If (2-MPM) is true of $Q_0 = \langle P_0, T_0, \bar{s}_0 \rangle$, then $\sum_{T'}^P$ has a solution with $m(p) > 0$, for all $p \in P$. If, further, a condition like (CT) is satisfied for *all* mass ratios, then this solution is unique up to a constant multiple, i.e. all mass ratios are uniquely determined. Thus, the values of all ratios of m-function values which make Q_0 a model for MPM are uniquely determined by the $\bar{s}$ and $D\bar{s}$ values in some finite sub-set $T' \subset T$. Hence, we have a method of determining mass ratios that requires less than a *complete* description of the paths and velocities of the particles.

In the case that $\eta(P) = 2$ and $\eta(T') = 1$, the condition (CT) will be satisfied whenever a solution to $\sum_{T'}^P$ exists. Whenever we have a two-particle system at a single instant which can be extended to a model for MPM, the ratio of the masses required to produce this extension is uniquely determined. This fact is the basis of Mach's ([23], pp. 180–185) well known "definition of mass". As we have already mentioned (Chapter III, p. 59) it is more appropriate to regard Mach's achievement as the discovery of a way of *observing* mass ratios, rather than as a *definition* of mass. Condition (CT) shows that "Mach's method" need not be restricted to two-particle systems. If the paths are appropriate, more complex systems could be used. Pendse ([32], p. 59) noted this.

It should be noted that a method of determining mass ratios from a finite number of positions and velocities at these positions introduces the following possibility. We can "determine" mass ratios from information about some fragment of the paths of particles and check to see if these mass ratios can be retained in making the remaining (perhaps, yet to be observed) portion of the paths a model for MPM. This is one way of providing an empirical check for (2-MPM). Intuitively, we observe the mass ratios in one part of the motion and check to see if the remainder of the motion is a model for MPM with these mass ratios.

Both (CS) and (CT) show that unique determinations of mass ratios are possible in the sense of not being ruled out by the mathematical structure of the theory ((MPM) or equivalently (NCPM)). Whether the possibility is realized depends upon whether (CS) or (CT) are in fact satisfied by particular PK's in the range of intended applications. That is, it depends on just exactly what physical systems we claim to be models for MPM. In particular, if we claim only that some physical systems are "approximately" models for MPM – neglecting, say, "external forces" – then we have only means of determining "approximate" values of mass ratios. It should also be clear that a determination of absolute mass values, rather than mass ratios, is never possible using this method since the solution of $\sum_T^P$ is never unique.

Having determined that it is, at least in principle, possible to determine mass ratios uniquely without assuming the values of the force function to be given, let us raise the following question. Suppose the values of the mass function m_0 are given. Under what circumstances are the values of the force function needed to extend $\langle P_0, T_0, \bar{s}_0, m_0 \rangle$ to a model for NCPM uniquely determined by values of $\bar{s}_0$, $D\bar{s}_0$ and m_0. This is equivalent to the question of when the system of equations:

$$m(p)D^2\bar{s}(p, t) - \sum_{i \in I} \bar{f}(p, t, i) = \bar{0}$$

$$\bar{f}(p, t, i) + \bar{f}(q, t, j) = \bar{0}$$

$$[\bar{s}(p, t) - \bar{s}(q, t)] \times \bar{f}(p, t, i) = \bar{0}$$

$p \in P$, $t \in T$, $i \in I$, where $\langle q, j \rangle = C \langle p, i \rangle$ and $m(p)$ is fixed, has a unique solution for $\bar{f}(p, t, i)$.

Clearly, such uniqueness never obtains. For given any solution $\bar{f}(p, t, i)$, the function

$$\bar{g}(p, t, k(q)) = \sum_i \bar{f}(p, t, i),$$

$$\bar{g}(p, t, l) = 0, \quad l \neq k(q)$$

where,

$$i \in \{i (\exists_j)(\langle p, i \rangle = C(\langle q, j \rangle))\}$$

and k is a one–one function from P into $I_{n(P)}$, is also a solution, no matter what $\bar{s}_0$ and m_0 are like. This is to say that, given any solution, we obtain another solution by taking the vector sum of all the forces that q exerts on p to be the only non-zero force that q exerts on p. Obviously we could "resolve" this force into "components" in different ways and produce still other new solutions. This fact rules out the possibility of determining force function values uniquely from values of the position function, even when the mass function values have already been determined, so long as we confine our attention to models for NCPM.

Intuitively, the reason why we cannot find a means of determining the values of the force function is this. Our mathematical formalism allows us to distinguish between *different* forces that the particle q exerts on the particle p, yet in no PK is the configuration of these different forces, required to produce a model for NCPM, uniquely determined by the paths of the particles. However, it is the case that in some PK's the *resultant* of all the forces that q exerts on p is uniquely determined by values of the position function.

CLASSICAL PARTICLE MECHANICS 137

To see this consider:

(D18) If $\langle P, T, \bar{s}, m, \bar{f}\rangle$ is an NCPM then $\bar{h}$ is a function from $P \times T \times P$ into the set of ordered triples of real numbers such that, for all $p, q \in P$ and $t \in T$,
$$\bar{h}(p, t, q) = \sum_i \bar{f}(p, t, i)$$
$$i \in \{i (\exists i)(\langle p, i\rangle = C(\langle q, j\rangle))\}.$$

Intuitively, $\bar{h}(p, t, q)$ is the resultant of all the forces that q exerts on p at t. We have the following theorem.

(T2) If $\langle P, T, \bar{s}, m, \bar{f}\rangle$ is an NCPM, then for all $p, q \in P$, $t \in T$,
(i) $\bar{h}(p, t, q) = -\bar{h}(q, t, p)$
(ii) $\bar{h}(p, t, q) \times [\bar{s}(p, t) - \bar{s}(q, t)] = \bar{O}$
(iii) $\bar{h}(p, t, p) = \bar{O}$.[1]

We can now ask when the values of the $\bar{h}$-function, determined by an $\bar{f}$-function which extends $\langle P, T, \bar{s}_0, m_0\rangle$ to a model for NCPM, are uniquely determined by values of $\bar{s}_0$ and m_0. This is equivalent to the question of when the solution to the system of equations:

$$m(p) D^2 \bar{s}(p, t) - \sum_{q \in P} \bar{h}(p, t, q) = \bar{O}$$
$$\lambda \quad \bar{h}(p, t, q) + \bar{h}(q, t, p) = \bar{O}$$
$$\bar{h}(p, t, q) \times [\bar{s}(p, t) - \bar{s}(q, t)] = \bar{O}$$

$p, q \in P$, $m(p)$ fixed, has a unique solution $\bar{h}(p, t, q)$ for each $t \in T$. If a solution exists, Pendse ([31], p. 1019) has shown that for $\eta(P) \leq 4$ it may be unique, but for $\eta(P) > 4$ it is never unique. Thus there does exist a possibility of determining the value of the resultant force that one particle exerts on another, given that the values of the mass function have already been determined.

It is important to understand clearly the status of these results in the view of classical particle mechanics being expounded here, since it is somewhat different from that usually given to them. Both Pendse [31] and Simon [37], [38] seem to take something like (2-NCPM) to express the empirical content of the claim that classical particle mechanics with third-law central forces can be applied to explain the motion of some particular physical system. They recognize that (2-NCPM) is an empirically significant claim regardless of whether the mass and force functions are uniquely determined for this particular application. Nevertheless, both seem to think that the *possibility* of unique determination of these values, in at least *some* applications, is a necessary condition for the epistemological respectability ('operational meaningfulness' is Simon's term) of these concepts. This is never explicitly stated, but it seems clear that something like this motivates their discussion of the circumstances in which m and $\bar{h}$ are uniquely determined. It is fairly clear that they would claim that, without assuming more specialized force laws, the notion of a particle's exerting more than one force on another particle is epistemologically suspect, since $\bar{f}$ can never be uniquely determined by the position function. Yet, they would find the net force exerted by q on p, $\bar{h}(p, t, q)$, to be acceptable, since it can be, at least in principle, uniquely determined by the position function in some applications.

So long as we confine our attention to (2-NCPM), it is difficult to understand this view. What is it that makes a function which admits the possibility of uniquely determining its values preferable to one that does not? There is, of course, the possibility of checking (2-NCPM) for a particular physical system by "observing" mass ratios on a finite sub-set of T, and checking to see if the same ratios "work" for other members of T. But even for systems where mass ratios are not uniquely determined by the values of the position function on a finite sub-set of T, checking (2-NCPM) for consistency with any finite set of data is a straight-forward matter. Unless we have some way of putting the "observed" values of the mass ratios and forces to a less trivial use than this, it seems that there is little justification for making a distinction between those functions which do and do not admit the possibility of such "observation".

On the other hand, the view of classical particle mechanics we are investigating allows us to understand why this distinction between (in principle) "observable" and "non-observable" functions is interesting. Suppose we regard the empirical content of classical particle mechanics to consist of a large number of claims of form (2), some made perhaps with 'is a CPM', some made with 'is an NCPM', and some made with even more severe restrictions of 'is a CPM', all "tied together" by certain constraints. Then we can understand why the force function $\bar{f}$ would be defective if its values were *never* uniquely determined by the position function, i.e. if there were no claims of form (2) in this collection that provided a unique determination. Were this the case, then we could never use the force function to construct crucial tests of the theory's claims involving only a small number of intended applications; we could never use the force function to make predictions about the behavior of one system on the basis of information about the behavior of another. At least, we could not use it in a direct way exemplified by the way the t-function was employed in (E–3) (Chapter IV). This does not, of course, show that theoretical functions whose values do not admit of (in principle) "observation" can never legitimately be employed for *any* purpose. It does, however, show that they would not have many of the features that theoretical functions like mass and force do appear to have. In this way, one can understand the concern of Pendse and Simon to show that mass and force do in fact admit of this sort of observation.

To explore further possibilities for determining mass and force values, consider the following restriction of 'is an NCPM', 'is a UNCPM' (is a unary Newtonian classical particle mechanics).

(D19) $x = \langle P, T, \bar{s}, m, \bar{f} \rangle$ is a UNCPM if and only if:
   (1) x is an NCPM
   (2) If $P \times I$ is balanceable by $\langle A_1, A_2, C \rangle$, $p, q \in P$, $\langle q, i \rangle = C(\langle p, j \rangle)$, $\langle q, k \rangle = C(\langle p, e \rangle)$, then $k = i$ and $l = j$.

Intuitively, a UNCPM is an NCPM in which each particle "exerts" at most one non-zero force on any other particle.

It is easy to show that a claim of form (2-UNCPM) and a claim of form (2-NCPM) are equivalent to each other and to a claim of form (2-MPM), in the sense that they determine the same class of PK's. This is to say that exactly the same PK's can be extended respectively to model for MPM, NCPM, and UNCPM, or that the $\bar{f}$-function is Ramsey eliminable from 2-UNCPM as well as 2-NCPM. Clearly, this means that no new possibilities for determining mass function values are introduced by (2-UNCPM).

However, (2-UNCPM) does introduce new possibilities for determining force function values. The question of when the $\bar{f}$-function satisfying (2-UNCPM), for a given m-function, is uniquely determined by the paths of the particles is equivalent to the question of when the system of equations:

$$\lambda_t^p \quad \begin{array}{l} m(p) D^2 \bar{s}(p, t) - \sum_{i \in I\ \eta(P)} \bar{f}(p, t, i) = \bar{O} \\ \bar{f}(p, t, i) + \bar{f}(q, t, j) = \bar{O} \\ \bar{f}(p, t, i) \times [\bar{s}(p, t) - \bar{s}(q, t)] = \bar{O} \\ \text{where } i, \text{ and } j \text{ are such that } \langle q, j \rangle = C(\langle q, i \rangle) \end{array}$$

has a unique solution, for all $t \in T$, for given $m(p)$, $D^2\bar{s}(p, t)$ and $\bar{s}(p, t)$. Clearly λ_t^p is the same system of equations as λ and thus has a unique solu-

140 THE LOGICAL STRUCTURE OF MATHEMATICAL PHYSICS

tion only if $\eta(P) \leq 4$. This is not surprising since (D19) effectively takes the $\bar{h}$-function, which we *defined* for NCPM, to be an undefined primitive. Note that here the force values, not just ratios, are uniquely determined once the masses are given. However, if only mass-ratios are given, then only ratios of force values are uniquely determined.

In addition to asking when the $\bar{f}$-function required to satisfy (2-UNCPM) for given values of the *m*-function, is uniquely determined by the paths of the particles, we can also ask when *both* the masses and forces required to satisfy (2-UNCPM) are uniquely determined by the paths. This is equivalent to the question of when the system of equations λ_T^p composed of λ_t^p for all $t \in T$ has a unique solution for $m(p)$ and $\bar{f}(p, t, i)$, given $D^2\bar{s}(p, t)$ and $\bar{s}(p, t)$. Pendse ([32], p. 58) has shown that this solution is unique, up to a constant multiple, only if $\eta(P) \leq 7$. That is *both* masses and forces *may be* determined, up to a constant, in systems containing less than seven particles but not in systems containing more. For a two-particle system it is always the case that the masses and forces required to produce a model for UNCPM are determined, up to a constant, by the accelerations. It should be noted that this does not introduce any essentially new possibilities for determining mass-ratios. Any PK for which the masses and forces required to satisfy (2-UNCPM) are determined up to a constant multiple is also one for which the masses required to satisfy (2-MPM) are determined up to a constant multiple.

Thus (2-UNCPM) provides us with essentially the same possibilities for determining mass-ratios as does (2-MPM). It provides some means of determining force function values, given values of the mass function and some means of determining, up to a constant multiple, both mass and force function values. Let us now consider how these possibilities might be extended by employing claims of form (2) made with various restrictions of 'is a UNCPM'.

First we define "is a DNCPM" (distance, Newtonian classical particle mechanics).

(D20) $x = \langle P, T, \bar{s}, m, \bar{f} \rangle$ is a DNCPM if and only if:
(1) x is a CPM;
(2) There exist A_1, A_2, C, and $(\eta(P))^2$ functions $O_{ij}(i, j \in I_{\eta(P)})$ from the real numbers into the real numbers such that:
(i) $P \times I$ is balanced by $\langle A_1, A_2, C \rangle$;
(ii) If $\langle q, i \rangle = C(\langle p, i \rangle)$ then for all $t \in T$

(a) $\bar{f}(p, t, i) = -\bar{f}(q, t, i)$
(b) $s(p, t) \times \bar{f}(p, t, i) = -\bar{s}(q, t) \times \bar{f}(q, t, i)$
(c) if $i \leq \eta(P)$ then
$|\bar{f}(p, t, i)| = O_{ij}(|\bar{s}(p, t) - \bar{s}(q, t)|)$
(d) if $i > \eta(P) \bar{f}(p, t, i) = \bar{O}$.

Intuitively, a DNCPM is an NCPM in which each particle "exerts" at most one non-zero force on any other and the magnitude of this force at any time is a function of the distance between the two particles at that time. Note that this function *may* differ for different pairs of particles.

Next, define "is an HNCPM" (Hooke's Law, Newtonian classical particle mechanics) by a restriction of (D20) produced by adding the following axiom:

(D21–2–iii) For all $i, j \in I_{\eta(P)}$, there exist real numbers K_{ij} and d_{ij} such that, for all real numbers x,
$O_{ij}(x) = K_{ij}(x - d_{ij})$.

(D21) characterizes a physical system in which all the forces of mutual interaction between the particles are "Hooke's Law" forces. That is, they are proportional to the degree to which the distance between two particles departs from an "equilibrium" distance d_{ij}. An example of this sort of system would be a number of particles in "free space" all connected to each other by coil springs.

We can also restrict (D20) to produce a definition of "is an INCPM" (inverse-square, Newtonian classical particle mechanics) by adding:

(D22–2–iii) For all $i, j \in I_{\eta(P)}$, there exists a real number K_{ij} such that for all real numbers x,
$O_{ij}(x) = K_{ij} x^{-1-2}$.

We can further restrict (D22) to produce a definition of "is a GNCPM" (gravitational Newtonian classical particle mechanics) by adding:

(D23–2–iv) For all $i, j \in I_{\eta(P)}$, if $\langle q, j \rangle = C(\langle p, i \rangle)$, there exists a real number G such that
$K_{ij} = Gm(p)m(q)$.

(D22) characterizes a physical system in which forces of mutual interaction between particles are inversely proportional to the square of the distance between particles, while (D23) requires that, in addition, these forces be proportional to the product of the masses of the particles – i.e. it characterizes "gravitational forces".

First, note that claims of form (2) for none of these restrictions of "is a UNCPM" allow us to determine more than mass-ratios. It is not difficult to show the following:

(T3) If $\langle P, T, \bar{s}, m, \bar{f} \rangle$ is a DNCPM (HNCPM, INCPM, GNCPM) and α is any positive real number then $\langle P, T, \bar{s}, \alpha m, \alpha \bar{f} \rangle$ is a DNCPM (HNCPM, INCPM, GNCPM).

Thus mass and force functions are determined, at best, up to a constant multiple by the position function. Of course, if we arbitrarily pick one suitable mass function, we may then find the force function to be uniquely determined by the position function. It should also be noted that (T3) would not be true of H, I, GNCPM if particular values of the constants K_{ij}, d_{ij}, and G were specified. If we did this we could obtain determinations of mass values, instead of just ratios.

Next, note that (2-MPM) is not logically equivalent to 2-D (H, I, G) (NCPM), in the sense that there may be models for MPM which are not extendible to models for $D(H, I, G)$ NCPM. Any two particle system whose accelerations are not proportional to the distance between the particles provides an example. This means that $\bar{f}$ is not Ramsey eliminable from 2-D (H, I, G) (NCPM) by (2-MPM), though it may be Ramsey eliminable in some other way.

Further, this raises the possibility that 2-D (H, I, G) (NCPM) provides means of determining mass-ratios, in addition to those provided by (2-MPM) (or equivalently (2-NCPM) and (2-UNCPM)). There are two ways this might happen. First, there might be a PK, extendible to a model for MPM with either of two m-functions which are *not* constant multiples of each other, and yet only *one* of these admitting of a further extension to produce a model for $D(G, I, H)$ (NCPM). Second, there might be a PK, extendible to a model for MPM only with m-functions which are constant multiples of each other, and yet it might require less information about the position function to determine the mass-ratios necessary to make this PK a model for $D(H, I, G)$ MCPM than to determine the mass-ratios necessary to make it a model for MPM.

The first possibility arises, for each of these restrictions of 'is a UNCPM', if and only if the relevant theorem of the following form is false.

(TF1) If $\langle P, T, \bar{s}, m, \bar{f} \rangle$ is a $D(H, I, G)$ NCPM and $\langle P, T, \bar{s}, m \rangle$ is an MPM, then there exists an $\bar{f}$ such that $\langle P, T, \bar{s}, m, \bar{f}' \rangle$ is a $D(H, I, G)$ NCPM.

At present, I am able neither to prove the theorem of form (TF1), nor produce a counterexample to it, for any of $D(H, I, G)$ NCPM. To do either apparently requires a consideration of the consequences of assuming that condition (CS) does not hold.

The second possibility may arise in the following way. Let $\lambda_{T'}^P$ be the system of equations obtained from λ_T^P by taking only the equations for $t \in T' \subset T$. Suppose that T' is finite. To $\lambda_{T'}^P$ add the equations that express the additional restrictions on the force functions in the various restrictions of UNCPM. For example, DNCPM requires that we add equations of the form

$$\bar{f}(p, t, i) + \bar{f}(q, t, i) = \bar{O}$$

whenever there is a j such that $\langle q, i \rangle = C \langle p, i \rangle$ and $\bar{s}(p, t) - \bar{s}(q, t) = \bar{s}(p, t') - \bar{s}(q, t')$. Call these systems of equations $\lambda D(H, I, G)_{T'}^P$. Suppose now that $\sum_{T'}^P$ has solutions for $m(p)$ which are not unique up to a constant multiple. Might it be that nevertheless $\lambda D(H, I, G)_{T'}^P$ has solutions for $m(p)$ and $\bar{f}(p, t, i)$ which *are* unique up to a constant multiple. This would be the case if and only if the relevant theorem of the following form is false.

(TF2)   If $\{m(p), \bar{f}(p, t, i)\}, p \in P, t \in T, i \in I$ is a solution to $\lambda D(H, I, G)_{T'}^P$ and $\{m'(p)\} p \in P$ is a solution to $\sum_{T'}^P$, then there exists on $\{\bar{f}(p, t, i)\} p \in P, t \in T, i \in I$ such that $\{m'(p), \bar{f}'(p, t, e)\}$ is a solution to $\lambda D(H, I, G)_{T'}^P$.

Again, I am unable to offer a proof for any theorems of form (TF2), nor am I able to offer counterexamples. An investigation of this problem would require a detailed understanding of the conditions under which the condition (CT) failed to hold. For example, Narlikar ([30], p. 36) has shown that the mass-ratios for a system of n particles *may be* uniquely determined by their accelerations at $n-1$ times, provided we assume that motion is a model for GNCPM. There will clearly be situations in which $\eta(P) - 1 = \eta(T')$ and (CT) fails to hold. But whether these will also be situations in which Narlikar's system of equation fails to have a unique (to a constant multiple) solution is not apparent. However, I conjecture that, for GNCPM at least, (TF2) is false.

It is clear that $2\text{-}D(H, I, G)$ (NCPM) do introduce new possibilities for determining force function values, assuming that the mass values are given. The systems of equations $\lambda D(H, I, G)_{T'}^P$, where $m(p)$ is considered given, will generally have unique solutions for $\bar{f}(p, t, i)$ in situations where $\lambda_{T'}^{p'}$, $m(p)$ given, will not.

As an example of this consider a system of three particles with known masses all of whose paths lie in the same straight line.

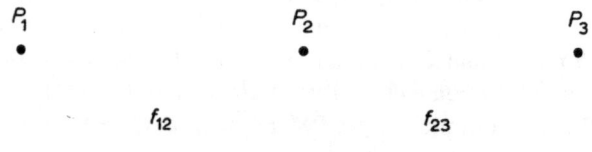

Fig. 2

The forces required to make this system a model for UNCPM are not uniquely determined since the system of equations:

$$m_1 a_1 = f_{12} + f_{13}$$
$$m_2 a_2 = -f_{12} + f_{23}$$
$$m_3 a_3 = -f_{23} - f_{13}$$

does not have a unique solution for the f_{ij}'s. (To see this, note that the determinate of coefficients is equal to zero.) However, the system of equations:

$$m_i a_i(t) = k_{ij}(x_i(t) - d_{ij}) + k_{ik}(x_i(t) - d_{ik})$$
$$m_i a_i(t') = K_{ij}(x_i(t') - d_{ij}) + k_{ik}(x_i(t') - d_{ik})$$
$$i, j, k, \in \{1, 2, 3\} i \neq j \neq k$$

will generally have a unique solution for the k_{ij}'s and d_{ij}'s, if it has a solution at all. This means that, if it is possible to make this system a model for HNCPM, then information about the positions and accelerations of the particles at two times is sufficient to uniquely determine the forces which will produce this model. These forces are determined by determining the values of the constants K_{ij} and d_{ij}.

Obviously, "is an NCPM" and the restrictions of it we have examined do not exhaust all possibilities for mass and force determinations. Indeed, they do not even constitute the most practically significant. Determinations of mass-ratios using an Atwood's machine, for example, appear to require that we suppose that this apparatus is a model for a set-theoretic structure in which some of the forces are not directed along the line connecting the particles, and others are proportional to the masses. The same is true of determinations using an analytical balance. Indeed, perhaps the most satisfactory way of dealing with these methods of mass-ratios determinations is to regard them, not as applications of classical particle mechanics, but as applications of rigid body mechanics. A precise understanding of the relation between rigid mechanics and particle mechanics would then be required, to make clear how the mass-ratios thus determined are relevant to applications of particle mechanics.

There are also other means of determining force values, more practically relevant than those we have mentioned. For example, we suppose that a body near the earth suspended from a coil spring is acted upon by a 'Hooke's Law' force and a constant force, equal to the weight of the body. We then determine the spring constant by observing the path of the body. Another example is a static structure, e.g. a bridge. We assume the "particles" in this structure to be part of a model for HNCPM such that for particles not connected by structural members the K_{ij}'s are zero. We further assume certain external forces. From this information, if the structure is "statically determinate" we can calculate the forces acting on any "particle".

To produce a full-blown logical reconstruction of classical particle mechanics, it would be necessary to examine, in detail, all of the various possibilities for mass and force determinations, or at least all the practically significant ones. We shall not do this here. This is another respect in which the account in this chapter falls short of complete logical reconstruction. However, the examples we have considered are sufficient to illustrate how one would proceed in analyzing any account of mass or force determinations found in the existing expositions of classical particle mechanics. To complete our account of the measurement of mass and force function values we must see how it is that the values determined in the ways we have described (as well as in other ways) are 'applicable' outside the application in which they are determined. That is, we must understand how the assignments of mass and force functions in various intended applications of classical particle mechanics are constrained.

In connection with the discussion of the claim (3-CPM), we have already examined some possibilities for constraining both the mass and force functions. In the case of the mass function, it does not appear that further discussion is required. I know of no use that is made of calculated values of the mass functions which can not be accounted for by the constraints that particles have the same mass in all applications in which they appear and that mass values be extensive with respect to concatenation of particles. In the case of the force function, the $\langle =, = \rangle$ constraint mentioned in this discussion does not appear to be sufficient to account for many of the ways we actually use calculated values of the force function. To obtain some idea of the sort of constraints that are required to make these uses intelligible, it is instructive to look at some examples.

Consider a situation in which we have a supply of coil springs which we may use to "tie together" a large number of particles of known mass in such a way that the resulting system is a model for HNCPM. We assume that we are constructing the system in "free space" so that only the Hooke's Law forces due to the springs act on the particles. Suppose we are interested in predicting the paths of the particles in the system we construct. An obvious way to proceed would be the following. Take each spring in our supply, connect two arbitrarily chosen particles with it, and observe the paths of the particles. If these observations are consistent with the claim that this system is a model for HNCPM, then we can calculate the value of K_{ij} and d_{ij} required to produce this model. Then assume that these constants are "intrinsic properties" of the spring. That is, assume that for *any* physical system constructed using this spring *these* values of K_{ij} and d_{ij} must be used, for the particles connected by *this* spring, to make this system a model for HNCPM. Having determined all the K_{ij}'s and d_{ij}'s in this way we would then predict the motion of any system that we constructed from our supply of springs. Intuitively, the constraint on the assignment of force function values to the physical systems considered here is that the K_{ij}'s and d_{ij}'s are intrinsic properties of the springs which do not change as the springs appear in different applications.

As another example, consider the way in which we employ "gravitational forces". There are some applications in which we claim that *only* gravitational forces act on the particles. That is, we claim the systems in these applications are models for GNCPM. Examples are: the bodies in the solar system, taken all together, as well as various sub-sets of these bodies, e.g. Jupiter and its moons; "freely falling" bodies near the surface of the earth. On the other hand, there are also systems in which we claim that gravitational forces as well as other kinds of forces act on the same particles, for example, a body suspended from a spring near the surface of the earth. In these situations we apparently make roughly the following claim. If a set of particles is a part of some intended application that is a model for GNCPM, then the forces required to make it a model for GNCPM must still act on the particles when they appear in some *other* "related" application. That is, the gravitational force required to make the freely falling body a model for GNCPM must still act on the body when it is suspended from the spring. In our terminology, the claim is a constraint on the assignments of force functions to intended applications which may be used to satisfy some claim of form (4).

It is perhaps worthwhile to consider further the system consisting of a body suspended from a spring near the surface of the earth. The most natural way to treat this system appears to be the following. Consider it to be a system consisting of three particles; the support, p_1, the body p_2, and the earth, p_3. The observed paths of the particles are such that they move in the same straight line and the distance between p_1 and p_3 remains constant. The claim that we make for this system is that it can be made a model for UNCPM. But the masses and forces employed to do this may not be chosen arbitrarily. Some of the particles in this system appear in other intended applications of classical particle mechanics which are claimed to be models for things like MPM, HNCPM, and GNCPM. Further, the system we are considering is "related" to these systems in such a way that the masses and forces used to make these systems models for the appropriate structure must be used to make this system a model for UNCPM. In particular, f_{12} is constrained to be a Hooke's Law force with K_{12} and d_{12} determined by the paths of particles connected by this spring in applications which are models for HNCPM. Likewise, f_{23} is constrained to be a "gravitational force", or equivalently, to be equal to $m(p_2)g$, where g is the acceleration of a freely falling body near the earth. If we know K_{12} and d_{12}, the claim (2-UNCPM) for this system, together with these constraints, justifies a trivial calculation of $m(p_2)g$. If g is known, we can further calculate $m(p_2)$. These calculations are trivial, but the role of constraints on the force functions in them is seldom explicitly recognized.

The general features of constraints on force function assignments illuminated by these examples are roughly the following. Particles may be acted upon by various special sorts of forces such as Hooke's Law and other elastic forces, gravitational forces, electromagnetic forces, viscous and frictional forces. Typically, these forces are characterized by special 'force laws' like (D21–2–iii), (D22–2–iii), (D23–2–iv). In addition, they are usually regarded as being "produced by" or "caused by" some, more or less identifiable objects in the physical systems – like springs, electric charges, or fluid media. Gravitational forces are an exception in that they are regarded as being "caused' by the particles themselves, and not by other objects. More explicitly, there are certain constants appearing in the 'force laws' – like elasticity coefficients, electric charge, coefficients of friction and viscosity – which are regarded as "intrinsic properties" of these objects. That is to say, whenever these objects (or other objects of the same kind) appear in applications of classical particle mechanics, we claim that they "produce" forces characterized by the same law and the same values of the appropriate constants. For gravitational forces, we simply claim that they are *always* present (though we sometimes "neglect" them). Thus, there appear to be a wide variety of constraints on force function assignments – one for each special kind of force. The exact nature of these constraints – how the constants in the 'force laws' are related to the appropriate objects – appear to differ widely. To formulate any of these constraints precisely would require a detailed exposition of the "theory" of this special sort of force, for example, the theory of the elastic properties of materials. This we shall not do, thus leaving still another lacunae in our sketch of a logical reconstruction.

Even though we do not examine, in detail, any of these constraints on the force function assignments, it is not difficult to see, in general, how they operate and that this supports the account of theoretical functions offered in Chapter IV, and V. In many applications of classical particles mechanics we simply claim that the system is a model for UNCPM, or perhaps even NCPM, and the paths of the particles do not determine the value of the force function.

Nevertheless, we claim also that one or more of these special kinds of forces act on the particles in this system, and perhaps other forces as well, whose 'force laws' are not explicitly specified. Indeed, we might even claim to know the values of these forces. How do we justify these latter claims?

The answer is roughly this. There are applications of classical particle mechanics in which the effects of one of these special kinds of forces may be (to a good approximation) "isolated". That is, there are applications in which we have enough information to allow us to calculate the values of the constants in the relevant 'force law'. One way this happens is the following. There may be applications in which we claim that *only* forces of this special kind act on the particles, e.g. we might claim that the application is a model for HNCPM. In some of these situations, though not in all, the paths of the particles will uniquely determine the values of the constants in the 'force law'. Once these constants are determined, the constraint that the same constants must work in all applications in which "objects" having these constants as "intrinsic properties" produce forces allows us to make claims about the values of the force function in these other applications. Indeed, we may even be able to so determine the forces in these other applications that further means of "isolating" the effects of special kinds of forces are provided. In this way it might be that not all methods of determining constants in 'force laws' require systems in which *only* forces of this kind are claimed to act.

This discussion suggests strongly that the appropriate way to depict the logical structure of classical particle mechanics is with a claim of form (5). This claim is constructed with various restrictions of 'is a CPM' characterizing special kinds of force laws. We have seen that restrictions can be produced which allow us to calculate both mass ratios and force values. Moreover, these restrictions are recognizable as characterizing 'force laws' which actually play a role in existing expositions of the theory. Further, we have seen that constraints on both the mass and force function assignments can be found which apparently justify the practice of employing these calculated values outside the application used in their calculation. In the case of the mass function these were few in number and quite simple. For the force function, however, there appear to be numerous and rather complicated constraints whose exact formulation appears to require a detailed account of the physical processes which "produce" forces. Indeed, it is likely that there is a certain amount of vagueness in these constraints as they are actually used. Just as there is probably a certain amount of vagueness in the specification of the range of intended applications in this claim of form (5). This vagueness allows for a certain amount of "cut and try" in attempting to get the experimental data to be consistent with the claim of form (5). In this sense then there is not one single claim settled for all time that captures the empirical content of classical particle mechanics. Rather one might envision physicists as being committed to using some claim of form (5) made with restrictions of 'is a CPM' to tell us how the world is, but continually changing their minds about just which one to use.

At this point something should be said about the notion of "invariance" in classical particle mechanics. It is frequently maintained that the laws of mechanics are, or ought to be, invariant under Galilean transformations. What does this mean? According to our view of classical particle mechanics, it means roughly this. Whether or not mass and force functions can be found to extend a given PK to a model for some restriction of 'is a

CPM' is, or should be, independent of the coordinate system used in the PK to describe the position of the particles. A bit more precisely, if a claim of form (2) made with some restriction of 'is a CPM' about some PK is true, and if PK' is obtained from PK by a Galilean transformation then the claim of form (2) is true of PK'. Moreover, the mass-functions satisfying both these claims are the same and the force functions differ only in that the force vectors are represented in different coordinate systems. The notion of a Galilean transformation on PK's and PM's can be precisely defined (see [27]) and with it this rough formulation of the claim that the laws of mechanics are, or ought to be, invariant under such transformations can be made precise. Doing this, however, would take us far afield. The rough formulation suffices for the following remarks about the claim. One may take it to be simply a claim about the special force laws that have, in fact, been postulated in the course of "doing" classical particle mechanics. All the force laws we have considered in this discussion are Galilean invariant. On the other hand, one might take the claim as a restriction on the special force laws that *might be* employed in "doing" classical particle mechanics. In this case the claim amounts to a commitment to make empirical claims with only Galilean invariant force laws. As such, it has the same status as the commitment to seek some tenable claim of form (5) made with 'is a CPM' and restrictions of it. We simply add the requirement that these restrictions be Galilean invariant. The second interpretation of the claim appears to be more in accord with the historical development of classical particle mechanics. The inability to find Galilean invariant laws for electromagnetic forces occasioned our giving up attempts to make tenable claims with 'is a CPM' and its restrictions.

Several much discussed questions about classical particle mechanics are illuminated by the sketch of the logical structure of this theory presented here. One of these is the question of the epistemological status of Newton's second law. One common way of raising this question is this. Is the second law to be understood as a definition of force or as a straightforward empirical truth? (See for example [29], p. 160 ff.) On the view sketched here neither of these alternatives gives an accurate account of the role of the second law. In all the set-theoretic predicates we employ in our logical reconstruction, both mass and force appear as primitive symbols. In none of them can the second law (D8–2), be taken as a definition of the $\bar{f}$-function. Indeed trivial applications of Padoa's principle will show that in all the predicates we employ neither the m nor $\bar{f}$-function is definable in terms of the remaining primitive symbols. (See [26], pp. 268–272.) On the other hand, the second law is not a straight-forward empirical truth either. We do not "independently" measure values of the mass force and position functions and then discover that they happen to satisfy the second law. Rather, we maintain that mass and force functions can be found for all applications which satisfy the second law. In most applications we claim that the force function satisfies other laws as well. Moreover, all these mass and force functions must satisfy certain constraints which operate "across" applications.

It is only in the content of this rather complicated nexus of claims that we can understand the role of the second law. It is true that the claim that the second law holds in any one application has some aspects of a non-empirical claim. It is difficult to conceive of anything that would count as conclusively falsifying it. The reason is roughly this. The basic empirical claim of classical particle mechanics is some claim of form (5) – made with restrictions of 'is a CPM'. This claim is vague enough to allow for modifications in the range of intended applications, in the constraints on the force functions, and perhaps even in the form of the force function claimed to hold in certain applications. Indeed, one might say that practitioners of classical particle mechanics are committed to the project of finding *some* claim of form (5) that they can maintain in the face of observational data about paths. That is, they are committed to seeking some way of carving up the world into intended applications and some way of adding special force laws to the second law so that a claim like (5) come out to be true. In a sense the second law is the "core" of this enterprise since it is alleged to hold in all applications. Recalcitrant data might force us to give up the claim that some special force law held in some application and yet we might try to find some other force law to maintain the truth of (5). But only repeated failures to make a claim like (5) fit the data would convince us to give up the second law. For, in doing this, we would, in effect, be admitting that mass and force functions satisfying the second law were simply not effective conceptual tools for understanding the motion of particles. In this sense the second law is a "core feature" of the concepts of force and mass. Yet it does not entirely determine the extension of these concepts. Each set-theoretic predicate appearing in (5) adds something to these concepts in that it typically introduces new possibilities or determining values of one or both the functions.

Another question that has been the source of much discussion is whether or not the concept of force is essential to classical particle mechanics. The preceding discussion suggests, but by no means proves, that the concept of force is essential to the Newtonian formulation classical particle mechanics. The $\bar{f}$-function appears as a primitive symbol in all our set-theoretic predicates except 'is an MPM'. In no case where it appears can it be eliminated by explicit definition. However, it was noted that $\bar{f}$ is Ramsey eliminable from (2-NCPM) and (2-UNCPM) by (2-MPM). But, though no proof was given that it cannot be done, there do not appear to be any obvious ways to Ramsey eliminate $\bar{f}$ from claims of (form 2) made with predicates which put further restrictions on this function. This strongly suggests that no way could be found to Ramsey eliminate $\bar{f}$ from a claim of form (5) made with these predicates. Intuitively, this means that possible configurations of paths that we rule out by stipulating force laws stronger than Newton's third law cannot be ruled out by adding to the definition of 'is an MPM' conditions solely in terms of the mass and position function. This, of course, does not mean that a completely new set-theoretic predicate involving neither mass nor force could have the same empirical content as the initial claim of form (5). But such a claim could hardly be counted as a logical reconstruction of the *Newtonian formulation* of classical particle mechanics. In this connection, we also noted that it is not likely that the mass function can be Ramsey eliminated from *any* claim of classical particle mechanics.

One final question should be mentioned. The fact that all known ways of measuring forces appear to presuppose that some force law is true has frequently raised doubts about the epistemological respectability of the concept of force. Indeed this appears to have been the principle motivation for wondering whether mechanics could be done without the concept of force. The legitimacy of using laws about theoretical functions to determine values of these functions was discussed in general in Chapter V (p. 104). The situation with the force function provides a concrete illustration to supplement that discussion.

Doubts about the legitimacy of force measurements can be raised in the following way. How is it that we can have reason to believe that a certain force law, used to calculate values of the force function, actually holds in the relevant application? Must not this reason ultimately be that certain relations among *observed values* of forces have been found to hold, either in this application, or in others like it? But then, how was the calculation of *these* force values justified? By appealing to some other force law, holding in some other application? But what reason do we have for believing that *this* law holds? Reasoning in this way leads to the conclusion that a claim about observing force values can only be justified by arguing in a circle.

On our account of the role of the force function in classical particle mechanics, the error in this line of reasoning is roughly this. It regards each claim that a special force law holds to be logically independent of other such claims. It envisions them as being subjected to empirical verification independently of one another. On our account, when we make a claim of form (5), we claim that force functions, having various special forms in various different applications, exist which make the paths of particles in these applications models for CPM. Moreover, these force functions satisfy certain constraints operating "across" applications. Thus, *all* our claims about force laws holding in various applications stand or fall together with this claim of form (5). In attempting to put this claim to an empirical test, or use it to make predictions, we are free to use any consequences of the claim. In particular, we can exploit situations in which the values of the force function satisfying this claim, in a particular application, are uniquely determined by the observed paths of the particles in this application. We *may* then infer from the constraints that these same, or related force values must appear in other applications. In this way, calculated or measured force values may be used in testing the claim of form (5) or in making predictions based on it. In doing this we assume, at least provisionally, that the claim is true. But there is nothing illegitimate about this. The claim may be accepted, rejected, or modified in the light of observations in the usual ways. It is simply that it is a rather complicated claim and typically the only way we can test it is by comparing some rather remote consequences with observations.

NOTE

[1] (T2-iii) is a direct consequence of (D13-1) and justifies our earlier comment that (D13-1) rules out the possibility of a particle's exerting forces on itself.

CHAPTER VII

IDENTITY, EQUIVALENCE AND REDUCTION

In this chapter we will attempt to use our understanding of the logical structure of the empirical claims in theories of mathematical physics – the account developed in the first five chapters – to clarify some other questions about these theories. First, we will attempt to say, as precisely as we can, just what a theory of mathematical physics *is*. That is, we will attempt to give some general, and precise characterization of theories of mathematical physics. Once we have developed this characterization, we will employ it to investigate the properties of two relations – equivalence and reduction – that are commonly alleged to hold between some theories of mathematical physics. In the course of this discussion, we will have occasion to examine the Lagrangian and Hamiltonian formulations of particle mechanics as examples of theories of mathematical physics that are, in some sense, equivalent to the Newtonian formulation of particle mechanics. We shall also examine rigid body mechanics as an example of a theory which reduces to particle mechanics.

The first questions we want to consider are roughly these: (a) what sort of thing is a theory of mathematical physics; and (b) how does one determine that there are two distinct theories of mathematical physics being considered, rather than just one? Thus far, we have been provisionally committed to the view that scientific theories are sets of statements (cf. Claim (A), Chapter I). Further, we have seen that, for theories of mathematical physics, there appears to be *one* statement that plays a central role in the theory. We say that sentences of form (5) appeared to be adequate, in general, for making such statements, though for some theories one might be able to use sentences of form (3), (2) and even (1). These statements were central to the theories under consideration in that, roughly speaking, they expressed the entire empirical content of the theory. All empirical claims of the theory could be regarded as consequences of these statements. Following our provisional commitment, it might seem natural to *identify* each theory of mathematical physics with the statement of this sort associated with it. There are however, implicit in the preceding discussion, several reasons for believing that this is not an accurate account of our intuitive understanding of 'theory of mathematical physics'.

To make these reasons explicit, and to suggest a plausible alternative view, consider the sentence (5). We may distinguish three entities associated with this sentence: (i) the sentence itself; (ii) the mathematical structure referred to in the sentence; and (iii) the statement made by the sentence. The mathematical structure referred to in the sentence is essentially the extensions of the predicates appearing in the sentence. We may think of this structure as being *characterized* by the predicate appearing in the sentence and as being *used* to make the statement made by the sentence. This can be made more precise and will be shortly. The statement made by sentence (5) is roughly this: a certain set – the set of intended applications – is alleged to be a sub-set of some distinguished sub-set of the set of all sub-sets of the class of models for the predicate 'is a P_0'. This sub-set is distinguished as the set of all sets of models for 'is a P_0' which can be extended, in an appropriate way, to models for 'is an S'. 'In an appropriate way' means that some of the extensions must satisfy certain restrictions of 'is an S' characterizing "theoretical laws" and all, taken together, must satisfy certain constraints. These entities may likewise be distinguished for sentences of forms (1), (2), and (3). There are certain relations among the entities (i), (ii), and (iii) which are relevant to the question at hand.

First, it is clear that we may, in general, modify (i) and obtain another sentence, of the same form as (5), in which (ii) and (iii) are the same as in the initial sentence. That is we can find other predicates with the same extensions – the same class of models – as the initial predicates and use them to make the same statement as the initial sentence. It is also true that we may modify *both* (i) and (ii) and obtain another sentence, either of the same form as (5) or of the form of (1), (2), or (3), in which (iii) is the same. For example, we might, in some cases, simply find a predicate (perhaps a restriction of 'is a P_0') whose class of models is such that the sub-set distinguished by the initial sentence is the set of all sub-sets of this class. For another example, we might use another sentence of form (5) in which the predicate 'is an R' plays the same role as 'is an S' in the initial sentence but, intuitively, has different *theoretical* functions than 'is an S'. That is, the set of possible partial models for 'is an R' and 'is an S' are the same, but the set of models is different. We might be able to find constraints and theoretical laws which would allow us to make the same statement as the initial sentence, using this new sentence containing 'is an R'. The possibilities of doing such things as this are discussed in detail below. On the other hand, it seems clear that (iii) can not be modified without changing (ii), and that (ii) can not be modified without changing (i).

In our previous discussion we have suggested that axiomatizations of physical theories which are definitions of set-theoretic predicates are relevant to providing a logical reconstruction of these theories because the empirical content of these theories may be expressed by sentences in which the predicate defined by the axiomatization appears. We have examined four sentence forms in which such predicates could be used and concluded that sentences of all these forms might be used to make empirical claims with this predicate, but that sentences of form (5) seemed to be required by some kinds of theories. It is useful to think of the results of this discussion in the following way. Sentences of the forms (1), (2), (3), and (5) which contain predicates characterizing the appropriate sort of mathematical structures, and whose range of intended applications is appropriate, correspond to possible theories of mathematical physics. That is, we have shown that there is *some* correspondence between sentences of a certain kind and (possible) theories of mathematical physics, but the exact nature of this correspondence has not yet been made explicit. The kind of sentences that correspond to such theories are sentences of the forms we have considered in which, roughly speaking, the predicates characterize mathematical structures and the members of the range of intended applications are "physical systems". This will be made more explicit soon.

It is fairly evident that this correspondence is at least many – one from the set of sentences of the appropriate kind onto the set of (possible) theories of mathematical physics. That is, each sentence of this kind corresponds to *exactly* one (possible) theory of mathematical physics, though more than one such sentence *may* correspond to the same theory. Equivalently each sentence of form (1), (2), (3), or (5) corresponds to *at most* one (possible) theory of mathematical physics, though several such sentences may correspond to one theory. Further, it is very plausible to expect that the explicit nature of this correspondence is going to be determined, in some way, by the three entities we have just mentioned as being associated with sentences of form (5). 'Determined' in the sense that theories of mathematical physics and the set of these determining entities correspond one–one and all sentences of form (5) associated with the same determining entity correspond to the same theory. Were this the case, it would be quite natural to *identify* the theory with the determining entity. It is possible to regard the present discussion as an attempt to describe this correspondence explicitly and to identify theories of mathematical physics in the manner just suggested.

Now, it is certainly plausible to think that the statement made with a sentence of form (5) has *something* to do with the theory corresponding to this sentence. But there are sound intuitive reasons for thinking that within certain limits, the statement made by the sentence (and *a fortiori* (ii) and (i)) might change, and yet we would not want to say the theory corresponding to the new sentence was a different one. On the other hand, it appears that, even if the statement made by the sentence remained the same while (ii) (and *a fortiori* (i)) changed, intuitively, we might want to say that the theory corresponding to the new sentence was different. It is less clear that simply changing (i), keeping (ii) and (iii) fixed, would ever motivate us to say that the new sentence corresponded to a different theory. These are the kinds of reasons that can be given for not identifying theories of mathematical physics with statements made by sentences of form (5). Let us examine them in more detail.

First, consider the claim that different statements made with sentences of the forms we are considering might correspond to the same theory. In Chapter IV we noted that the truth values of statements made with sentences of form (3) might be sensitive to slight changes in the range of intended applications, both as to what this range included and as to how it was "carved-up" into different intended applications. We noted there that this feature of such statements allowed us to account for the putative fact that theories of mathematical physics sometimes appear to be tenaciously maintained in the face of recalcitrant data. We suggested there that one might understand 'having a theory' to mean 'being committed to use a certain mathematical structure, together with certain constraints on theoretical functions to account for the behavior of a, not too precisely specified, range of phenomena'. Typically, the range of intended applications might be specified as a class of physical systems for which some claim of form (3), employing this mathematical structure and constraints, was well confirmed, together with other systems "like this in relevant respects". But, just exactly what these "relevant respects" are may not be spelled out explicitly. This leads to a situation in which the range of intended applications in all statements associated with the theory contain a certain core set of intended applications, together with other intended applications that are "like these in relevant respects". Roughly, these statements represent (perhaps successive) attempts to enlarge the range of application of the theory to include more physical systems. This seems to indicate that the mathematical structure referred to in a sentence of form (3) is much more characteristic of the theory involved than the statement it makes. On the other hand, it is surely not the case that *all* sentences of this form, using the same mathematical structure, correspond to the same theory. There is some, but not unlimited, imprecision in what counts as a range of intended applications for a particular theory.

In Chapter V, where we considered statements in which theoretical laws were "postulated" to hold in certain intended applications it was noted that, roughly speaking, different theoretical laws might be postulated on different occasions in an effort to make the same "theory" apply to an imprecisely specified range of intended applications. This is just to say that sentences of form (5), containing different restrictions of the basic predicate and making different statements, correspond to the same theory. They simply represent different attempts to use the central mathematical structure associated with the theory to make tenable empirical claims. This central mathematical structure consists of a basic mathematical structure alleged to be common to all intended applications, together with constraints on the theoretical functions. Predicates characterizing it appear in every sentence of form (5) corresponding to the theory. There are however other mathematical structures used in these sentences – those characterized by restrictions of the predicate characterizing the basic structure and identifiable as theoretical laws. These structures may differ and yet the sentences containing references to them correspond to the same theory. In Chapter VI we saw this account exemplified by classical particle mechanics. In addition it was noted there that certain constraints on the theoretical functions might be intimately associated with the postulation of theoretical laws and consequently might be expected to change when different theoretical laws were postulated. This means that not all the constraints need be a part of the central mathematical structure characteristic of every sentence of form (5) corresponding to a particular theory.

These considerations suggest that the correspondence between sentences of form (5) (and the related forms (3), (2), and (1)) and theories of mathematical physics has the following properties. Sentences of these forms used to make different statements may correspond to the same theory. The sentences corresponding to the same theory are related in two ways. First, the range of intended applications, though not identical, overlap to a large extent. That is, there is a core of intended applications that is common to all empirical claims of the theory. Further, the particular way in which other applications may be added to this core is also characteristic of the theory. Second, there is some central, core mathematical structure that is referred to in all these sentences, though there are typically other mathematical structures referred to in some of these sentences, but not in all. We might summarize the view suggested here in the following way. That a sentence of form (5) (or related forms) contains reference to a core mathematical structure characteristic of a theory and makes a statement about some range of intended applications characteristic of the theory is *sufficient* to assure that the sentence corresponds to this theory.

Is there reason to believe that this is also a *necessary* condition for a sentence of this form corresponding to a given theory? That the sentence is "about" some range of intended applications characteristic of the theory seems clearly to be necessary. It is less clear that reference to some particular mathematical structure is necessary. We have noted that sentences in which reference is made to different mathematical structures may be used to make the same statement about a given range of intended applications. Are these sentences to be taken to correspond to the same, or to different theories? In answer to this, two considerations are relevant.

First, axiomatizations of physical theories which are definitions of set-theoretic predicates are usually thought to characterize some essential feature of the theory in question. This means, at least, that the mathematical structure characterized by the predicate is regarded intuitively as an essential feature of the theory. It *may* even mean that the particular syntactical form of the predicate is essentially connected with the theory. This suggests that we should regard reference to a particular mathematical structure as a necessary condition for a sentence corresponding to a particular theory. However, there is a difficulty here. It is clear that one can make "trivial" modifications in these axiomatizations without thereby rendering them intuitively unacceptable as axiomatizations of the theory in question. For example, one may simply permute the order of the members of the ordered quintuples that are CPM's. But, these "trivial" modifications produce predicates which, strictly speaking, characterize different mathematical structures. Yet we would not intuitively want to say that these "trivially different" structures correspond to different theories. There are two things one could say about this. One could simply say that there are "trivially different" and "significantly different" theories. "Trivially different" axiomatizations correspond to "trivially different" theories. Alternatively, one could say that theories really correspond to equivalence classes of mathematical structures under the "trivially different" equivalence relation, rather than to mathematical structures. Common usage does not seem to significantly favor either of these views. I shall opt for the first, primarily for reasons of systematic convenience. It should be noted that both views rest on an unexplicated notion of "trivially different". The task of explicating this is left to those readers with a flair for such problems.

Further support for the view that different mathematical structures correspond to different theories, though they are used to make the same statements, is afforded by our discussion at the beginning of Chapter VI about what we intuitively recognize as "equivalent" formulations of particle mechanics. Each of these formulations may be regarded as being associated with a different core mathematical structure which is used to make statements about essentially the same range of intended applications. Roughly speaking, given any statement about this range made with one of these structures, the same statement can be made using either of the other structures. Do we intuitively regard these different structures as being characteristic of different theories of mathematical physics? It does not appear that ordinary usage provides any decisive guidance on this point. I do, however, think that one can give a systematic and precise account of the relation between intuitively equivalent formulations of particle mechanics (and other similar situations like the Schrödinger and Heisenberg formulations of quantum mechanics) by agreeing that the mathematical structures involved are characteristic of numerically distinct, but equivalent (in a sense to be explicated), theories of mathematical physics. Admittedly, this involves some arbitrariness in choice of the meaning of 'theory of mathematical physics', but I do not think that this choice is inconsistent with any large part of our common usage of this term. That it has the advantages just claimed will be argued shortly.

Thus far, our discussion seems to suggest roughly the following account of the correspondence between sentences of form (5) (and related forms) and theories of mathematical physics. The theory to which a given sentence corresponds is completely determined by a certain part of the mathematical structure referred to in the sentence and the properties of the range of intended applications – namely that it be related to the appropriate way to the core of applications characteristic of the theory. This, in turn, suggests a rough account of the identity criteria for theories of mathematical physics. Such theories may be identified as ordered pairs consisting of a certain mathematical structure and a certain core set of intended applications.

We have not, so far, considered the possibility that the syntactical properties of a sentence of form (5) (and related forms) are relevant in determining the theory to which it corresponds and thus to the identity criteria for theories. This is a significant possibility and can not be summarily dismissed. However, we shall defer consideration of it until after we have formulated more precisely the view of identity criteria mentioned here.

To make this view of the identity criteria for theories of mathematical physics precise, we need some precise, general way of characterizing the mathematical structure referred to in those sentences of form (5) (and related forms) which correspond to theories of mathematical physics. Roughly, what we want is a completely general characterization of the mathematical structures that *could be* used to make the empirical claims of theories of mathematical physics. We want to characterize these structures in such a way that the core structure which is characteristic of a particular theory can be distinguished conveniently from the rest of the structure. Further, we want the characterization to be such that, for any given range of intended applications, we can say precisely what statement is made about this range by a structure of this sort when expressions referring to both appear in the appropriate way in sentences of form (5) (and related forms).

To begin, consider the predicates 'is an S' and 'is a CPM' that appear in our examples of sentences used to make statements of form (5). These predicates, we have said, characterize the basic mathematical structure of the respective theories. 'Characterize' in the sense that they determine a set of entities such that all and only members of this set "have" this structure. It *may*, of course, be possible to characterize this same set of entities with other, different set-theoretic predicates. For this reason, we should *identify* this basic mathematical structure as the set of models for these predicates, rather than with the predicates themselves. In each of our examples we found it necessary to speak of possible models for the predicates which characterized the basic mathematical structure. These were respectively models for 'is an S_0' and models for 'is a PM'. Thus, associated with each of our examples is a set of possible models for the predicate characterizing the basic mathematical structure. Roughly, this set of possible models consists of all the entities which might possibly have the basic mathematical structure, provided the numerical functions in them satisfied certain relations. This is to say, the basic mathematical structure is a sub-set of the set of possible models. Let us call this set of possible models 'the matrix of the theory'. Intuitively, the matrix for the theory "determines" the sort of mathematical entities the theory is going to employ, without specifying any relations among these entities. The basic mathematical structure is a sub-set of the matrix in which these mathematical entities – numerical, vector, or other functions – stand in certain specific relations.

We can make an attempt to give a general characterization of the things that *could be* matrices for theories of mathematical physics. This is done in (D24). It is convenient to define an *n*-ary matrix, where *n* is the number of functions appearing in the members of the matrix.

(D24) X is an *n-ary matrix for a theory of mathematical physics* if and only if

(1) X is a set,
(2) there exist sets $R_1, R_2, \ldots, R_n$ (not necessarily distinct) such that, $y \in X$ if and only if there exist $D, f_1, f_2, \ldots, f_n$ such that
   (i) $y = (D, f_1, f_2, \ldots, f_n)$;
   (ii) D is a set;
   (iii) for all i, $1 \leq i \leq n$, f_i is a function from D into R_i.

According to (D24), an *n*-ary matrix for a theory of mathematical physics is a set of ordered $n+1$-tuples, the first member of which is a set D and the $1+i$-th member of which is a function f_i with domain D and range R_i. Moreover, for any given D, *every* function from D into R_i appears as the *i*-th member of some member of the matrix. Intuitively, the R_i are to be thought of as things like the set of real numbers, the set of *n*-dimensional vectors over the field of real or complex numbers, or some set of linear operators on these vector spaces. They remain the same for all members of the matrix.

This definition of a matrix is deficient in at least three respects. First, it should be the case that the set of all models for 'is a PM '((D7), p. 114) is a tertiary matrix for a theory of mathematical physics. According to our definition, it is not, simply because there is not one, single domain of individuals in the models for 'is a PM' on which all the functions are defined. There is yet a second reason why this set is not a matrix by our definition. To be a PM the functions in $(P, T, \bar{s}, m, \bar{f})$ must satisfy certain conditions, e.g. $\bar{s}$ must be twice differentiable, and $\bar{f}$ must satisfy a convergence condition. To be a matrix the set of models would have to include members with *all* functions from $P \times T$ into the 3-vectors over the field of real numbers, not just the twice differentiable or convergent ones.

Now, it is not clear that these difficulties are decisive. One might argue that it would be a simple matter to reformulate our definition of 'is a PM' so that all the functions were defined on $P \times T$ – the time argument in the mass function being vacuous. One might also claim that we could leave the requirements of continuity, differentiability and convergence out of the definition of 'is a PM' and put them in the definition of 'is a CPM'. By doing these two things we could bring the definition of 'is a PM' into line with our definition of a matrix. The first modification may sound plausible; however, the second rather suggests that it is our definition of a matrix that ought to be modified. Intuitively, what we want in the matrix is just the underlying mathematical apparatus of the theory out of which the rest of it is constructed. We need not include *all* functions from the domains into the range associated with the i-th function place. We just need to include enough of them to supply what is needed for the rest of the theory. But, how to draw this line, in complete generality, is not easy to see. Indeed, this just might be a feature of theories of mathematical physics about which no tenable and interesting generalizations can be made.

Further examples of difficulties of this same sort are provided by the axiomatizations of the Lagrangian and Hamiltonian formulations of particle mechanics (see below (D45), p. 208 and (D50), p. 216). In these axiomatizations, the problem of finding some way of recasting them so that all the functions in each model are defined on the same domain does not admit of any obvious and natural solution. There seems to be no plausible way to bring them into accord with our definition of a matrix.

These difficulties with (D24) both tend to show that it is too restrictive. Some things that we would like to admit as matrices for theories of mathematical physics get ruled out by our definition. On the other hand, there is some reason to think that this definition would also let in some things that we would not intuitively recognize as things that *could be* associated with a theory of mathematical physics. For example, we have not attempted to characterize precisely what are suitable ranges for the functions in the members of the matrix. We have not attempted to restrict the R_i's to be sets of things like numbers, vectors, tensors, etc. that commonly appear in theories of mathematical physics. It thus appears almost certain that some things will satisfy our definition which could never be regarded as part of even a "possible" theory of mathematical physics.

In the light of these difficulties, it would be nice to have an account of the nature of a theory of mathematical physics that did not depend too heavily on (D24), or any similar attempt to characterize the matrices for such theories. This can be provided, and very soon we shall effectively cut our account loose from this definition. However, the definition does serve a useful, and perhaps essential, expository purpose in motivating the crucial definition in our account. Indeed, it continues to serve some heuristic purposes throughout the development of our account. However, it is not essential to the account, in the sense that none of the definitions and theorems depend upon it. For the moment, though, we shall continue to talk as if (D24) were an adequate characterization of a matrix for a theory of mathematical physics.

Among the functions that appear in members of the matrix of a theory of mathematical physics, a distinction can be made in the way that these functions are employed in applying the mathematical formalism of the theory to make empirical claims. This is the distinction between those functions that are used as theoretical functions and those which are used as non-theoretical functions. The significance of this distinction will, of course, not become apparent until we come to say precisely how the mathematical formalism is used to make empirical claims. However, we can foresee the need for such a distinction now and introduce the apparatus for making it. Clearly, the order in which the functions appear in members of the matrix is irrelevant, so that we simply agree that the last m-n functions in these members will be treated as theoretical functions. In general, the values of the integers m and n will depend on the theory we are dealing with. In the case that $m=n$, the theory simply has no theoretical functions. At this point, we shall also introduce the sub-set of the matrix which we are to identify as the basic mathematical structure of the theory. These features of the theory: the matrix; the distinction between theoretical and non-theoretical functions; and the basic mathematical structure; are among those aspects of the mathematical structure of the theory which remain constant throughout all its applications. We attempt to make these ideas precise with the following definition:

(D25) x is a *frame for a theory of mathematical physics* if and only if there exist M_0, N_0, r, M, and integers m and n, $n \leq m$ such that:

(1) $x = \langle M_0, N_0, r, M \rangle$;
(2) M_0 is an m-ary matrix for a theory of mathematical physics;
(3) $N_0 = \{y \mid \text{there is a } \langle D, f_1, f_2, \ldots, f_m \rangle \in M_0 \text{ such that } y = \langle D, f_1, f_2, \ldots, f_n \rangle \}$
(4) r is a function whose domain is M_0 and whose range is N_0 such that, if $\langle D, f_1, f_2, \ldots, f_m \rangle \in M_0$ then
$r(\langle D, f_1, f_2, \ldots, f_m \rangle) = \langle D, f_1, f_2, \ldots, f_n \rangle$.
(5) $M \subseteq M_0$.

In (D25) the distinction between theoretical and non-theoretical functions is made by specifying the set N_0. Intuitively, N_0 determines the kinds of mathematical functions that will be used in describing the intended applications of the theory. Since all possible functions of this kind appear in some member of N_0, it is natural to call N_0 'the set of possible applications of the theory'. In doing this we should remember that many of the members of N_0 will not even be what we would call 'physical systems'. They are simply things that have the same mathematical structure as the physical systems to which we shall apply our theory. It should also be noted that N_0 is an n-ary matrix for a theory of mathematical physics. In our previous terminology, N_0 is the set of possible *partial* models for the predicate characterizing the basic mathematical structure of the theory. In our previous examples, it was respectively the set of all models for 'is a P_0' and the set of all models for 'is a PK'. By choosing to employ N_0 in the way that we subsequently do, we effectively make the functions $f_{n+1}, f_{n+2}, \ldots, f_m$ theoretical functions by virtue of the way they are used in applying the mathematical structure of the theory. The set M is intended to be the set of all models for the predicate characterizing the basic mathematical structure of the theory. Roughly, it is the mathematical structure common to *all* physical systems to which the theory applies. In our examples, M was respectively the set of all models for 'is an S' and the set of all models for 'is a CPM'.

It should be noted that, strictly speaking, the distinction drawn in (D25) is one between theoretical and non-theoretical *function places* rather than functions. That is, we agree that the functions appearing in the last m-n places of the ordered $m+1$-tuples in M_0 are going to be used as theoretical functions. It is, however, rather awkward to continually speak of theoretical and non-theoretical function places. Thus we choose to speak of theoretical and non-theoretical functions, both when we refer to the last m-n functions in a particular member of M_0, and when we refer to the last m-n function places. In most instances, the intended reference will be clear from the context. Where there might be some doubt, the reference will be made explicit.

Let us now pause for a moment and ask what is really essential about the concept of a frame. Clearly, one must have some set of models to count as the matrix and some sub-set of the matrix to count as the basic mathematical structure. Further, there must be some set to be identified as the set of intended applications of the theory, and the members of this set must be related to the members of the matrix in an appropriate way. Roughly speaking, the members of the set of possible applications must be obtainable from members of the matrix by simply "lopping off" the theoretical functions. In our definition this relation is characterized by (D25-3), and this characterization depends on the way we have characterized the members of the matrix. However, it seems clear that what is essential is only that every member of the matrix correspond to exactly one member of the set of possible applications – its "restriction". We can always intuitively regard this "restriction" operation as a "lopping off" of the theoretical functions in the members of the matrix, but what it amounts to formally is simply a function from the matrix onto the set of possible applications. Once we realize this, then the possibility arises of characterizing *some* of the essential properties of a frame for a theory of mathematical physics without being very explicit about the properties a matrix for such a theory. Consider the following definition.

(D26) x is a *frame for a theory of mathematical physics* if and only if there exists an M_0, N_0, M and r such that
(1) $x = \langle M_0, N_0, r, M \rangle$;
(2) M_0 and N_0 and M are sets;
(3) $M \subseteq M_0$;
(4) $N_0 \cap M_0 \neq \Lambda$ only if $N_0 = M_0$;
(5) r is a function whose domain is M_0 and whose range is N_0 such that, if $N_0 = M_0$ then $y = r(x)$ if and only if $y = x$.

Note that (D26-4) allows us to include theories in which there are no theoretical functions. In such theories $N_0 = M_0$. Except in these theories,

IDENTITY, EQUIVALENCE AND REDUCTION

we require that the matrix and the set of possible applications be disjoint. (D26–5) assures us that for theories with no theoretical functions the restriction function r is the identity function.

In (D26), we simply characterize, the "set-theoretic structure" of a frame without paying any attention to what the members of M_0 and N_0 must be like, or to the explicit nature of the restriction function r. This definition actually expresses all that we need to know about frames for our account of theories of mathematical physics. However, it is obvious that there are going to be things that are frames, according to this definition, which are not, and could not plausibly be, related to any such theory. What one needs to avoid this is something like (D24) where some attempt is made to rule out unwanted matrices, and, corresponding to this, something like (D25–3) and (D25–4) where inappropriate sets of possible applications and inappropriate restriction functions are ruled out. Unfortunately, there appears to be no *simple* way of doing this that does not rule out too much, e.g. our account to the mathematical structure of Lagrangian mechanics. So, rather than rule out too much, we choose to let in too much and take (D26) as the definition of a frame for a theory of mathematical physics. This is perhaps not so objectionable as it appears at first glance. After all, what we are primarily concerned about is saying something "interesting" about *all* the things that we, in fact, regard as theories of mathematical physics. That what we say "happens" to be true about some other rather outré entities should not disturb us unduly. No pernicious confusion would result if we simply agreed to call these things 'theories of mathematical physics'. Alternatively, one could take the requirements of (D26) to be necessary conditions for a frame. This would put all our subsequent account into the form of necessary conditions for theories of mathematical physics. I would not object to this way of taking the account. It is primarily for reasons of expository elegance that the first alternative is adopted.

At this point it is convenient to introduce some of the conceptual apparatus necessary to describe how the mathematical structures we are specifying are used to make statements. Consider a sentence of form (2). This sentence may be used to make a statement of the following sort. It "picks out" some sub-set of the set of possible applications, N_0, and "claims" that the set of intended applications is a sub-set of this distinguished sub-set. This sub-set of N_0 is distinguished as the set of all possible applications that are extendible, by adding theoretical functions, to produce members of M. In general, for any sub-set of N_0 it is distinguished as the set of all possible applications that are extendible, by adding theoretical functions, to produce members of M. In general, for any sub-set X of M_0, we may define $R(X)$ (D27-1) as the set of all members of N_0 that can be extended, by adding theoretical functions, to produce members of X. To state this definition, we make use of the notion of the restriction $r(x)$ of a member x of M_0. $r(x)$ is simply the member of N_0 one gets by "lopping off" the theoretical functions from x. Likewise, for any set of sub-sets of M_0, $\bar{X}$, we may define $\bar{R}(\bar{X})$ to be the set of all sub-sets of N_0 that can be extended, by adding theoretical functions to their members, to produce members of $\bar{X}$ (D27-2). It is also convenient to speak of the sets of extensions of members of sub-sets of N_0. The set of extensions of $x \in N_0$, $e(x)$, is simply all the members of M_0 which have the same configuration of non-theoretical functions as x (D27-3). The set of extensions of sub-sets and sets of sub-sets of N_0 are defined respectively by (D27-4) and (D27-5). In (D27-6) the notion of *an* extension is defined in the obvious way. Note that $\mathscr{E}(X)$, $X \in \mathscr{P}(N_0)$ is a *set* of sub-sets of M_0. If $Y \in \mathscr{E}(X)$ then, by "lopping off" the theoretical functions of every member of Y, one obtains X. We also need some notation for the *sub-set* of M_0 whose members have as their "non-theoretical parts" all and only the configurations of non-theoretical functions in X. $E(X)$ may also be thought of as the sub-set of M_0 obtained by adding all possible configurations of theoretical functions to each member of X. Likewise, $\bar{E}(\bar{X})$ (D27-8) is the sub-set of $\mathscr{P}(M_0)$ obtained by "extending" each *set* in $\bar{X}$ in all possible ways by adding theoretical functions to its members.

(D27) If $\langle M_0, N_0, r, M \rangle$ is a frame for a theory of mathematical physics then:
(1) R is a function whose domain is $\mathscr{P}(M_0)$ and whose range is $\mathscr{P}(N_0)$ such that, for all $X \in \mathscr{P}(M_0)$,
$R(X) = \{y \mid y \in N_0 \text{ and there is an } x \in X \text{ such that } y = r(x)\}$.
(2) $\bar{R}$ is a function whose domain is $\mathscr{P}(\mathscr{P}(M_0))$ and whose range is $\mathscr{P}(\mathscr{P}(N_0))$ such that, for all $\bar{X} \in \mathscr{P}(\mathscr{P}(M_0))$,
$\bar{R}(\bar{X}) = \{Y \mid Y \in \mathscr{P}(N_0) \text{ and there is an } X \in \bar{X} \text{ such that } Y = R(X)\}$.
(3) e is a function whose domain is N_0 and whose range is a sub-set of $\mathscr{P}(M_0)$ such that, for all $x \in N_0$,
$e(x) = \{y \mid y \in M_0 \text{ and } x = r(y)\}$.

IDENTITY, EQUIVALENCE AND REDUCTION 169

(4) $\mathscr{E}$ is a function whose domain is $\mathscr{P}(N_0)$ and whose range is a sub-set of $\mathscr{P}(\mathscr{P}(M_0))$ such that, for all $X \in \mathscr{P}(N_0)$,
$\mathscr{E}(X) = \{Y \mid Y \in \mathscr{P}(M_0) \text{ and } X = R(Y)\}$.

(5) $\bar{\mathscr{E}}$ is a function whose domain is $\mathscr{P}(\mathscr{P}(N_0))$ and whose range is a sub-set of $\mathscr{P}(\mathscr{P}(\mathscr{P}(N_0)))$ such that, for all $\bar{X} \in \mathscr{P}(\mathscr{P}(N_0))$,
$\bar{\mathscr{E}}(\bar{X}) = \{\bar{Y} \mid \bar{Y} \in \mathscr{P}(\mathscr{P}(M_0)) \text{ and } \bar{X} = \bar{R}(Y)\}$.

(6) For all $x \in N_0 (X \in \mathscr{P}(N_0))$, $\bar{X} \in \mathscr{P}(\mathscr{P}(N_0))$, $y(Y, \bar{Y})$ is *an extension* of $x(X, \bar{X})$ if and only if $y \in e(x)(Y \in \mathscr{E}(X), \bar{Y} \in \bar{\mathscr{E}}(\bar{X}))$.

(7) E is a function whose domain is $\mathscr{P}(N_0)$ and whose range is a sub-set of $\mathscr{P}(M_0)$ such that, for all $X \in \mathscr{P}(N_0)$,
$E(X) = \{y \mid y \in M_0 \text{ and } r(y) \in X\}$.

(8) $\bar{E}$ is a function whose domain is $\mathscr{P}(\mathscr{P}(N_0))$ and whose range is a sub-set of $\mathscr{P}(\mathscr{P}(M_0))$ such that, for all $\bar{X} \in \mathscr{P}(\mathscr{P}(N_0))$,
$\bar{E}(\bar{X}) = \{Y \mid Y \in \mathscr{P}(M_0) \text{ and } R(Y) \in \bar{X}\}$.

Some useful consequences of (D27) are enumerated in (T4). The proofs for these are, without exception, trivial and will not be given here. Particular note should be taken of (T4-4), (T4-5), and (T4-8). Much use will be made of these facts in what follows.

(T4) If $\langle M_0, N_0, r, M \rangle$ is a frame for a theory of mathematical physics then:
(1) For all $x \in M_0$, $R(\{x\}) = \{r(x)\}$, and, for all $X \in \mathscr{P}(M_0)$, $\bar{R}(\{X\}) = \{R(X)\}$;
(2) For all $X \in \mathscr{P}(M_0)$ if $X \neq \Lambda$ then $R(X) \neq \Lambda$, and, for all $X \in \mathscr{P}(\mathscr{P}(M_0))$, if $\bar{X} \neq \Lambda$ then $\bar{R}(\bar{X}) \neq \Lambda$;
(3) For all $X, Y \in \mathscr{P}(M_0)$, if $X \subset Y$ then $R(X) \subseteq R(Y)$, and, for all $\bar{X}, \bar{Y} \in \mathscr{P}(\mathscr{P}(M_0))$, if $\bar{X} \subset \bar{Y}$ then $\bar{R}(\bar{X}) \subseteq \bar{R}(\bar{Y})$;
(4) For all $X, Y \in \mathscr{P}(M_0)$, $R(X \cap Y) \subseteq R(X) \cap R(Y)$, and, for all $\bar{X}, \bar{Y} \in \mathscr{P}(\mathscr{P}(M_0))$, $\bar{R}(\bar{X} \cap \bar{Y}) \subseteq \bar{R}(\bar{X}) \cap \bar{R}(\bar{Y})$;
(5) For all $X \in \mathscr{P}(M_0)$, $\mathscr{P}(R(X)) = \bar{R}(\mathscr{P}(X))$;
(6) For all $x \in M_0$, $x \in e(r(x))$, for all $X \in \mathscr{P}(M_0)$, $X \in \mathscr{E}(R(X))$, and, for all $\bar{X} \in \mathscr{P}(\mathscr{P}(M_0))$, $\bar{X} \in \bar{\mathscr{E}}(\bar{R}(\bar{X}))$;
(7) For all $x \in N_0 (X \in \mathscr{P}(N_0), \bar{X} \in \mathscr{P}(\mathscr{P}(N_0))$, if $y \in e(x)(Y \in \mathscr{E}(X), \bar{Y} \in \bar{\mathscr{E}}(\bar{X}))$ then $r(y) = x(R(Y) = X, \bar{R}(\bar{Y}) = \bar{X})$;

(8) For all $X \in \mathcal{P}(N_0)$, $(\bar{X} \in \mathcal{P}(\mathcal{P}(N_0)))$, if
$y \in E(X)$, $(Y \in \bar{E}(\bar{X}))$ then
$r(y) \in X$, $(R(Y) \in \bar{X})$;

(9) For all $x \in N_0$, $(X \in \mathcal{P}(N_0))$, $R(e(x)) = \{x\}$, $(\bar{R}(\mathscr{E}(X)) = \{X\})$;

(10) For all $X \in \mathcal{P}(N_0)$, $(\bar{X} \in \mathcal{P}(\mathcal{P}(N_0)))$, $R(E(X)) = X$, $(\bar{R}(\bar{E}(\bar{X})) = \bar{X})$;

(11) If $M_0 = N_0$, then for all $x \in M_0$, $(X \in \mathcal{P}(M_0)$, $\bar{X} \in \mathcal{P}(\mathcal{P}(M_0)))$, $r(x) = x$, $(R(X) = X$, $\bar{R}(\bar{X}) = \bar{X})$;

(12) For all $x \in N_0$, $(X \in \mathcal{P}(N_0)$ and $Y \in \mathcal{P}(M_0)$, $(\bar{Y} \in \mathcal{P}(\mathcal{P}(M_0)))$; $x \in R(Y)$, $(X \in \bar{R}(\bar{Y}))$ if and only if $e(x) \cap Y \neq \Lambda (\mathscr{E}(X) \cap \bar{Y} \neq \Lambda)$.

(13) For all $X \in \mathcal{P}(N_0)$, $(\bar{X} \in (\mathcal{P}(\mathcal{P}(N_0)))$, if $y \in X$, $(Y \in \bar{X})$ then $e(y) \subseteq E(X) (\mathscr{E}(Y) \subseteq \bar{E}(\bar{X}))$.

We want now to introduce an additional element of the mathematical structure of the theory – the constraints on the theoretical functions. The intuitive idea of constraints is this. Not all possible sets of theoretical functions may be used to extend the set of intended applications of the theory to models for the predicate characterizing the basic mathematical structure. Some are ruled out by constraints on these functions. We want to characterize, in general, the notion of a constraint, whether it be on the theoretical functions, the non-theoretical functions, or both. We do this by singling out a set of sub-sets of M_0; call it C. If this set C is to be a constraint on certain functions, its properties must reflect two intuitive ideas about constraints. We must be able to recognize that C is a constraint on *only* these functions, and that it is a constraint on *sets* of these functions, i.e. no way of filling these places will be ruled out. We begin by defining a constraint for *any set* in such a way that the second intuitive feature is secured. (D28–2) assures us that every member of the set X will occur in at least one member of the set of sub-sets of X in the constraint C.

(D28)   If X is a set then C is a constraint for X if and only if
(1) $C \subseteq \mathcal{P}(X)$,
(2) For all $x \in X$, $\{x\} \in C$.

In (T5) some useful properties of constraints are listed. Again the proofs are trivial.

(T5)   If X is a set and C and C' are constraints for X then:
(1) For all $Y \subseteq X$, $C \cap \mathcal{P}(Y)$ is a constraint for Y;

(2) There is no $Y \subset X$ such that $C \subseteq \mathscr{P}(Y)$;
(3) For all $Y \neq \Lambda$, $Y \subseteq X$, $\mathscr{P}(Y) \cap C \neq \Lambda$;
(4) For all Y such that $X \subset Y$ there is a C^* such that C^* is a constraint for Y and $C = \mathscr{P}(X) \cap C^*$;
(5) $\mathscr{P}(X)$ is a constraint for X;
(6) $C \cap C'$ is a constraint for X.

Next, we define (D29) the notion of a core for a theory of mathematical physics. This is just a frame together with a constraint for the matrix M_0 in the frame. This definition is broad enough to allow that the constraint in the core may be on only theoretical functions, on only non-theoretical functions, or on both. These cases will be distinguished precisely in (D30), thus accounting for the first intuitive feature of constraints that we mentioned. Since we have, thus far, only spoken of constraints on theoretical functions, the intuitive motivation for permitting such generality in our definition of a core may not be apparent. This will be discussed in some detail below.

(D29) x is a *core for a theory of mathematical physics* if and only if there exist M_0, r, N_0, M and C such that
(1) $x = \langle M_0, N_0, r, M, C \rangle$;
(2) $\langle M_0, N_0, r, M \rangle$ is a frame for a theory of mathematical physics;
(3) C is a constraint for M_0.

Some properties of cores are listed in (T6). (T6-1) means intuitively that the effect of imposing constraints can not be achieved by stipulating additional requirements that must be satisfied in *all* the applications of the theory. Any such stipulation would exclude sub-sets of M_0 which were not excluded by the constraints. Intuitively, constraints are a very sensitive way of singling out sub-sets of M_0. (T6-1) is a trivial consequence of (T5-2). (T6-2) is a trivial consequence of (T6-1) it assures us that no constraint fails to be satisfied by some sub-sets of any set of models that we take to be the basic mathematical structure of the theory. (T6-3) notes that if $\mathscr{P}(M)$ is a sub-set of C, then C rules out no sub-sets of M. That is, C is effectively vacuous so far as *this* basic structure is concerned. However, were the basic structure to be changed then C might serve to rule out some sets of models for the new basic structure. (T6-4) notes that only in the case where $C = \mathscr{P}(M_0)$ can it be that both the restriction of C is all of

$\mathscr{P}(N_0)$ and the extension of this restriction lets in no more than was initially in C.

(T6) If $H=\langle M_0, N_0, r, M, C\rangle$ is a core for a theory of mathematical physics then:
(1) There is no $M'\subset M_0$ such that $C\subseteq\mathscr{P}(M')$;
(2) There is no non-empty $M'\subset M_0$ such that $\mathscr{P}(M')\cap C=\Lambda$;
(3) If $\mathscr{P}(M)\subset C$ then C is not a constraint for M;
(4) If $C\neq\mathscr{P}(M_0)$, then $\bar{R}(C)=\mathscr{P}(N_0)$ only if $\bar{E}(\bar{R}(C))\neq C$.

Intuitively, the constraint C in a core $H=\langle M_0, N_0, r, M, C\rangle$ constrains only the theoretical function places in M_0 if all possible *sets* of configurations of non-theoretical functions appear in members of C. But $\mathscr{P}(N_0)$ is essentially just the set of all sets of possible configurations of non-theoretical functions – i.e. ways of filling in the non-theoretical function places. And $\bar{R}(C)$ is the set of all sets of configurations of non-theoretical functions that appear in C. Thus, we can say that the core H is only theoretically constrained if and only if $\bar{R}(C)=\mathscr{P}(N_0)$. Similarly, the constraint C constrains only non-theoretical functions if it contains *all* the members of $\mathscr{P}(M_0)$ that can be obtained by extending any configuration of non-theoretical functions that appears in it, that is if $\bar{E}(\bar{R}(c))=C$. Thus, we say that it is only non-theoretically constrained if and only if $\bar{E}(\bar{R}(C))=C$. If $C=\mathscr{P}(M_0)$ then the constraint rules out nothing. In this case we say that H is vacuously constrained. In the case that H is not vacuously constrained (T4–4) says that H can not both be only theoretically and only non-theoretically constrained. However, it can be neither; in which case, we say that H is heterogeneously constrained. Intuitively this means that the way both theoretical and non-theoretical functions places are filled is constrained by C. In the case that $C=\mathscr{P}(M)$ and C rules out no sub-sets of models for M, we shall say that H is effectively vacuously constrained. This terminology is defined precisely in (D30).

(D30) If $H=\langle M_0, N_0, r, M, C\rangle$ is a core for a theory of mathematical physics:
(1) H is *only theoretically constrained* if and only if $\bar{R}(C)=\mathscr{P}(N_0)$;
(2) H is *only non-theoretically constrained* if and only if $\bar{E}(\bar{R}(C))=C$;
(3) H is *vacuously constrained* if and only if $C=\mathscr{P}(M_0)$;

(4) H is *heterogeneously constrained* if and only if x is not vacuously constrained and neither only theoretically, nor only non-theoretically constrained.

(5) H is *effectively vacuously constrained* if and only if $C = \mathscr{P}(M)$.

Note that (D30) forces us to say that vacuously constrained cores are both only theoretically and only non-theoretically constrained. In particular $\langle N_0, N_0, r, M, \mathscr{P}(N_0) \rangle$ is only theoretically constrained even though, intuitively, it has no theoretical functions.

Analogous to the way that a frame determines a sub-set $R(M)$ of the set of all possible applications N_0, a core determines a sub-set $\bar{R}(\mathscr{P}(M) \cap C)$ of the set of all sub-sets of possible applications. Intuitively, in the case where the constraints are on only theoretical functions, $\bar{R}(\mathscr{P}(M) \cap C)$ is just the set of all sets of possible applications that can be extended, by adding theoretical functions, to produce sub-sets of M in which the theoretical functions satisfy the constraints. Let us examine some properties of $\bar{R}(\mathscr{P}(M) \cap C)$ in the general case where constraints may also operate on non-theoretical functions. These properties are summarized in (T7).

(T7) If $\langle M_0, N_0, r, M, C \rangle$ is a core for a theory of mathematical physics then:
(1) $\bar{R}(\mathscr{P}(M) \cap C) \subseteq \bar{R}(\mathscr{P}(M)) \cap \bar{R}(C)$;
(2) $\bar{R}(\mathscr{P}(M) \cap C)$ is a constraint for $R(M)$;
(3) $\bar{R}(C)$ is a constraint for N_0;
(4) There is a C' such that:
   (a) C' is a constraint for N_0,
   (b) $\bar{R}(\mathscr{P}(M) \cap C) = \mathscr{P}(R(M)) \cap C'$;
(5) There is an $N \subseteq N_0$ such that $\mathscr{P}(N) = \bar{R}(\mathscr{P}(M) \cap C)$ if and only if $\bar{R}(\mathscr{P}(M) \cap C) = \bar{R}(\mathscr{P}(M))$;
(6) If $\mathscr{P}(M) \subseteq C$ then $\bar{R}(\mathscr{P}(M) \cap C) = \bar{R}(\mathscr{P}(M))$.

(T7–1) follows from (T4–4). From (T7–1) and (T4–5) we have $\bar{R}(\mathscr{P}(M) \cap C) \subseteq \mathscr{P}(R(M))$. Further if $x \in R(M)$ then there is some $y \in M$ such that $x = r(y)$. But $\{y\} \in \mathscr{P}(M) \cap C$, since $\mathscr{P}(M) \cap C$ is a constraint for M ((T5–1) and (D27–2)), and hence $\{r(y)\} = \{x\} \in \bar{R}(\mathscr{P}(M) \cap C)$. Thus we have (T7–2). In a similar manner we get (T7–3). (T7–5) follows from (T7–2) and (T5–4). Intuitively, the content of (T7–2) and (T7–5) is this. The basic mathematical structure M and the constraints C determine

174 THE LOGICAL STRUCTURE OF MATHEMATICAL PHYSICS

some sub-set $\bar{R}(\mathscr{P}(M) \cap C)$ of $\mathscr{P}(N_0)$. This determined sub-set is a constraint on the set of possible applications $R(M)$. That is, it determines some set of sub-sets of $R(M)$. Moreover, this determined sub-set can be seen as the intersection of $\mathscr{P}(R(M))$ and some constraint C' for the set of possible applications N_0.

It is of some interest to know when this set $\bar{R}(\mathscr{P}(M) \cap C)$ is the power of some sub-set of N_0 – i.e. when it can be determined without a constraint for N_0. (T7-5) says that this is the case if and only if $\bar{R}(\mathscr{P}(M) \cap C) = R(\mathscr{P}(M)) = \mathscr{P}(R(M))$, that is, if and only if $R(\mathscr{P}(M) \cap C)$ is a vacuous constraint for $R(M)$. In this situation, the constraint C rules out no more sub-sets of N_0 than could be ruled out using M alone. The 'if' part of (T7-5) is obvious. To see the "only if" part, suppose $\bar{R}(\mathscr{P}(M) \cap C) = \mathscr{P}(N)$ and $N \neq R(M)$. By (T7-1), $\mathscr{P}(N) \subseteq \mathscr{P}(R(M))$, thus there must be an x in $R(M)$ and not in N. But, by (T7-2) $\{x\} \in \bar{R}(\mathscr{P}(M) \cap O)$ and thus $\{x\} \in \mathscr{P}(N)$ and hence $x \in N$. Thus it must be that $N = R(M)$. (T7-6) follows immediately from (T7-1). It says that if C is a vacuous constraint for M then $\bar{R}(\mathscr{P}(M) \cap C)$ is a vacuous constraint for $R(M)$. It should be noted that $\bar{R}(\mathscr{P}(M) \cap C)$ *may* turn out to be a vacuous constraint for $R(M)$ even if C is not a vacuous constraint for M. This situation is illuminated by the following simple example. (Note that in this example we employ the notion of a matrix defined in (D24).)

(E5) Let M_0 be any binary matrix for a theory of mathematical physics such that $R_1 = R_2 =$ the real numbers. Let N_0 be a unary matrix for a theory of mathematical physics such that $R_1 =$ the real numbers. Let D be any set and x, y, z, a, b, c be functions from D into the real numbers. Denote $\langle D, i, j \rangle$, $i \in \{x, y, z\}, j \in \{a, b, c\}$ by 'ij', and $\langle D, i \rangle$ by 'i'.

(a) Let $M = \{xa, yb, zc\}$ and
$C = \mathscr{P}(M_0) - \{x \mid x \in \mathscr{P}(M_0)$ and there are $y, z \in M_0$ such that $x = \{y, z\}\}$.
Then $R(M) = \{x, y, z\}$,
$$\mathscr{P}(M) \cap C = \begin{Bmatrix} \{xa\}, \{yb\}, \{zc\} \\ \{xa, yb, zc\} \end{Bmatrix}$$
and $\bar{R}(\mathscr{P}(M) \cap C) = \{\{x\}, \{y\}, \{z\}, \{x, y, z\}\}$.

(b) Let $M = \{xa, yb, ya\}$ and
$C = \mathscr{P}(M_0) - \{x \mid x \in \mathscr{P}(M_0)$ and there are $y, z, w \in M_0$ such that $x = \{y, z, w\}\}$.

Then $R(M)=\{x, y\}$ and
$$\bar{R}(\mathscr{P}(M) \cap C) = \{\{x\}, \{y\}, \{x, y\}\}$$
$$= \mathscr{P}(\{x, y\})$$
$$= \mathscr{P}(R(M)).$$

In (E5–a) there is clearly no sub-set of N_0 whose power set is $\bar{R}(\mathscr{P}(M) \cap C)$. The constraint C has the effect of excluding all two-element sub-sets of $R(M)$, while M rules out everything but members of N_0 consisting of D and the functions x, y, and z. In (E5–b), $\bar{R}(\mathscr{P}(M) \cap C) = \mathscr{P}(R(M))$, even though $\mathscr{P}(M)$ is not a sub-set of C, i.e. C is not a vacuous constraint for M.

Let us now look at how $\bar{R}(\mathscr{P}(M) \cap C)$ differs when the core $H = \langle M_0, N_0, r, M, C \rangle$ is: (i) only theoretically constrained; (ii) only non-theoretically constrained; and (iii) heterogeneously constrained. We may summarize these cases as follows:

(i) only theoretical
 (a) $\bar{R}(C) = \mathscr{P}(N_0)$
 (b) $\bar{E}(\bar{R}(C)) = \mathscr{P}(M_0)$
 (c) $\bar{R}(\mathscr{P}(M) \cap C) \subseteq \bar{R}(\mathscr{P}(M)) \subseteq \bar{R}(C)$,
(ii) only non-theoretical
 (a) $\bar{R}(C) \subseteq \mathscr{P}(N_0)$
 (b) $\bar{E}(\bar{R}(C)) = C$,
 (c) There is an $N = N_0$ such that
  $\mathscr{P}(N) = \bar{R}(\mathscr{P}(M) \cap C)$ only if $\mathscr{P}(M) \subseteq C$.
(iii) heterogeneously
 (a) $\bar{R}(C) \subseteq \mathscr{P}(N_0)$
 (b) $C \subseteq \bar{E}(\bar{R}(C))$.

For the case where H is not effectively vacuously constrained, these situations are depicted in Figure 3. Note that (ii–c) means intuitively that $\bar{R}(\mathscr{P}(M) \cap C)$ is a vacuous constraint for $R(M)$ only if H is effectively vacuously constrained. This should not be surprising since the constraints are all on non-theoretical functions.

In addition to illuminating some of the properties of $\bar{R}(\mathscr{P}(M) \cap C)$, the preceding discussion should convince us that there is no purely formal reason why one should insist that constraints operate only on theoretical functions. Whatever functions happen to be constrained in a core, the core picks out some set of sets of possible applications. It singles out some sets of possible "observable" states of affairs. So far as the preceding discussion goes, there is no apparent reason to distinguish the case in which the

176 THE LOGICAL STRUCTURE OF MATHEMATICAL PHYSICS

constraints are on only theoretical functions from the other cases. This seems to justify our decision to allow the constraints on cores to operate on non-theoretical, as well as theoretical, functions. However, this is not the whole story. We shall return to this point shortly.

Roughly, the view being investigated here is this. The core of a theory of mathematical physics is one of the aspects of the theory that remains

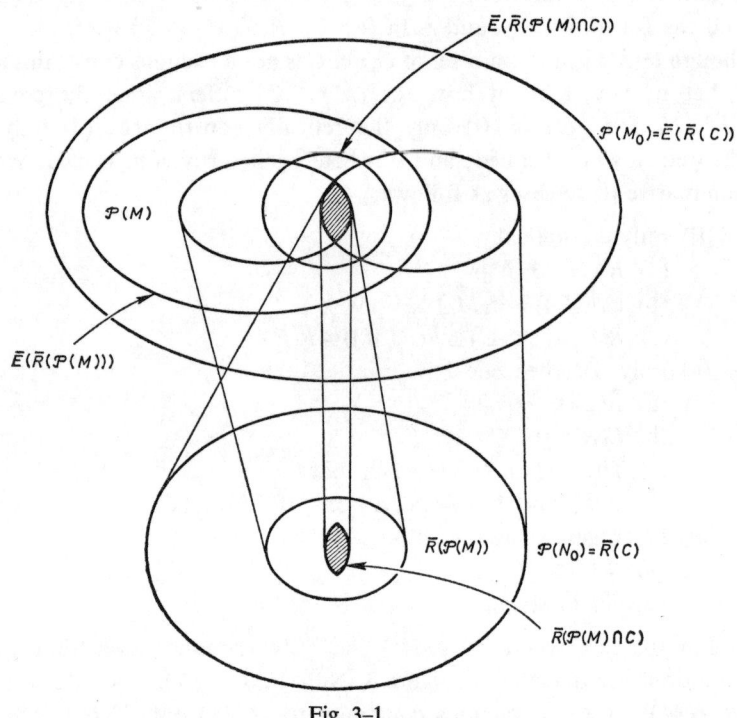

Fig. 3–1

unchanged throughout various different attempts to use the mathematical structure of the theory to make empirical claims. It is not the only thing that remains unchanged. Some "core" set of intended applications also seems relevant to specifying the identity conditions for theories of mathematical physics. But, what we have defined as the core of the theory is the part of the formal, mathematical apparatus of the theory that remains fixed through various different attempts to make tenable empirical claims with the mathematical structure associated with the theory. There may, however, be some additional formal apparatus that differs in various different attempts to use the mathematical structure of the theory. It was suggested in Chapter V that the theoretical laws postulated to hold in some intended applications might differ in different attempts to "carve up" the world in such a way that a tenable empirical claim might be made using a given core mathematical structure. In addition, we saw in Chapter VI

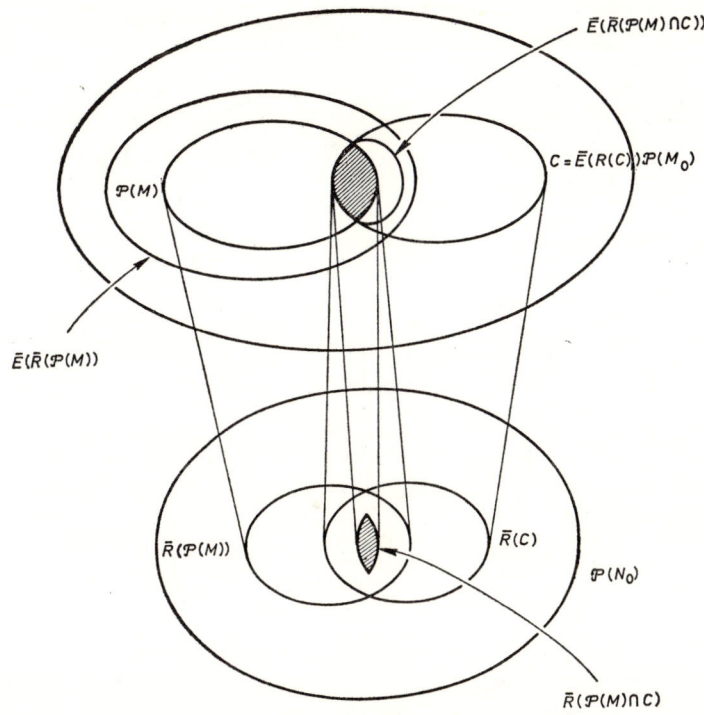

Fig. 3-2

that there might be special constraints on theoretical functions associated with particular theoretical laws that are postulated. That is, there may be two kinds of constraints: those like the ones on the mass function which appear in every attempt to use the core structure; and those like the ones on the force function which are tied intimately to special force laws and would not appear were we not to postulate the relevant force law.

178 THE LOGICAL STRUCTURE OF MATHEMATICAL PHYSICS

To capture these ideas, we first define the notion of a law for a frame $F = \langle M_0, N_0, r, M \rangle$. A law for F is simply a sub-set of M_0. A law is a non-theoretical law if it only rules out certain ways of filling the non-theoretical function places; that is, if it contains all extensions of the configurations of non-theoretical functions that it contains. This is stated precisely in (D31–2). Note the analogy to the definition of a non-theoretical constraint

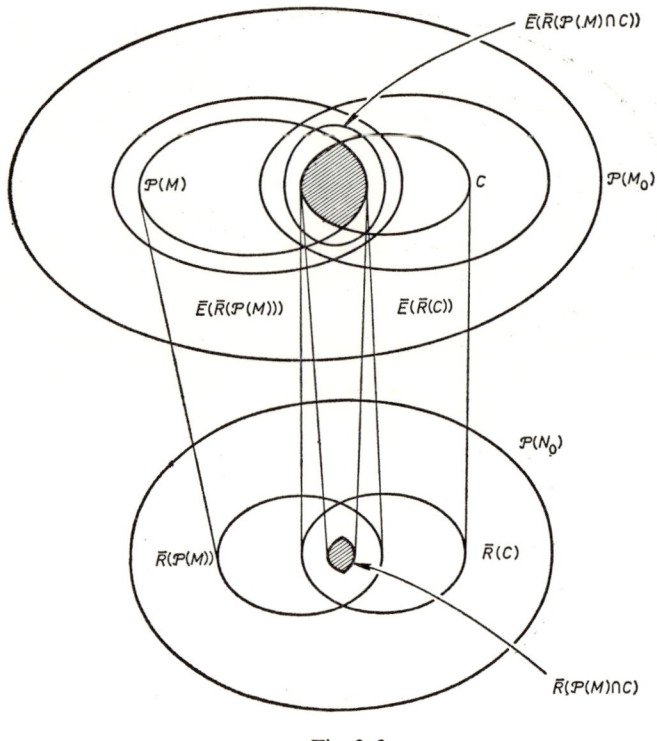

Fig. 3-3

(D30–2). We say that a law is theoretical if and only if it is not non-theoretical. This permits a theoretical law to rule out configuration of non-theoretical functions, as well as configurations of theoretical functions. Intuitively, this means that theoretical laws can "connect" the values of theoretical functions with the values of non-theoretical functions as in the case of Newton's second law.

(D31) If $F = \langle M_0, N_0, r, M \rangle$ is a frame for a theory of mathematical physics then:

(1) X is *a law for F* if and only if $X \subseteq M_0$;
(2) X is *a non-theoretical law for F* if and only if X is a law for F and $E(R(X)) = X$;
(3) X is *a theoretical law for F* if and only if X is a law for F and X is not non-theoretical.

Next, we define an expanded core for a theory of mathematical physics. Roughly, this consists of a core, a set of laws L, a set of constraints C_L associated with the functions "involved" in the laws in L, and a relation α which assigns members of N_0 to laws in L. Intuitively, we are going to claim that law $1 \in L$ "holds" in just those intended applications that are assigned to 1 by the relation α. Since we want to leave open the possibility of claiming that, in some intended applications, no functions have special forms, we include M in L (D32-3-ii). Note that we do not require that all the laws in L be theoretical laws. But we do require that if they all are, then C_L must constrain only theoretical functions (D32-4). This is an effort to capture the intuitive requirement that the constraints in C_L have something to do with the laws in L. The requirement that (D32-5) places on α is essentially this. If law x is a special case of law y (i.e. x is a sub-set of y) and we allege x to hold in z, then we must also allege y to hold in z. It should be intuitively clear that "application relations" lacking this property would not be very interesting. An expanded core is the sort of mathematical structure that is referred to in sentences of form (5).

(D32) x is *an expanded core for a theory of mathematical physics* if and only if there exist $M_0, x, N_0, \alpha, C, L, C_L$, or such that:
(1) $x = \langle M_0, N_0, r, M, C, L, C_L, \alpha \rangle$;
(2) $\langle M_0, N_0, r, m, C \rangle$ is a core for a theory of mathematical physics;
(3) L is a set such that:
   (i) for all $x \in L$, x is a law for $\langle M_0, N_0, r, M \rangle$;
   (ii) $M \in L$;
(4) C_L is a constraint for M_0 such that, if, for all $x \in L$, $x \neq M$, x is a theoretical law for $\langle M_0, N_0, r, M \rangle$, then $\bar{R}(C_L) = \mathscr{P}(N_0)$;

180 THE LOGICAL STRUCTURE OF MATHEMATICAL PHYSICS

>(5) α is a relation whose domain is N_0 and whose range is L such that, for all $z \in N_0$, $x, y \in L$, if $(z, x) \in \alpha$ and $x \subseteq y$, then $\langle z, y \rangle \in \alpha$.

Some useful terminology is introduced in (D33). An expansion of a core H is simply any expanded core which contains H. The null-expansion of H is the expansion obtained by adding no new laws and no new constraints.

>(D33) If $H = \langle M_0, N_0, r, M, C \rangle$ is a core for a theory of mathematical physics then:
>(1) x is *an expansion of H* if and only if there exist L, C_L and α such that $x = \langle H, L, C_L, \alpha \rangle$ and x is an expanded core for a theory of mathematical physics;
>(2) x is *the null-expansion of H* if and only if
>$$x = \langle H, \{M\}, \{\mathscr{P}(M_0)\}, N_0 x \{M\} \rangle;$$
>(3) For all $z \in N_0$, $\alpha(z) = \cap \{x \mid \langle z, x \rangle \in \alpha\}$.

It is also convenient to define the notion of a theoretically expanded core. This is roughly a core which has been expanded by adding theoretical laws and constraints on the *theoretical* functions "involved" in these laws. Note that any core, not just theoretically constrained cores, may be theoretically expanded. In the example of classical particle mechanics, we dealt with a theoretically constrained core and found it necessary to consider only theoretically constrained expansions of this core. One might naturally ask if, in this respect, classical particle mechanics is typical of all theories of mathematical physics. We will consider this question below. It should only be noted now that there appear to be no purely formal reasons why we should restrict ourselves to using only theoretical expansions of a theoretically constrained core to pick out a certain sub-set of sets of intended applications.

>(D34) x is *a theoretically expanded core for a theory of mathematical physics* if and only if there exist $M_0, r, N_0, M, C, L, C_L, \alpha$ such that:
>(1) $x = \langle M_0, N_0, r, M, C, L, C_L, \alpha \rangle$ is an expanded core for a theory of mathematical physics;
>(2) For all $x \in L$, $x \neq M$, x is a theoretical law for $\langle M_0, N_0, r, M \rangle$.

Analogous to the way that a core determines a set of sets of intended applications $\bar{R}(\mathscr{P}(M) \cap C)$, an expanded core also determines a set of sets of intended applications. Roughly, these are the sets of intended applications which can be extended to models for the basic structure M in such a way that both the constraints C and C_L are satisfied and such that the extension of each member of the set is a member of the member of L assigned to it by C, i.e. the extension of x satisfies the law $\alpha(x)$. This is made more precise in (D35):

(D35) If $\mathscr{E} = \langle M_0, N_0, r, M, C, C_L, \alpha \rangle$ is an expanded core for a theory of mathematical physics then

$$N_{\mathscr{E}} = \left\{ X \middle| \begin{array}{l} X \subseteq N_0 \text{ and there is a } Y \text{ such that:} \\ \text{(i) } Y \in (\mathscr{P}(M) \cap C \cap C_L) \\ \text{(ii) } X = R(Y) \\ \text{(iii) for all } y \in Y, \text{ if there is an } x \in X \text{ such that} \\ \quad x = r(y) \text{ then } y \in \alpha(x). \end{array} \right\}$$

To an expanded core for a theory of mathematical physics we add a set of intended applications, I, to obtain an applied core for a theory of mathematical physics. We shall say that the core is correctly applied if the set of intended applications is a sub-set of the set of possible applications. Intuitively, this just means that we are applying the theory to the *sort* of thing to which it might be expected to apply, though we still might not be applying it "successfully". We shall say that the core is successfully applied if I is in the set $N_{\mathscr{E}}$ determined by the expanded core. Roughly speaking, applied cores correspond to the mathematical structure which is referred to in sentences of form (5), together with the statement made by the sentence using this structure. Speaking a bit more precisely, this is the case only if the core is correctly applied. This is because we require that the $\bar{Q}$ referred to in sentences of form (5) be partial possible models for the predicate 'is an S'. Our reasons for doing this are discussed in Chapter II (pp. 16 ff.). Successfully applied cores correspond to true statements.

(D36) x is an *applied core for a theory of mathematical physics* if and only if there exist $M_0, N_0, r, M, C, C_L, I, \alpha$ such that:
(1) $x = \langle M_0, N_0, r, M, C, L, C_L, I, \alpha \rangle$;
(2) $\langle M_0, N_0, r, M, C, L, C_L, \alpha \rangle$ is an expanded core for a theory of mathematical physics;
(3) I is a non-empty set.

(D37) If $\mathscr{E}$ is an applied core for a theory of mathematical physics then:

(1) $\mathscr{E}$ is *correctly applied* if and only if $I \subseteq N_0$;
(2) $\mathscr{E}$ is *successfully applied* if and only if $I \in N_{\mathscr{E}}$.

The following special cases should be noted. In the case that L contains only M and $C_L = C$, then the condition that I be in $N_{\mathscr{E}}$ reduces to the condition that I be in $\bar{R}(\mathscr{P}(M) \cap C)$. In this case, the applied core corresponds to the mathematical structure referred to in sentences of form (3), together with the statement made by the sentence using this mathematical structure. In the case where, in addition, $\mathscr{P}(M) \subseteq C$, the constraints are vacuous and the condition that I be in $\bar{R}(\mathscr{P}(M) \cap C)$ reduces to the condition that I be a sub-set of $R(M)$. In this case, the applied core corresponds to the mathematical structure and statements associated with sentences of form (2). In the case where $N_0 = M_0$ and there are no constraints on non-theoretical functions and no non-theoretical laws, the condition that I be a sub-set of $R(M)$ reduces to the condition that I be a sub-set of M. In this case, the applied core corresponds to the mathematical structure and statements associated with sentences whose form is a generalization of form (1). Thus, an applied core may correspond to the mathematical structure and statement associated with any of the sentence forms we have considered.

We may think of applied cores for theories of mathematical physics in the following way. Any given applied core determines three things: a statement, a mathematical structure, and one of the sentence forms (1), (2), (3), and (5). The relation between these things is this. Some sentences of the determined form may make the determined statement by referring to the determined mathematical structure. There will, in general, be more than one sentence of this form which may do this. These sentences will differ in the predicates they employ to characterize the mathematical structure. Applied cores are, roughly, statements plus the mathematical structure used to make them, abstracted from any particular way of referring to this structure.

The apparatus we have constructed allows us to formulate more precisely the view about the identity criteria for theories of mathematical physics suggested earlier. An applied core for a theory of mathematical physics consists roughly of a core for a theory of mathematical physics, some theoretical laws, perhaps with constraints peculiar to them, and a set of intended applications. According to the view suggested earlier, two applied cores correspond to, or are applications of, the same theory if and only if the cores are identical and both sets of intended applications are included in the set of intended applications that is characteristic of the theory. Alternatively, we would say that theories of mathematical physics are to be identified as ordered parts consisting of a core for a theory of mathematical physics, together with some characteristic set of intended applications. This is still not quite an adequate formulation of the view, for we have said nothing about the properties that the characteristic set of intended applications of the theory must have. We have suggested, in an intuitive way, that this set may be characterized by some core set of intended applications, perhaps together with some stipulation about the acceptable ways of enlarging this core. We have not, however, attempted to say just what sort of things are admissible as intended interpretations of theories of mathematical physics – what sort of things physical systems are – nor have we been very explicit about the account of the various sets of intended applications appearing in applications of the same theory are related. About the first point, I have virtually nothing to say; about the second, what I have to say is most expediently dealt with in the next chapter where the "dynamic" aspect of physical theorizing is discussed. At this point, we may, however, give a necessary condition for something's being a theory of mathematical physics which is entailed by the view of identity criteria we are considering.

(D38) x is a *theory of mathematical physics* only if there exists H and I such that:

(1) $x = \langle H, I \rangle$;
(2) $H = \langle M_0, N_0, r, M, C \rangle$ is a core for a theory of mathematical physics;
(3) I is a set such that $I \subseteq N_0$.

The necessary condition (D38) should be understood in the following way. In order to make (D38) into a necessary and sufficient condition it will certainly be necessary to specify additional conditions to be satisfied by the characteristic set of intended applications I. It *may be* that other conditions on the core need to be added also. We have already suggested the possibility that *only* theoretically constrained cores are associated with theories of mathematical physics. The intuitive significance of (D38-3) is that the intended applications of the theory must, at least, have the same mathematical structure as the members of the set of possible applications. In conjunction with (D39) (below) this requires that every application of a theory of mathematical physics be a correctly applied core for a theory of mathematical physics.

(D39) If $T=\langle H, I\rangle$ is a theory of mathematical physics and $S=\langle M_0, N_0, r, M, C, C_L, L, \alpha, I_i\rangle$ is an applied core for a theory of mathematical physics, then *S is an application of T* if and only if $H=\langle M_0, N_0, r, M, C\rangle$ and $I_i \subseteq I$.

(D39) makes explicit the relation between theories of mathematical physics and applied cores that we have already described intuitively. Note that it allows that *any* expansion of a core may be used to produce an application, not just theoretical expansions. We have already mentioned the possibility that *only* theoretically expanded cores are used in producing applications of theories of mathematical physics. It should also be noted that there is an ambiguity in our usage of 'application'. We refer both to certain applied cores and to members of N_0 as 'applications'. Both these usages seem quite natural and the context will always make it clear which is intended.

The necessary condition (D38) allows us to define the relation of *formal identity* among theories of mathematical physics. Roughly speaking theories of mathematical physics are formally identical if and only if they have the same core mathematical structure, regardless of their range of intended applications. We shall find that the relation of formal identity is sufficient for discussing some other interesting ways in which theories of mathematical physics might be related. This is to say, these relations are independent of the explicit nature of the range of intended applications. They depend only on the core mathematical structures associated with the theories.

(D40) If $T=\langle H, I\rangle$ and $T'=\langle H', I'\rangle$ are theories of mathematical physics then *T is formally identical to T'* if and only if $H=H'$.

Intuitively, the view being considered here is that two theories of mathematical physics are formally identical if and only if they have the same basic mathematical structure, the same distinction between theoretical and non-theoretical functions, and the same core of constraints. This view may be contrasted with two other possible views about formal identity. One is the view that mere identity of the basic mathematical structure is sufficient for formal identity of two theories. This "weak" view of formal identity appears to have been held by Adams ([1], p. 6 ff.). It is a view that one would be naturally inclined to if no attention had been given to the question of how the basic formalism is related to the empirical content of the theory. The discussion in Chapters II–IV, however, suggests that, when attention is given to this question, the distinction between theoretical and non-theoretical functions and at least some of the constraints on the theoretical functions are as essential to the theory as the basic mathematical structure. (The question of whether *only* constraints on the theoretical functions may appear in the core will be considered shortly.) I can think of no example of a real theory in which the functions that are treated as theoretical functions differ in what we would regard intuitively as different attempts to apply the same theory. The reason for this appears very roughly to be that the mathematical structure of the intended applications is "given" as a part of the problem facing the theoretician. He is constrained by the formulation of his problem to treat the functions appearing in the description of the "observed data" as non-theoretical. He may theorize using only these functions, or he may find it expedient to employ more sophisticated mathematical structures with theoretical functions. It also seems that at least some constraints on theoretical functions, like those on the mass function in particle mechanics, are such that, were they to change, we would say intuitively that we were dealing with a different physical quantity and thus with a different physical theory. It appears likely that this would be our reaction to a non-extensive mass function. Just exactly what determines the constraints that practitioners of a theory regard as essential, and the more general question of how any constraints at all come to be imposed on theoretical functions will be considered in the next chapter when we consider the dynamic aspect of scientific theorizing. All these considerations appear to count against the "weak" view of formal identity. I can, however, think of no explicit counterexample to it where we, for example, use the same basic mathematical structure with different functions treated as theoretical functions and do, in fact, intuitively think of this as applying two different theories.

In contrast to this "weak" view of formal identity, a "strong" view would be one which required, in addition to identity of the basic mathematical structure and the theoretical, non-theoretical distinction, identity of the laws postulated and *all* the constraints, i.e. one which required identity of expanded cores. That this view is intuitively unsatisfactory can be seen by referring to the discussion of classical particle mechanics in the preceding chapter. It is clear that we would not regard ourselves as dealing with a different theory were we to discover that, contrary to our previous conjectures, an inverse-cube force law gave a better account of the paths of some planets. We ordinarily regard the "trial and error" postulating of different theoretical laws, and perhaps constraints connected with them, to be a part of the activity connected with trying to successfully apply *one* theory to some class of phenomena. Of course, there is usage of 'theory' that runs counter to this view – as when we speak of "the theory of gravitation" or of "the theory of harmonic oscillators". But there seems to be good reason to regard the entities referred to by these expressions as "sub-theories" of particle mechanics. The reason is this. The only way we can account for the way theoretical terms function in these "sub-theories" is to regard them as a "part" of a claim like (5). Attempting to regard them as theories in their own right by virtue of a "strong" criteria of formal identity will simply not allow us to explain why it is that the values of theoretical functions in these various "sub-theories" are not independent of one another. To explain this, they must be "tied together" in a claim of form (5).

It is instructive, at this point, to relate this discussion of the identity conditions for theories of mathematical physics to our attempt, in Chapter VI, to provide a logical reconstruction of classical particle mechanics. From the present point of view, at least one of the things we attempted to do in that discussion was provide a description of the *theory* itself, i.e. the core of the theory and the set of admissible sets of intended applications. In addition, we attempted to say something about the expansions of this core that were actually used by practitioners of the theory to make empirical claims. This second endeavor may be regarded as pursuing the ultimate end of logical reconstruction, and the description of the theory as a means to that end.

In our attempt at describing the core of classical particle mechanics, we were considerably more successful than in our attempt to describe the intended applications. The first three elements of the core are the following:

M_0 = the set of all models for 'is a PM'
N_0 = the set of all models for 'is a PK'
M = the set of all models for 'is a CPM'
r = the function such that $r(\langle P, T, \bar{s}, m, \bar{f}\rangle) = \langle P, T, \bar{s}\rangle$.

It should be clear that, identified in this way, $\langle M_0, N_0, r, M \rangle$ satisfies (D26). Regarding the fourth element, the constraints, we were not able to be as precise. It seems clear that the "identity" and "extensitivity with respect to concatenation" constraints on the mass-function are to be regarded as core constraints. With respect to the force function, the "identity" constraints surely belongs in the core, but it seems likely that other constraints on the force function are associated with special force laws and thus do not properly belong in the core.

This understanding of the identity criteria for theories of mathematical physics allows us to formulate precisely a notion of equivalence for theories of mathematical physics which appears to be adequate for dealing with the examples of intuitively equivalent theories we have mentioned. To this end, we begin by defining four senses of equivalence for cores for theories of mathematical physics. Cores are *application equivalent* if and only if they have the same set of possible applications (D41–A). They are *initially equivalent* if and only if they are application equivalent and the same set of possible applications is "picked out" by both cores (D41–C). Intuitively, this means that both cores, before they are expanded by adding special laws, may be used to say exactly the same thing about any given range of intended applications. Cores are *effect-equivalent*, roughly speaking, if anything that can be said using one can be said using the other, and conversely. More precisely, any pair of cores are effect equivalent if and only if they are application equivalent and, for any way of expanding one core, $\mathscr{E}$, which determines a set $N_\mathscr{E}$, there is some way of expanding the other core to determine the same set. *Theoretical effect-equivalence* is just effect equivalence when we are restricted to the use of theoretically expanded cores in "picking out" the set N.

(D41) If $H = \langle M_0, N_0, r, M, C \rangle$ and $H' = \langle M'_0, N'_0, r', M', C' \rangle$ are cores for theories of mathematical physics then:
(A) H is *application equivalent* to H' if and only if $N_0 = N'_0$.
(B) H' *initially dominates* H if and only if:
   (1) $N_0 = N'_0$
   (2) $\bar{R}'(\mathscr{P}(M') \cap C') \subseteq \bar{R}(\mathscr{P}(M) \cap C)$
(C) H is *initially equivalent* to H' if and only if
   H initially dominates H' and H' initially dominates H.
(D) H' *effect-dominates* H if and only if
   (1) $N_0 = N'_0$

(2) For all L, C_L and α such that $\mathscr{E} = \langle M_0, N_0, r, M, C, L, C_L, \alpha \rangle$ is an expanded core for a theory of mathematical physics, there is an L', $C_{L'}$, and α' such that $\mathscr{E}' = \langle M'_0, N'_0, r', M', C', L', C_{L'}, \alpha' \rangle$ is an expanded core for a theory of mathematical physics and $N_\mathscr{E} = N_{\mathscr{E}'}$.

(E) H is *effect-equivalent* to H' if and only if H effect-dominates H' and H' effect-dominates H.

(F) H' *theoretically effect-dominates* H if and only if:
(1) $N_0 = N'_0$
(2) For all L, C_L, and α such that $\mathscr{E} = \langle M_0, N_0, r, M, C, L, C_L, \alpha \rangle$ is a theoretically expanded core for a theory of mathematical physics, there exists L', $C_{L'}$, and α' such that $\mathscr{E}' = \langle M'_0, N'_0, r', M', C', L', C_{L'}, \alpha' \rangle$ is theoretically expanded core for a theory of mathematical physics and $N_\mathscr{E} = N_{\mathscr{E}'}$.

(G) H is *theoretically effect-equivalent* to H' if and only if H theoretically effect-dominates H' and H' theoretically effect-dominates H.

Some properties of the relations defined in (D41) should be noted. First, we should note that we have, in fact, defined equivalence relations.

(T8)   The relations 'is application equivalent to', 'is effect-equivalent to' and 'is theoretically effect-equivalent to' are equivalence relations.

Next, a necessary condition for the initial equivalence of application equivalent cores is to be noted.

(T9)   If $H = \langle M_0, N_0, r, M, C \rangle$ and $H' = \langle M'_0, N_0, r', M'_0, C' \rangle$ are cores for theories of mathematical physics then H is initially equivalent to H' only if $R(M) = R'(M')$.

To see this, suppose $\bar{R}(\mathscr{P}(M) \cap C) = \bar{R}'(\mathscr{P}(M') \cap C')$ and $R(\mathscr{P}(M)) \neq R'(\mathscr{P}(M'))$. Suppose there is an x in $R(M)$ but not in $R'(M')$. Then $\{x\} \in \bar{R}(\mathscr{P}(M) \cap C)$ and $\{x\} \notin \bar{R}'(\mathscr{P}(M') \cap C')$ (by (T7-2)). Thus $\bar{R}(\mathscr{P}(M) \cap C) \neq \bar{R}'(\mathscr{P}'(M') \cap C')$.

(T10) and (T11) are trivial consequences of (T9). They say respectively that, for application equivalent cores with vacuous constraints, $R(M) = R'(M')$ is both necessary and sufficient for initial equivalence, and that, for application equivalent cores without theoretical functions, identity for the

IDENTITY, EQUIVALENCE AND REDUCTION

basic mathematical structure is both necessary and sufficient for initial equivalence. The proof of (T11) requires (T4–12).

(T10) If $H = \langle M_0, N_0, r, M, \mathcal{P}(M_0) \rangle$ and $H' = \langle M'_0, N_0, r', M', \mathcal{P}(M'_0) \rangle$ are cores for theories of mathematical physics then H is initially equivalent to H' if and only if $R(M) = R'(M')$.

(T11) If $H = \langle N_0, N_0, r, N, C \rangle$ and $H' = \langle N_0, N_0, r, N', C' \rangle$ are cores for theories of mathematical physics then H is initially equivalent to H' only if $N = N'$.

(T13) states that initial equivalence and effect equivalence are coextensive relations. Intuitively, this means the following. If every set of sets of possible applications that can be picked out using one expanded core can also be "picked out" using another, and conversely, then the sets that are "picked out" by both cores, without the addition of special laws and constraints, must be identical. Also, if the sets that are "picked out" by the unexpanded cores are identical, then every set that we can "pick out" using an expansion of one core, we can also "pick out" using some expansion of the other core and conversely. To establish (T13) it is expedient to first prove (T12) which says that H effect-dominates H' if and only if H' initially dominates H. The asymmetry of H and H' in this result "washes out" to yield (T13).

(T12) If $H = \langle M_0, N_0, r, M, C \rangle$ and $H' = \langle M'_0, N_0, r', M', C' \rangle$ are cores for theories of mathematical physics then H effect-dominates H' if and only if H' initially dominates H.

To see this, suppose H effect-dominates H' and consider the expansion of H',

$$\mathcal{E}' = \langle M'_0, N_0, r', M', C', \{M'\}, \mathcal{P}(M'), N_0 \times \{M'\} \rangle.$$

Note that $N_{\mathcal{E}'} = \bar{R}'(\mathcal{P}(M') \cap C')$. Let

$$\mathcal{E} = \langle M_0, N_0, r, M, C, L, C_L, \alpha \rangle$$

be any expansion of H such that $N_{\mathcal{E}} = N_{\mathcal{E}'}$. Note that

$$N_{\mathcal{E}} \subseteq \bar{R}(\mathcal{P}(M) \cap C \cap C_L)$$

and thus

$$\bar{R}'(\mathcal{P}(M') \cap C') \subseteq \bar{R}(\mathcal{P}(M) \cap C \cap C_L)$$
$$\subseteq \bar{R}(\mathcal{P}(M) \cap C) \cap \bar{R}(C_L).$$

Thus $\bar{R}'(\mathcal{P}(M') \cap C') \subseteq \bar{R}(\mathcal{P}(M) \cap C)$ and H' initially dominates H.

Now suppose that H' initially dominates H. Then $\bar{R}'(\mathcal{P}(M') \cap C') \subseteq$

$\bar{R}(\mathscr{P}(M) \cap C)$ and thus $\bar{R}'(\mathscr{P}(M')) \subseteq \bar{R}(\mathscr{P}(M))$ (by an argument analogous to that in (T9) and (T4–5). Let

$$\mathscr{E}' = \langle M'_0, N_0, r, M', C', L', C_{L'}, \alpha' \rangle$$

be any expansion of H' and construct an expansion of H in the following way. Let

$$C_L = \bar{E}(\bar{R}'(\mathscr{P}(M') \cap C' \cap C_{L'})).$$

For any $x \in N_0$, let

$$\alpha(x) = E(\bar{R}\alpha'(x))$$

and let

$$L = \{M\} \cap \{y \mid y \in \mathscr{P}(M_0) \text{ and there is an } x \in N_0 \text{ such that } y = \alpha(x)\}.$$

Note first that

$$\bar{R}(\mathscr{P}(M) \cap C \cap C_L) = \bar{R}(\mathscr{P}(M) \cap C \cap \bar{E}(\bar{R}'(\mathscr{P}(M') \cap C' \cap C_{L'})))$$
$$\subseteq \bar{R}(\mathscr{P}(M) \cap C) \cap \bar{R}(\bar{E}(\bar{R}'(\mathscr{P}(M') \cap C' \cap C_{L'})))$$
$$\subseteq \bar{R}(\mathscr{P}(M) \cap C) \cap \bar{R}'(\mathscr{P}(M') \cap C' \cap C_{L'})$$
$$\subseteq \bar{R}'(\mathscr{P}(M') \cap C' \cap C_{L'}).$$

Now suppose there is an $X \in \bar{R}'(\mathscr{P}(M') \cap C' \cap C_{L'})$ and $\notin \bar{R}(\mathscr{P}(M) \cap C \cap C_L)$. Then (by (T4–12)),

$$\mathscr{E}(X) \cap (\mathscr{P}(M) \cap C \cap C_L) = \Lambda.$$

But since $X \in \bar{R}'(\mathscr{P}(M') \cap C' \cap C_{L'})$, (by (T4–13))

$$\mathscr{E}(X) \subseteq \bar{E}(\bar{R}'(\mathscr{P}(M') \cap C' \cap C_{L'})) = C_L.$$

Thus

$$\mathscr{E}(X) \cap \mathscr{P}(M) = \Lambda$$

and (by (T4–12)), $X \notin \bar{R}(\mathscr{P}(M))$. But, since $X \in \bar{R}'(\mathscr{P}(M') \cap C' \cap C_{L'})$, $X \in \bar{R}'(\mathscr{P}(M'))$ and thus $\bar{R}(\mathscr{P}(M))$, $\bar{R}'(\mathscr{P}(M'))$ which contradicts the hypothesis. Thus $\bar{R}(\mathscr{P}(M) \cap C \cap C_L) = \bar{R}'(\mathscr{P}(M') \cap C' \cap C_{L'})$. In a precisely analogous way, it can be shown that, for all $x \in N_0$, $\bar{R}(\alpha(x)) = \bar{R}'(\alpha'(x))$. Further all laws in L are non-theoretical laws and C_L is a constraint on only non-theoretical functions. Thus $\mathscr{E}$ is an expansion of H and $N_\mathscr{E} = N_{\mathscr{E}'}$. Hence H effect-dominates H. As a trivial corollary of (T12), we have:

(T13) If H and H' are cores for theories of mathematical physics then H is effect-equivalent to H' if and only if H is initially equivalent to H'.

The relation between initial equivalence and theoretical effect-equivalence is somewhat different. It is still the case that theoretical effect-equivalence entails initial equivalence. This can be seen by noting that the null-expansion of H – the one which adds no laws and only vacuous constraints – is a theoretically expanded core. Then the proof follows that of (T12). We state this precisely as follows:

(T14)   If $H = \langle M_0, N_0, r, M, C \rangle$ and $H' = \langle M'_0, N_0, r', M', C' \rangle$ are cores for theories of mathematical physics and H is theoretically effect-equivalent to H' then H is initially equivalent to H'.

On the other hand, initial equivalence does not entail theoretical effect-equivalence. The basic idea of the proof that initial equivalence entails effect-equivalence was, roughly, this. If H' initially dominates H then, given any expansion of H' containing laws L' and constraints $C_{L'}$, we can construct a set of laws and constraints for H which has the same effect by simply ignoring the theoretical functions in H. That is, we simply take the constraints on the non-theoretical functions determined by $C_{L'}$, to be the new constraints C_L, and the non-theoretical laws determined by L' to be the new laws L. The formal devices for doing this are respectively the $\bar{R}$ and $\bar{E}$ functions. These functions simply fill out the configurations of non-theoretical functions determined by $C_{L'}$ and by adding all possible configurations of theoretical functions. It should be evident that this method of proof would not work if we required the expansion of H to be a theoretically expanded core, i.e. if the laws in L were required to be theoretical laws and the constraints in C_L constraints on theoretical functions. This suggests that cores may be initially equivalent without being theoretically-effect equivalent. The following is a simple example of such a situation. The notation is the same as that of (E5).

(E5–e)   Let $M'_0 = M_0$, $N'_0 = N_0$,
$M = \{xa, ya, za\}$, $M' = \{xa, yb, za\}$
$C = \mathscr{P}(M_0)$ and $C' = \mathscr{P}(M')$. Then
$\bar{R}(\mathscr{P}(M) \cap C) = \bar{R}(\mathscr{P}(M') \cap C') = \mathscr{P}(\{x, y, z\})$, so that
$H = \langle M_0, N_0, r, M, C \rangle$ and
$H' = \langle M'_0, N'_0, r', M', C' \rangle$ are initially equivalent.
Let $l = \{m \mid m \in M_0$ and there is an $i \in I$ and $j \in \{a, b, c\}$ such that $m = \langle D, i, j \rangle\}$
and $L' = \{M'_0, l\}$. Note that l is a theoretical law for $F' =$

$\langle M'_0, N'_0, r', M' \rangle$. Note also that $M' \subseteq l$ so that $M' \cap l = M'$, thus L' is a trivial addition to H'. Let
$C_{L'} = \mathcal{P}(M'_0) - \{m \mid \text{there are } i, j \in I \text{ such that } m = \{ia, jb\}\}$.
Note that $C_{L'}$ is a theoretical constraint for F' and hence $\mathscr{E}' = \langle M'_0, N'_0, r', M', C', L', C_{L'}, \alpha \rangle$, for any α, is a theoretically expanded core. Further,

$$N_{0'} = \bar{R}'(\mathcal{P}(M') \cap C_{L'} \cap C') = \begin{cases} \{x\}, \{y\}, \{z\} \\ \{x, z\}, \{y, z\} \\ \{x, y, z\} \end{cases}.$$

Now, it should be clear that no theoretical expansion, $\mathscr{E}$, of H can be found such that $N_{\mathscr{E}} = N_{\mathscr{E}'}$. Since the theoretical function place in all members of M is filled in the same way any *theoretical* law of constraint which rules out more than $M^2 \cap C$ will rule out everything.

(T15) states that any core is initially equivalent (and thus effect-equivalent) to a core with no theoretical functions. This is a trivial consequence of (T7–4).

(T15) If $H = \langle M_0, N_0, r, M, C \rangle$ is a core for a theory of mathematical physics then there exist a C' such that $H' = \langle N_0, N_0, r', R(M), C' \rangle$ is a core for a theory of mathematical physics and H' is initially equivalent to H.

It should be noted that an analogous situation does not obtain for theoretical effect-equivalence. The reason is simple. There are no theoretical expansions of cores without theoretical functions. Thus, if H has theoretical functions, i.e. $M_0 \neq N_0$, then no H' without theoretical functions is theoretically effect-equivalent to H. (T16) makes more explicit one instance of the situation described in (T14).

(T16) If $H = \langle M_0, N_0, r, M, \mathcal{P}(M_0) \rangle$ is a core for a theory of mathematical physics then $H' = \langle N_0, N_0, r', R(M), \mathcal{P}(N_0) \rangle$ is a core for a theory of mathematical physics and H' is initially equivalent to H.

(T17) says that constraints (possibly on theoretical functions) which produce non-trivial constraints on $R(M)$ can only be reproduced in a theory without theoretical functions by non-trivial constraints on the non-theoretical functions. That is, one may "eliminate" theoretical functions, but one can not "eliminate" constraints.

(T17) If $H=\langle M_0, N_0, r, M, C\rangle$ and $H'=\langle N_0, N_0, r', R(M), C'\rangle$ are theories of mathematical physics such that H is initially equivalent to H' and $\bar{R}(\mathscr{P}(M) \cap C) \neq \bar{R}(\mathscr{P}(M))$ then $C' \neq \mathscr{P}(N_0)$.

(T18) states that every core is initially equivalent to an only theoretically constrained core. Intuitively, this means that anything we can say using constraints on non-theoretical functions we can also say using constraints on only theoretical functions.

(T18) If $H=\langle M_0, N_0, r, M, C\rangle$ is a core for a theory of mathematical physics, then there exists an $H'=\langle M_0, N_0, r', M', C'\rangle$ such that H' is an only theoretically constrained core for a theory of mathematical physics and H' is initially equivalent to H.

To see this, construct H' in the following way. Extend each member of $R(M)$ to a member of M_0 by adding arbitrarily chosen theoretical functions to it. That is, for every $x \in R(M)$ choose arbitrarily some $y \in e(x)$. Call this y_x. Then let $M'=\{z \mid \text{there is an } x \in R(M) \text{ and } z=y_x\}$. Construct C' in the following way. Let C' contain: (1) all members of $\mathscr{P}(M')$ that are produced by the above mentioned, arbitrarily chosen extensions of the members of the sets in $\bar{R}(\mathscr{P}(M) \cap C)$; (2) no other members of $\mathscr{P}(M')$; (3) all members of $\mathscr{P}(M_0) - \mathscr{P}(M')$. It is easy to see that $\bar{R}'(\mathscr{P}(M') \cap C') = \bar{R}(\mathscr{P}(M) \cap C)$. Further, C' is a constraint for M_0 simply because it contains all singletons in $\mathscr{P}(M_0)$. It is a constraint on the theoretical functions because any configuration of theoretical functions that appears in some member of C' appears in C' with all possible configurations of non-theoretical functions. It should be noted that the essential idea of the above construction is this. The members of $R(M)$ are tied to some arbitrarily chosen configurations of theoretical functions to produce M'. Intuitively, this amounts to connecting the theoretical and non-theoretical functions by laws involving both. Then constraints on the theoretical functions are introduced which, when confined to $\mathscr{P}(M')$, "pick out" just the sets of non-theoretical functions that $\bar{R}(\mathscr{P}(M) \cap C)$ picks out.

We now use the notions of equivalence for cores defined in (D41) to define a notion of formal equivalence for theories of mathematical physics. Intuitively, two theories are formally equivalent if and only if the core mathematical structures associated with the theories admit the same possibilities for making statements about any given set of intended applications. That is, if and only if anything that can be said about a given range of intended applications using one core can also be said using the other, and conversely. This seems to indicate that we want to demand that the cores be effect-equivalent. In the light of (T13), however, we had just as well demand that they be initially equivalent. This is what we do in (D42–A).

We have seen that there is some reason to think that, in reality, only theoretically expanded cores are used to make the empirical claims of theories of mathematical physics. Were this the case, then initial equivalence would no longer entail the intuitive property of cores we are trying to capture (see the discussion surrounding (E5–C)). This property is, however, in this case, captured by theoretical effect-equivalence. In (D42–B) we define a notion of strong formal equivalence which requires that the cores be theoretically effect-equivalent. This is the notion of formal equivalence between theories of mathematical physics that would be appropriate were we convinced that we should restrict ourselves to theoretically expanded cores. It should be noted that, by (T14), strong formal equivalence also requires that the cores be initially equivalent so that strong formal equivalence entails weak formal equivalence.

(D42) If $T = \langle H, I \rangle$ and $T' = \langle H', I' \rangle$ are theories of mathematical physics then:
(A) T is *formally equivalent to T' in the weak sense* if and only if H is initially equivalent to H'.
(B) T is *formally equivalent to T' in the strong sense* if and only if H is theoretically effect-equivalent to H'.

Before we proceed to argue that the notion of formal equivalence defined in (D42) gives an adequate account of our intuitive ideas about equivalent theories, there are some lingering issues that need to be cleared up. Up to now, we have deferred discussing these questions because it seemed more expedient to deal with them in the light of a precise and explicit statement of one possible view about the identity conditions for theories of mathematical physics and an account of the relation of equivalence among such theories based on this view. We have now provided (in (D40) and (D42)) an account of *formal* identity and *formal* equivalence for theories of mathematical physics. We can now consider, in some detail, three possible objections to this account.

First, let us consider the possibility, mentioned earlier, that syntactical properties of sentences of the forms we are considering are relevant to determining the theories to which they correspond. Were this the case, then formal identity could not be defined in terms of the mathematical structure alone. We would have to take account of the ways in which this mathematical structure was, in fact, characterized in sentences in which it was used to make the empirical claims of the theory. Intuitively, the main reason for thinking that syntactical properties might be relevant is this. A great deal of mathematical structure can appear "implicitly" in a predicate without appearing "explicitly" as a part of the entities which satisfy the predicate. For example, when we define a function as the solution of a certain differential equation, there may be functions referred to in the definition that are not a "part" of the entity defined. In this way, some of the mathematical structure that we intuitively regard as characteristic of the theory might get "buried" in the predicates used in a certain formulation of the theory and not actually appear in the models for these predicates. Our criterion of formal identity would then lead us to decide that we had two formally distinct theories, when intuitively we would want to say that there was only one.

To understand this more explicitly consider the two applied cores:

$$S = \langle M_0, N_0, r, M, \mathscr{P}(M_0), \{M\}, \mathscr{P}(M_0), N_0 \times \{M\}, I \rangle$$
$$S' = \langle N_0, N_0, r', R(M), \mathscr{P}(N_0), \{R(M)\}, \mathscr{P}(N_0), N_0 \times \{R(M)\}, I \rangle.$$

Both of these applied cores correspond to the same statement: that $I \subseteq R(M)$. In the case of S, this statement could be made using a sentence of form (2) which referred to the core $H = \langle M_0, N_0, r, M, \mathscr{P}(M_0) \rangle$. In the case of S', the statement could be made using a generalization of a sentence of form (1) which referred to the core $H' = \langle N_0, N_0, r', R(M), \mathscr{P}(N_0) \rangle$. According to our account of formal identity, these sentences correspond to formally equivalent, but nevertheless distinct, theories of mathematical physics since the cores H and H' are initially equivalent (T16), but distinct.

However, it *might be* that we would, intuitively, want to say that these two sentences correspond to the *same* theory. If the statement associated with S', were made using a sentence in which $R(M)$ were characterized by a predicate which was a restriction (see Chapter III, pp. 41 ff.) of the predicate characterizing N_0 in the sentence used to make the statement associated with S, we would say that the theoretical functions had been Ramsey eliminated from the latter sentence.[1]

In this case, we probably *would* want to say that these two sentences correspond to numerically distinct, but equivalent theories. But, if $R(M)$ were characterized by a predicate which "involved" the theoretical functions appearing in the second sentence, then the theoretical functions have not been Ramsey eliminated, and it seems likely that we might say that these sentences correspond to the *same* theory. The reason that we might say this is that, roughly speaking, the same mathematical structure that appears in S gets incorporated into the predicate used to characterize $R(M)$. We have only relocated it, rather than eliminated it.

Admittedly, these syntactical notions like "involving" and "restriction" are rather vague and imprecise. Nevertheless, it appears that they are sometimes relevant in determining whether or not sentences of forms (1) and (2) are, intuitively, regarded as corresponding to the same theory. This suggests that syntactical features of the predicates that are, in fact, used to characterize the mathematical structures associated with theories are relevant to identifying these theories. That is, H and H' might be different cores and yet the predicates used to characterize them, in certain situations might be such that we would want to say that they were, in these situations, associated with the *same* theory. Were this the case, then the requirement of our proposed definition of 'formal identity' (D40) would not be a necessary condition for what we intuitively regard as formal identity.

One way to see the importance of these syntactical properties is to note the following. (T6) assures us that, for any theory of mathematical physics, there is another theory with *no theoretical functions* which is formally equivalent to the given theory. This result would trivially answer the question of whether we could "get on" without theoretical functions, were it not the case that we had to attend to syntactical matters. It might be that the cores without theoretical functions whose existence is guaranteed by (T16) do not admit of characterization by any syntactical apparatus that does not "involve" reference to the mathematical structure in the initially given core. In this case, we would not want to say that (T16) provided us with a theory, formally equivalent to the given theory, from which the theoretical functions had been "eliminated". "In principle" we can always get on without theoretical functions; whether we can "in practice" depends upon the properties of our syntactical apparatus.

Let us now consider the seriousness of this difficulty with our account of the identity condition for theories. Remember, that the job we want to do with these conditions is this. Given existing expositions of theories of mathematical physics, we want to be able to say whether or not they are expositions of the same theory. More precisely, given logical reconstructions of these expositions, using sentences of the forms we have discussed, we want to be able to tell whether these sentences correspond to the same or different theories. Now, it is clear that, if we permit unlimited latitude in the syntactical apparatus used in these sentences, our identity conditions will not be of much help in this task. We will never be certain that we have gotten all the relevant mathematical structure "out of the predicate" and into the entities satisfying the predicate. Unless we can do this, we can never be certain that we are not assigning some sentences to different theories that belong with the same theory. One way of solving this problem would be to provide some sort of canonical form for the syntactic apparatus used to characterize the mathematical structure which guaranteed that the maximal amount of mathematical structure appeared in the entities satisfying the predicates. This suggests that something like a formal language for defining the set-theoretic predicates used to characterize the mathematical structure might be a solution. In any event, the important thing to understand is that this is a serious difficulty only if we lack some means of determining when there is intuitively significant mathematical structure "buried" in the syntactical apparatus.

This way of viewing it suggests two further observations about this difficulty. First, it really does not count as an objection to the intuitive idea, reflected in our conditions for formal identity, that it is the mathematical structure that is "really" characteristic of the formal part of the theory. Nor does it count as an objection to our account of what this mathematical structure is like, i.e. that it is what we have called 'a core for a theory of mathematical physics'. This difficulty merely shows that the relevant mathematical structure *may not* be written on the face of our logical reconstructions of theories. Some of it may be buried in the syntax. Understood in this way, we have not really discovered a difficulty with our identity conditions, but only a difficulty in applying them. Second, it is very likely that this difficulty in application will be quite insignificant in "real life" situations when we are confronted with existing expositions of real theories. It seems likely to arise, if at all, in connection with attempts to provide theories formally equivalent to given theories with one or more of the theoretical functions "eliminated". This seems to indicate that this difficulty does not provide sufficient reason to think syntactical concepts must play an essential role in identifying theories.

The next objection to be considered is essentially an objection to our characterization of a core for a theory of mathematical physics (D28). We have already mentioned that there is some reason to think that none but only theoretically constrained cores (D30) actually appear in theories of mathematical physics. Were this the case, one might contend that we should either revise our definition of a core to admit only constraints on theoretical functions, or revise our definition of a theory (D38) to admit none but only theoretically constrained cores.

In dealing with this objection, it is important to understand that our definition of a theory of mathematical physics is being offered as a conceptual truth about the sorts of things that *could be* theories of mathematical physics. It is not being offered as simply an empirical claim about the things that have actually been offered, or recognized, as theories of mathematical physics. Thus, simply from the claim that all actually propounded theories employ constraints only on theoretical functions, it does not follow that our definition should be modified in the way suggested. To force such a modification in our definition, some compelling reasons must be given for thinking that, in some sense, theories of mathematical physics *could not* be formulated using constraints on non-theoretical functions.

To illustrate this point, consider the following question. Can theoretical functions always be eliminated from any given theory of mathematical physics. Putting aside syntactical questions, this is the same as asking whether there is a theory without theoretical functions that is formally equivalent to any given theory. The answer, given by (T16), is, yes. However, were we to decide that, for some reason, cores with constraints on non-theoretical functions *could not* be a part of a theory of mathematical physics, the answer might be no. For, in this case, no core without theoretical functions which contained non-trivial constraints would be a part of a theory, and thus, the core guaranteed by (T16) might not suffice to produce a formally equivalent theory. Indeed, (T17) shows that there are cases were no formally equivalent theory without non-trivial constraints on the non-theoretical functions can be produced. The point we note here is that the answer to this question is not to be determined by an examination of existing theories. It is a question about what *can* count as a theory of mathematical physics. Even if all theories that *have* actually been proposed contain theoretical functions and only constraints on these functions, one would not want to conclude that the answer to this question is that theoretical functions can not be eliminated. To do so would be simply to betray a misunderstanding of the question.

As to the empirical aspect of this issue, I conjecture that there are no recognized theories of mathematical physics in which the non-theoretical functions are constrained. More precisely, I conjecture that there are no theories of mathematical physics that have actually been propounded, whose logical reconstruction would require us to employ constraints on non-theoretical functions. I shall not give arguments in support of this conjecture here. The only sort of argument that would be conclusive is a case by case examination of theories. I can only invite the reader to examine the theories most familiar to him and make a tentative effort at formulating the claim of the theory along the lines that have been suggested. I believe that anyone who does this will find that constraints on non-theoretical functions do not appear to be required. I emphasize again that this is an empirical conjecture. A clear-cut counterexample would conclusively refute it.

Let us suppose that this conjecture is correct. Why do we not use constraints on non-theoretical functions? Is it simply an accidental fact that no theories happen to have been constructed using constraints on non-theoretical functions, or is there some deeper explanation for this fact? Is there any reason to believe that a theory of mathematical physics "could not" be formulated using constraints on non-theoretical functions? Is the use of constraints essentially tied to theoretical functions?

The preceding discussion should have convinced us that there is no *semantical* reason why constraints operate on only theoretical functions. Whatever functions happen to be constrained in a core, the core "picks out" some set of possible applications, $\bar{R}(\mathscr{P}(M) \cap C)$. It singles out some sets of "observable" states of affairs. So far as the preceding discussion goes, there is no apparent reason to distinguish the case when the constraints are on only theoretical functions from the other cases. Since we have chosen to characterize theories in purely semantical terms, it seems that any definition of a theory which rules out constraints on non-theoretical functions will appear to be arbitrary. At least, it appears that the most expedient line is to make the definition of a theory as wide as possible and then raise questions as to why some kinds of theories may be preferable to others, on other than semantical grounds. That is, as a terminological convenience, we will agree that entities with non-theoretically constrained cores shall be called 'theories', but we will then ask what reasons there are to think that such theories *can* never actually be propounded.

To understand what these reasons are, note that (T18) assures us that, for any given core, H, there is a theoretically constrained core, H' that is initially equivalent to it. Let us suppose that H contains constraints on non-theoretical functions and ask the following question. Is there any reason why a theory containing the core H' is preferable to a theory containing H?

One reason why one might prefer to use H' rather than H is that the syntactical apparatus necessary to characterize H' is, in some sense, simpler or more tractable than that necessary to characterize H. We saw an example of this in Chapter IV (p. 74) in considering the claim (3I). In this example, the effect of rather simply describable constraints on the theoretical functions could be reproduced only by employing constraints on the non-theoretical functions with quite complex descriptions. However, if this is to be an argument for ruling out *all* constraints on non-theoretical functions from theories of mathematical physics, it must be shown that this situation occurs in general. That is, it must be shown, that constraints on theoretical functions *always* admit of description by simpler syntactical devices than equivalent constraints on non-theoretical functions. I know of no argument that shows this. Indeed, though it is intuitively clear in the example in Chapter IV, the concept of "simplicity" for syntactical apparatus is notoriously elusive. This makes it doubtful that any convincing argument to this effect could be produced.

It should be noted that the claim that there are syntactical reasons for preferring constraints on theoretical functions is very similar to the much more familiar claim that there are syntactical reasons for preferring theories with theoretical functions (or theoretical terms) over equivalent theories without any theoretical functions at all. The difficulty with both these lines of argument is that they appeal to a notion of syntactical simplicity which is at least as problematic as the phenomenon it is invoked to explain.

One might also offer a sort of "pragmatic" reason for preferring H' over H. It might be contended that, though H and H' are equivalent in what they say about N_0 – equivalent in "descriptive force" – they differ in "explanatory force". One might say that, for at least some of the known facts about N_0, H merely provides a description of these facts, while H' provides explanation of them. Equivalently, one might say that constraints on non-theoretical functions will always appear to be *ad hoc*. We will always be prompted to ask "why" values of these functions in different situations are related in ways described in the constraints. In some sense, H' provides an answer to this "why-question" by saying, roughly, "It is because there is a theoretical function, related in a certain way to this observable function, whose values are constrained in a certain way". But, if this is to serve as an account of our preference for constraints on theoretical functions, we must be told *why* this is accepted as a satisfactory explanation of facts about the non-theoretical function. Why are we not prompted to seek a further explanation of the constraints on the values of the theoretical function? Why don't they look just as *ad hoc* as the constraints on the non-theoretical functions?

One possible way to account for this alleged disparity in "explanatory force" is to say that constraints on theoretical functions "explain" the observed constraints on non-theoretical functions because they admit of a characterization by simpler syntactical apparatus. This, of course, amounts to explicating the pragmatic notion of explanatory force in terms of the syntactical notion of simplicity and has the effect of reducing the presently considered account of our preference for constraints on theoretical functions to the previous one. Is there any satisfactory way of accounting for this disparity in "explanatory force" that does not appeal to the notion of simplicity?

It appears that there may be. To see this, first, note the peculiar way that constraints on *any* function operate. For any particular application, just exactly which values of the constrained function are ruled out will depend on what else is in the range of intended applications and the values of this function in these other applications. Next, note that there are always going to be, at least, some applications in which the values of the *non-theoretical* functions will be determined in ways that are independent of the theory in question. Intuitively, this means that we would have to insist that these functions had these values, even at the cost of giving up the theory. Now suppose we describe what we know, at some given time, about the range of intended applications using a constraint on some non-theoretical function. If we do this, we may discover, at some later time, further members of the range of intended applications with values of non-theoretical functions which are incompatible with the constraint we had initially chosen and our total knowledge about the range of intended applications. In general, we will not be able to "give up" our beliefs about the values of the non-theoretical functions. Some of them are determined independently of *this* theory. So we must give up the constraint. In contrast, if we put constraints on a theoretical function which are consistent with our present knowledge, and subsequently discover an intended application that is incompatible with this constraint, it is sometimes open to us to maintain the constraint and change our beliefs about certain of the values of the theoretical function. This is because the values of the theoretical function are all determined in ways that are dependent on the theory at hand. More precisely, they are determined by a claim like (5) which can allow some latitude in the values that are assigned to them in particular applications. (For a detailed discussion of this see Chapter IV, pp. 89 ff. and Chapter VIII, p. 276).

Roughly, the situation is this. It may be possible to maintain constraints on a theoretical function in the face of recalcitrant data by, intuitively, admitting an "error" in determining some of the values of the theoretical function. Moreover, it may be possible to do this in situations where constraints on a non-theoretical function would have to be given up. For this reason, constraints on non-theoretical functions can never take on the aspect of a permanent feature of the theory – which we are confident may be maintained as new applications are added, or as more is discovered about old ones. They can never have the same immunity to new discoveries that constraints on theoretical functions have. This seems to go some way to explaining why they are regarded as *ad hoc*. They may be consistent with present knowledge, but there is little reason to expect they will continue to be so. In contrast, constraints on theoretical functions have "built in" to them a kind of invulnerability to newly discovered facts. They can even come to be regarded as "necessary" features of the concepts, rather than as empirical facts. This perhaps helps to explain why they are so acceptable as explanations. They can last long enough to become familiar; further research can be planned with their intuitive guidance; they are not likely to evaporate tomorrow in the face of one recalcitrant experiment. On the other hand, constraints on non-theoretical functions must always be regarded with reserve. As with fickle women, one should not stake too much on their forebearance through the vicissitudes of experience.

It should be noted that the facts just cited about *constraints* on non-theoretical functions do not mitigate against the use of non-theoretical *laws*, or against the use of theories with no theoretical functions. If we choose to describe the observed behavior of a *single* application using a mathematical structure that has no theoretical functions, or by adding a non-theoretical law to some structure with theoretical functions, we do not run the risk of later discovering further intended applications which would force us to give this up. Of course, the discovery of *new* facts about *this* application might force us to do this. However, we can keep on building up a stock of tenable non-theoretical laws without extensions of the theory's range of applications or the extent of our knowledge of it forcing us to give up some of them. It is only in the case of constraints on non-theoretical functions that such an extension might force us to "go back and start over again".

Is this "pragmatic" difference between constraints on theoretical and non-theoretical functions sufficient to allow us to conclude that there is some sense in which theories that contain constraints on non-theoretical functions *can not* actually be propounded? I suppose the answer depends greatly on just how strong a sense of 'can not' one demands. The difference certainly seems sufficient to explain why one would *prefer* to deal with constraints on theoretical functions and, perhaps, why only theories with such constraints are given serious attention. But it does not show that one *could not*, for some perverse reason, offer such a theory for serious consideration. It is however, at least sufficient to show that it is worth giving some special consideration to the properties of theories with theoretically constrained cores. There is at least some reason to think that these are the only "really interesting" theories.

The final objection we shall consider is this. We have noted that, in the case of classical particle mechanics, only theoretically expanded cores were required to account for the empirical claims of the theory. Were this generally the case, then one might think that we should allow *only* theoretically expanded cores to be employed when a core is used to make an empirical claim about some range of intended applications. This would mean that we should take theoretical effect-equivalence to be the appropriate way of capturing the intuitive idea that two cores allow us to say exactly the same things about a given range of intended applications, and that strong formal equivalence is the appropriate account of the intuitive notion of formal equivalence.

Again, I conjecture that it is the case that only theoretically expanded cores are used in theories of mathematical physics that are actually propounded. This conjecture is supported with the same sort of reasons that supported my conjecture that none but only theoretically constrained cores appear in theories that are actually propounded. Again, let us suppose that my conjecture is correct and ask why it is that one might prefer to use theoretically expanded cores. As with constraints on theoretical functions, there appears to be no semantical reason for such a preference. One could offer syntactical reasons of a similar sort that we noted in the discussion of our preference for constraints on theoretical functions. But they are open to the same objections. Are there any convincing pragmatic reasons for preferring theoretically expanded cores?

It is clear that the same considerations which explained our preference for theoretically constrained cores can not be invoked to explain our preference for theoretically expanded cores. We have already noted that these considerations do not mitigate against the use of non-theoretical laws. The most we can get from such considerations is that the constraints that we add in the expansion, if they are non-trivial, must be constraints on theoretical functions only. But, we can not squeeze out a reason why we should not simply add a non-theoretical law and trivial constraints on the non-theoretical functions.

There is, however, a further consideration that suggests some grounds for preferring to add theoretical laws. It is only by adding such laws, and accompanying constraints on the theoretical functions, that we can enlarge the possibilities for measuring values of theoretical functions. In effect, when we add a theoretical law, we are more specifically delineating the "theoretical concept" associated with the theoretical function involved. We are making it more concrete, more accessible to empirical detection, and less of an *ad hoc* device. The more we are able to account for with theoretical laws, the more confident we become that our theoretical functions "represent" real properties of the individuals, rather than mere "theoretical fictions". Indeed, there is some reason to think that, once we have settled on a core, with theoretical functions, any account of the behavior of the non-theoretical functions which does not involve these theoretical functions will be regarded as unsatisfactory. Once we have succeeded in explaining *something* with these functions, we believe that we should be able to explain everything with them. For, with every new success with these functions, our previous successes become more significant and less *ad hoc* – their "explanatory force" is enhanced.

Again, as in the case of theoretically constrained cores, I do not think that these considerations are absolutely compelling. Someone *might* seriously propose a theory in which the central claim was not made with a theoretically expanded core. But, again, I do think they are compelling enough to motivate consideration of what would be the case, were we to limit ourselves to theoretically expanded cores. Indeed, in the light of our previous discussion, it is instructive to consider a situation in which we are limited to theoretical expansions of theoretically constrained cores. To this end, we define a preferred theory of mathematical physics as a theory

IDENTITY, EQUIVALENCE AND REDUCTION

with a theoretically constrained core (D43) and investigate the properties of the strong formal equivalence relation on such theories.

(D43) x is a *preferred theory of mathematical physics* if and only if:
   (1) $x = \langle H, I \rangle$ is a theory of mathematical physics;
   (2) H is a theoretically constrained core for a theory of mathematical physics.

The first thing to note about preferred theories is that the only theories without theoretical functions that are preferred are those with trivial constraints. (Note the remarks following (D30).) Next note that (T17) says that cores which put non-trivial constraints on $R(M)$ can only be initially equivalent to cores without theoretical functions which have non-trivial constraints. This means that preferred theories whose cores put non-trivial constraints on $R(M)$ can never be even weakly equivalent to a *preferred* theory with no theoretical functions. This is one sense in which theoretical functions may not be (semantically) eliminable from preferred theories. But it is not a very interesting sense. Intuitively, it simply says that, if we refuse to countenance constraints on non-theoretical functions, there, to say some things, we must resort to constraints on theoretical functions. Note that (T18) assures us that (leaving aside syntactical considerations) we can always do this.

A slightly more interesting sense in which theoretical functions may be (semantically) essential when we confine our attention to preferred theories and strong equivalence can be seen in the following way. Suppose we have a preferred theory with more than one theoretical function and we are concerned, not with eliminating all theoretical functions, but only some; say one, of them. That is, we want to find some preferred theory, with all the functions except the one to be eliminated, which is strongly equivalent to the initial theory. A bit more precisely, given the theoretically constrained core $H = \langle M_0, N_0, r, M, C \rangle$, we want to find a theoretically constrained core $H' = \langle M'_0, N_0, r', M', C' \rangle$ such that M'_0 differs from M_0 only in that its members have one less theoretical function place. It is not difficult to see that we can always find *some* theory, without this function, that is weakly equivalent to the initial theory. The proof is essentially the same as that for (T16). Moreover, (T17) assures us that we can go further and find a *preferred* theory that is weakly equivalent to the initial theory. But we have no guarantee that this theory will be *strongly* equivalent to the initial theory. Of course this does not prove that *no* preferred theory can be found that is strongly equivalent to the initial theory. However, it does suggest that there might be examples where this is the case.

At present, I am unable to provide a concrete example to show that there are theories from which one of the theoretical functions can not be eliminated, i.e. examples of the situation described in the last paragraph. However, I conjecture that the force function in classical particle mechanics might be an example of this sort of situation. We have already seen that the force function is Ramsey eliminable from a claim of the form (2-NCPM). However, I suspect that, were we to consider all the possibilities open to us for ruling out sets of intended applications by constraints on the force function, we would find that some of these could not be reproduced by constraints on the mass function. We would have to employ constraints directly on the position functions and thus the expanded core we used to do this would not be theoretically expanded. Were this to be the case, it would be a reason, other than syntactical simplicity, for employing the force function. However, it would fall short of showing, in some absolute sense, that the force function is essential, since it depends on our insistence on preferred theories and strong equivalence.

To provide a "real life" example of the notions of equivalence we have been discussing, we shall consider the Lagrangian and Hamiltonian formulations of classical particle mechanics. Earlier, in our discussion of classical particle mechanics we restricted our attention to the Newtonian formulation of this theory and suggested that other formulations of the theory could be accounted for in roughly the following way. These other formulations, e.g. the Lagrangian and Hamiltonian formulations, were to be regarded as distinct theories of mathematical physics which were "equivalent" to the Newtonian formulation. We should like to see now if our account of identity conditions for theories of mathematical physics and the relation of equivalence among such theories does allow us to account for the alternative formulations of classical particle mechanics in this way.

First, let us consider the Lagrangian formulation of particle mechanics. In accord with the view suggested earlier, I will attempt to provide a characterization of the mathematical structure of this formulation which allows us to regard it as a distinct theory of mathematical physics – call it 'Lagrangian generalized mechanics'. In particular, I will attempt to characterize the *theoretical* part of this theory in a way that is independent of the theoretical part of the Newtonian formulation. That is, I will take the generalized forces and the total kinetic energy of the system to be primitive concepts, rather than defining them, in the more customary way, in terms of Newtonian forces and masses. On the other hand, the non-theoretical parts of the mathematical structure of classical particle mechanics and of our formulation of Lagrangian generalized mechanics will be essentially the same. In Lagrangian mechanics, we shall count the generalized coordinates as theoretical functions and link them in the usual way to the non-theoretical position function.

We begin by defining generalized mechanics.[2] Roughly, this is just the mathematical apparatus of Lagrangian generalized mechanics without the requirement of Lagrange's equations. The set of all models for 'is a GM' is to be taken as the set M_0 in the core for Lagrangian generalized mechanics.

(D44) x is a GM (generalized mechanics) if and only if there exist $P, T, \bar{s}, \bar{X}, h, q, Q, K$ such that:
(1) $x = \langle P, T, \bar{s}, \bar{X}, h, q, Q, K \rangle$;
(2) $\langle P, T, \bar{s} \rangle$ is a PK;
(3) h is an integer;
(4) q is a function from $I_h \times T$ into the real numbers such that, for all $i \in I_h$, $t \in T$, $Dq(i, t)$ exists;
(5) Q is a function from $I_h \times R^h \times T$ into the real numbers;
(6) $\bar{X}$ is a function from $P \times R^h \times T$ into R^3 such that:
  (i) for all $i, j \in I_{h+1}$ and for all $p \in P$, $\langle x_1, \ldots, x_h \rangle \in R^h$, $t \in T$, $D_i D_j \bar{X}(p, x_1, \ldots, x_h, t)$ exists;
  (ii) for all $p \in P$, $t \in T$, $\bar{s}(p, t) = \bar{X}(p, q(1, t), \ldots, q(h, t), t)$;
(7) K is a function from $R^{2h} \times T$ into the real numbers such that, for all $i \in I_{2h+1}$ and for all $\langle x_1, \ldots, x_h, y_1, \ldots, y_h, t \rangle \in R^{2h} \times T$, $Dx_i K(x_1, \ldots, x_h, y_1, \ldots, y_h, t)$ exists.[3]

(D44–2) simply provides that we can take the set of all PK's to be the set of possible applications N_0 in the core of Lagrangian generalized mechanics. In doing this we are, intuitively, committed to regarding generalized coordinates, generalized forces and total kinetic energy of the system of particles as theoretical functions. (D44–3) stipulates the number of generalized coordinates in the system of generalized mechanics, while (D44–4) and (D44–5) characterize respectively the generalized coordinates $q(i, t)$ and the generalized forces $Q(i, q, t)$. (D44–6) specifies that $\bar{X}$ is the function that relates the values to the generalized coordinates to the positions of the particles in P. In the usual expositions of the Lagrangian formulation, the equations (D44-6-ii) are called 'the constraints.'[4]

(D44-7) characterizes the numerical function whose value for the argument $\langle q(1, t), \ldots, q(h, t), Dq(1, t), \ldots, Dq(h, t), t \rangle$ is to be interpreted as the total kinetic energy of the particles in P at time t.

In (D45) we define a Lagrangian generalized mechanics by adding an additional requirement to a generalized mechanics. This requirement is simply that the generalized coordinates satisfy the h second order partial differential equations given by (D45-2). In standard expositions of the Lagrangian formulation, these are called 'Lagrange's equations'.[5] The set of all models for 'is an LGM' plays the role of M in the core for Lagrangian generalized mechanics. It is the basic mathematical structure of this theory.

(D45) x *is an* LGM (Lagrangian generalized mechanics) if and only if there exist $P, T, \bar{S}, \bar{X}, h, q, Q,$ and K such that:
(1) $x = \langle P, T, \bar{S}, \bar{X}, h, q, Q, K \rangle$ is a GM;
(2) For all $t \in T$, $i \subset I_h$,
$$D_t D_{h+i} K(q(1, t), \ldots, q(h, t), Dq(1, t), \ldots, Dq(h, t), t)$$
$$- D_i K(q(1, t), \ldots, q(h, t), Dq(1, t), \ldots, Dq(h, t), t)$$
$$= Q(i, q(1, t), \ldots, q(h, t), t).$$

Having described all the elements of the core for Lagrangian generalized mechanics except the constraints, we can now say roughly what the empirical content of this theory is. Ignoring the constraints, the claim is this. Given any PK – any set of particles and their positions in a certain time interval – we can extend this PK, in some way, to produce an LGM. That is, we can find some generalized coordinates related to the given position function via some $\bar{X}$ that satisfies (D44-6-ii), and some generalized forces and kinetic energy function K such that these generalized coordinates satisfy Lagrange's equations. Now, it is pretty clear that this claim is trivially true, just as the corresponding claim for the basic mathematic structure of the Newtonian formulation of classical particle mechanics. (See Chapter VI, p. 119.) Whatever empirical content the theory has comes from the constraints in the core and the constraints and special theoretical laws that are added when the core is expanded. In the case of Lagrangian mechanics, one would naturally expect these to be constraints on the kinetic energy function and generalized forces and laws involving these functions.

IDENTITY, EQUIVALENCE AND REDUCTION   209

Apparently the most natural thing to do now would be to launch into a discussion of the constraints that appear in the core for Lagrangian generalized mechanics. However, it is expedient to consider these constraints in the light of some facts about how systems of generalized mechanics can be "constructed" out of systems of particle mechanics. In most expositions of the Lagrangian formulation of mechanics, this formulation has a sort of symbiotic relationship to the Newtonian formulation in that the kinetic energy functions is defined in terms of the Newtonian masses and velocities of the particles and the generalized forces are defined in terms of the Newtonian forces and velocities. Something of this sort can be done in our framework in the following way. We can define a mapping from the set of all PM's into the set of all GM's such that the GM that corresponds, under this mapping, to a given PM has a kinetic energy function and generalized forces that are related in the usual way to the functions in the PM. To this end, we define the F-relation.

(D46) F is a binary relation such that $\langle x, x' \rangle \in F$ if and only if there exist $P, T, \bar{s}, m, \bar{f}$, and $P', T', \bar{s}', \bar{X}, h, Q, K$ such that:
(1) $x = \langle P, T, \bar{x}, m, \bar{f} \rangle$ is a PM;
(2) $x' = \langle P', T', \bar{s}', \bar{X}, h, q, Q, K \rangle$ is a GM;
(3) $\langle P, T, \bar{S} \rangle = \langle P', T', \bar{S}' \rangle$;
(4) For all $\langle x_1, \ldots, x_h, y_1, \ldots, y_h, t \rangle \in R^{2h} \times T$,
$$K(x_1, \ldots, x_h, y_1, \ldots, y_h, t)$$
$$= \sum_{p \in P} \frac{m(p)}{2} \sum_{j=1}^{h} [D_j \bar{X}(p, x_1, \ldots, x_h, t) y_j$$
$$+ D_{h+1} \bar{X}(p, x_1, \ldots, x_h, t)]^2;$$
(5) For all $\langle i, x_1, \ldots, x_h, t \rangle \in I_h \times R^{2h} \times T$,
$$Q(i, x_1, \ldots, x_h, t) = \sum_{p \in P} \left[\sum_{j=1}^{h} \bar{f}(p, t, j) \right]$$
$$\times D_i \bar{X}(p, x_1, \ldots, x_h, t).$$

To see that (D46–4) is the appropriate definition for the kinetic energy function of a generalized mechanics that is "constructed" from a particle mechanics, note that

15

$$D\bar{s}(p, t) = D_i \bar{X}(p, q(1, t), \ldots, q(h, t), t)$$

$$= \sum_{j=1}^{h} D_j \bar{X}(p, q(1, t), \ldots, q(h, t), t) D_t q(i, t)$$

$$+ D_{h+1} \bar{X}(p, q(1, t), \ldots, q(h, t), t).$$

And to see that (D46–5) is the appropriate definition for the generalized forces, note that

$$\sum_{j=1}^{h} \bar{f}(p, t, j)$$

is the total resultant force on the particle p and that,

$$D_i \bar{X}(p, q(1, t), \ldots, q(h, t), t)$$

is intuitively the partial derivative of the position of p with respect to the i-th generalized coordinate.

The F-relation has the following properties. First, for any given PM, there is some GM that stands in the F-relation to it. This is simply a reflection of the fact that the position functions themselves can *always* be taken as the generalized coordinates of $\bar{X}$ as an identity function in one of its arguments. In general, there will also be other possible choices of generalized coordinates that will produce a GM that corresponds to the given PM. Some of these will be mere changes in coordinate systems, as from rectangular to polar coordinates; others will be occasioned by physical constraints on the positions of the particles. Second, it is easy to show that if a PM satisfies Newton's second law (D8–2), then a GM constructed from it satisfies Lagrange's equations. Finally, a GM constructed from a given PM may satisfy Lagrange's equations and yet the PM fail to satisfy Newton's second law. However, in this case, it will always be possible to add one additional force to each particle in this PM and thereby make it satisfy Newton's second law. Proofs of the first two facts about the F-relation can be found in standard texts. The third fact is less well known. A proof can, however, be found in ([19], Theorem 13, p. 19). These facts are summarized below.

(T19) If x is a PM then there exists an x' such that $\langle x, x' \rangle \in F$.

(T20) If $\langle x, x' \rangle \in F$ and x is a CPM then x' is an LGM.

(T21) If $x = \langle P, T, \bar{s}, m, \bar{f} \rangle$, $x' = \langle P, T, \bar{s}, \bar{X}, h, q, Q, K \rangle$, $\langle x, x' \rangle \in F$ and x' is an LGM then, if $\bar{h}$ is a function from $P \times T$ into R^3 such that, for all $p \in P$, $t \in T$,

$$\bar{h}(p, t) = m(p), D_t^2 \bar{X}(p, q(1, t), \ldots, q(h, t), t) - \sum_{j=1}^{h} \bar{f}(p, t, j)$$

and $\bar{g}$ is a function from $P \times T \times I$ into R^3 such that, for all $p \in P$, $t \in T$,
  (i) $\bar{g}(p, t, 1) = \bar{h}(p, t)$,
  (ii) for $i \neq 1$, $\bar{g}(p, t, i) = \bar{f}(p, t, i-1)$, then $\langle P, T, \bar{s}, m, \bar{g} \rangle$ is a CPM.

In a natural way F imposes a mapping $\bar{F}$. From $\{x \mid x \text{ is a PM}\}$ into $\{x' \mid x' \text{ is a GM}\}$, we define $\bar{F}$ in (D47).

(D47) $\bar{F}$ is a binary relation such that $\langle X, X' \rangle \in \bar{F}$ if and only if:
  (1) $X \subseteq \{x \mid x \text{ is a PM}\}$;
  (2) $X' \subseteq \{x' \mid x' \text{ is a GM}\}$;
  (3) For all $x \in X$, there exists an $x' \in X'$ such that $\langle x, x' \rangle \in F$;
  (4) For all $x' \in X'$, there exists an $x \in X$ such that $\langle x, x' \rangle \in F$.

The properties of $\bar{F}$ listed as (T22) and (T23) follow directly from (T20) and (T21). (T24) follows from (D46–3).

(T22) From all $X \subseteq \{x \mid x \text{ is a PM}\}$ there is an $X' \subseteq \{x' \mid x' \text{ is a GM}\}$ such that $\langle X, X' \rangle \in \bar{F}$.

(T23) If $\langle X, X' \rangle \in \bar{F}$ and $X \subseteq \{x \mid x \text{ is a CPM}\}$ then $X' \subseteq \{x' \mid x' \text{ is an LGM}\}$.

(T24) If $\langle X, X' \rangle \in F$, then $R(X) = R'(X')$.

Next, for any set of sets of PM's, $\bar{X}$, we define $G(F, \bar{X})$ to be the set of all sets of GM's which stand in the F-relation with some member of $\bar{X}$. More precisely.

(D48) For all $\bar{X} \subseteq \mathscr{P}\{x \mid x \text{ is a PM}\}$, $G(F, \bar{X}) = \{X' \mid \text{there is an } X \in \bar{X} \text{ and } \langle X, X' \rangle \in \bar{F}\}$.

As a trivial consequence of (T22) and (T24) we have

(T25) For all $\bar{X} \subseteq \mathscr{P}\{x \mid x \text{ is a PM}\}$, $\bar{R}(\bar{X}) = \bar{R}'(G(F, \bar{X}))$.

Now, let us return to the question of the constraints in the core for Lagrangian generalized mechanics. Roughly, what we can show is this. Whatever constraints appear in the core of the classical particle mechanics, there are some constraints that we *could* put in the core for Lagrangian generalized mechanics that "have the same effect". In particular if classical particle mechanics is constrained by C then, we can "produce the same effect" by constraining Lagrangian generalized mechanics with $G(F, C)$. We shall make this precise below. What this *suggests* is roughly the following. If one is concerned with formulating Lagrangian generalized mechanics as a theory of mathematical physics, distinct from classical particle mechanics, the core constraints for the former theory can be obtained by simply "cranking" the constraints in the core of classical particle mechanics through the F-relation. Admittedly, this sounds like cheating, for what guarantee is there that these *are* the constraints that actually appear in expositions of the theory we are trying to reconstruct? There is, of course, no guarantee. But the fact that most expositions of Lagrangian mechanics trade very heavily on its "symbiotic" relationship with the Newtonian formulation suggest, that if we insist on regarding it as a distinct theory, we are going to have to allow that the constraints that appear in its core are just those taken over directly from the Newtonian formulation. Note also that if C constrains only theoretical functions then $\bar{R}(C) = \mathscr{P}(N_0)$. But, by (T25) $\bar{R}(G(F, C)) = \mathscr{P}(N_0)$ and so $G(F, C)$ constrains only theoretical functions. Thus if the core for classical particle mechanics is a theoretically constrained core then the core we are proposing for Lagrangian generalized mechanics is also theoretically constrained. A precise statement of the way $G(F, C)$ reproduces the effect of C is the following theorem.

(T26) If $H = \langle M_0, N_0, r, M, C \rangle$ is the core of classical particle mechanics and $H' = \langle M'_0, N_0, r', M', G(F, C) \rangle$ is the core of Lagrangian generalized mechanics, then H' effect-dominates H'.

To see that $\bar{R}(\mathscr{P}(M) \cap C) \subseteq \bar{R}'(\mathscr{P}(M') \cap G(F, C))$, suppose $Y \in \bar{R}(\mathscr{P}(M) \cap C)$. Then there is an $X \in \mathscr{P}(M) \cap C$ and $Y = R(X)$. Since $X \in \mathscr{P}(M_0)$, by (T21), there is an X' such that $\langle X, X' \rangle \in \bar{F}$. Since $X \in C$, $X' \in G(F, C)$ by (D48). Since $X \in \mathscr{P}(M)$, by (T23), $X' \in \mathscr{P}(M')$. Thus $X' \in \mathscr{P}(M') \cap G(F, C)$, and since $\langle X, X' \rangle \in \bar{F}$, by (T24), $Y = R(X) = R(X')$. Thus $Y \in \bar{R}(\mathscr{P}(M') \cap \bar{F}(C))$.

Thus far, we have seen that, *whatever* constraints we suppose be in the core H of classical particle mechanics, we can produce some constraints to put in the core H' for Lagrangian generalized mechanics such that H' initially dominates H. There is even some reason to think that these are the constraints that *ought* to go in H'. It is natural to ask if the H' that we constructed in this way is also initially dominated by H so that we have initial equivalence between H' and H. This question can not be answered without being somewhat more explicit about the nature of the constraints C in the core of classical particle mechanics. However, there appears to be reason to think that these constraints are such that initial equivalence obtains.

To understand this situation more precisely, let us attempt to prove that $\bar{R}(\mathcal{P}(M') \cap G(F, C)) \subseteq \bar{R}(\mathcal{P}(M) \cap C)$. Suppose $Y' \in \bar{R}(\mathcal{P}(M') \cap G(F, C))$. Then there is an $X' \in \mathcal{P}(M') \cap G(F, C)$ such that $Y' = R'(X')$. Since $X' \in G(F, C)$, by (D48), there is an X such that $X \in C$ and $\langle X, X' \rangle \in \bar{F}$. Now, there is no guarantee that this X is in $\mathcal{P}(M)$. But, from (T20) we know the following. Any $x \in X$ can be assured to be a model for CPM by simply "adding" the new force $\bar{h}(p, t)$ to the forces already acting on the particle p in x. In this way we can produce an X^* which will be in $\mathcal{P}(M)$ and $R(X^*) = R(X) = R'(X') = Y$. But now, we have no *assurance* that X^* will still be in C. This depends on what C is like. If C constrains only the mass function, then clearly changing the forces in this way will still leave X^* in C. Indeed, even if C constrains the force function, in some not too stringent ways, like those mentioned in Chapter VI as possible core constraints on the forces, it is clear that "adding" $\bar{h}(p, t)$ to each particle in every x and X would still leave X^* in C. *If* C permits this, then X^* and $\mathcal{P}(M) \cap C$, $Y = R(X^*)$ and hence $Y \in \bar{R}(\mathcal{P}(M) \cap C)$.

Thus, if we are willing to make some rather plausible assumptions about the nature of the constraints in the core of classical particle mechanics, we have the following theorem.

(T27) If $H = \langle M_0, N_0, r, M, C \rangle$ is the core of classical particle mechanics and $H' = \langle M'_0, N_0, r', M', G(F, C) \rangle$ is the core of Lagrangian generalized mechanics then H' is initially equivalent to H.

And, as an immediate consequence of (T14), we have:

(T28) If $H = \langle M_0, N_0, r, M, C \rangle$ is the core of classical particle mechanics and $H' = \langle M'_0, N_0, r', M', G(F, C) \rangle$ is the core of Lagrangian generalized mechanics then H is effect-equivalent to H'.

Now let us consider the question of theoretical effect-equivalence. First, does H' theoretically effect-dominate H? Let $\mathscr{E} = \langle H, L, C_L, \alpha \rangle$ be any theoretically expansion of H. Can we find some theoretical expansion of

$H', \mathscr{E}' = \langle H', L', C_{L'}, \alpha' \rangle$ such that $N_{\mathscr{E}} = N_{\mathscr{E}'}$? The natural way to construct $\mathscr{E}'$ would appear to be, roughly, to let L', $C_{L'}$, and α' be the "images" of L, C_L, and α under the F-mapping. More precisely, for any $1 \in L$, let $1'$ be *any* sub-set of M_0' such that $\langle 1, 1' \rangle \in \bar{F}$, let $C_{L'} = G(F, C)$; and let α' be such that, for all $x \in N_0$, $\alpha'(x') = 1'$ if and only if $\alpha(x) = 1$. It is easy to show, in a way analogous to the proof of (T26), that $N_{\mathscr{E}} \subseteq N_{\mathscr{E}'}$. But, when we attempt to employ the argument for (T27) to show that $N_{\mathscr{E}'} \subseteq N_{\mathscr{E}}$ we run into trouble. Any $X' \in \mathscr{P}(M') \cap G(F, C) \cap G(F, C)$ stands in the $\bar{F}$-relation with some $X \in C$. Again, there is no assurance that $X \in \mathscr{P}(M)$. Moreover, now there is the additional requirement to be met that $x \in y$ satisfy the law $\alpha(x)$ and there is also no assurance of this. And, indeed, there is nothing like (T21) which assures us that we can always modify x in some "trivial" way to produce an x^* which satisfies $\alpha(x)$. Further, even if we could do this, there is no reason to expect that the modifications required would be so "trivial" as to let us be assured that $X^* \in C_L$.

Roughly, the situation is this. The "obvious" way of expanding H' to reproduce the effect of a given expansion of H, appears to be sure to let in too much, at least in some cases. Of course, this does not prove that there is not some other "non-obvious", theoretical expansion of H' which will do the job. At present, I do not see, in general, how to construct such an expansion. Nor do I see a way of producing a theoretical expansion of H such that it can be shown that there is *no* expansion of H' that reproduces its effect. Thus the question of whether H' theoretically effect-dominates H remains open.

Does H theoretically effect-dominate H'? Here the answer is, if anything, even less clear. For there is apparently no "natural" way to construct a theoretical expansion of H which one might intuitively expect to reproduce the effect of some given theoretical expansion of H'. Let $\mathscr{E}' = \langle H', L', C_{L'}, \alpha' \rangle$ be any theoretical expansion of H'. One might think of trying to run the F-mapping backward to produce an expansion of H reproducing the effect of $\mathscr{E}'$. But note, given $1' \in L'$, there is no guarantee that there is *any* $1 \in \mathscr{P}(M_0)$ such that $\langle 1, 1' \rangle \in \bar{F}$. Intuitively, $1'$ could, for example, require that K be of such a form that it could not be made to satisfy (D46-4) for any PM whatsoever. Yet, one might well find PK's which could be extended to LGM's with such a K-function so that $K_{\mathscr{E}'}$ for an $\mathscr{E}'$ with such a law would not be empty. In this situation, one hardly knows where to start looking for an expansion of H to reproduce the effect of $\mathscr{E}'$. On the other hand, I am not able to produce a theoretical expansion of H' such that it can be shown that no theoretical expansion of H can be found which reproduces its effect.

Thus, the most that I can offer in the way of enlightenment about theoretical effect-equivalence is the following theorem.

(T27) If $H = \langle M_0, N_0, r, M, C \rangle$ is the core of classical particle mechanics and $H' = \langle M'_0, N_0, r', M', G(F, C) \rangle$ is the core of Lagrangian generalized mechanics, then if $\mathscr{E} = \langle H, L, C_L, \alpha \rangle$ is a theoretically expanded core for a theory of mathematical physics, there is a theoretically expanded core for a theory of mathematical physics $\mathscr{E}' = \langle H', L', C_{L'}, \alpha' \rangle$ such that $N_\mathscr{E} \subseteq N_{\mathscr{E}'}$.

The preceding discussion may be briefly summarized in the following way. It is possible to regard the Lagrangian formulation of particle mechanics as a distinct theory of mathematical physics. If we do so, we can show that this theory is formally equivalent, in the weak sense at least, to the Newtonian formulation of particle mechanics. Whether it is formally equivalent in the strong sense remains an open question.

Let us now briefly consider the Hamiltonian formulation of particle mechanics. Analogous to (D44) the predicate 'is an SM' characterizes the mathematical structure of Hamiltonian mechanics except for the requirement of Hamilton's equations. The set of all SM's is M_0 in the core of Hamiltonian mechanics.

(D49) x *is an* SM (is a specialized mechanic) if and only if there exist $P, T, \bar{S}, \bar{X}, h, q, p$, and H such that:
(1) $x = \langle P, T, \bar{S}, \bar{X}, h, q, p, H \rangle$;
(2) $\langle P, T, \bar{S} \rangle$ is a PK;
(3) h is an integer;
(4) q and p are functions from $I_h \times T$ into the real numbers such that, for all $i \in I_h$ and $t \in T$, $Dq(i, t)$ and $Dp(i, t)$ exist;
(5) $\bar{X}$ is a function from $P \times R^h \times T$ into R^3 such that:
  (i) for all $i, j \in I_{h+1}$ and for all $z \in P$, $\langle x_1, \ldots, x_h \rangle \in R^k$, $t \in T$, $D_i D_j \bar{X}(x, x_1, \ldots, x_h, t)$ exists;
  (ii) for all $p \in P$, $t \in T$, $\bar{S}(p, t) = \bar{X}(p, q(1, t), \ldots, q(h, t), t)$;
(6) H is a function from $R^{2h} \times T$ into the real numbers such that for all $i \in I_{2h+1}$ and for all $\langle x_1, \ldots, x_h, y_1, \ldots, y_h, t \rangle \in R^{2h} \times T$, $D_i H(x_1, \ldots, x_h, y_1, \ldots, y_h, t)$ exists.

In (D49) $q(i, t)$ is the value of the i-th generalized coordinate at time t and $p(i, t)$ is the value of the i-th generalized momentum at time t.

$H(q(1, t), \ldots, q(h, t), p(1, t), \ldots, p(h, t), t)$ is the value of the Hamiltonian of the particles in P at time t.

We now introduce the requirement that the generalized positions and momenta satisfy Hamilton's equations.

(D50) x *is an* HSM (Hamiltonian specialized mechanics) if and only if there exist $P, T, \bar{s}, \bar{X}, h, q, p$, and H such that:
(1) $x = \langle P, T, \bar{s}, \bar{X}, h, q, p, H \rangle$ is an SM;
(2) for all $1 \in T, i \in I_h$:
  (i) $D_i H(q(1, t), \ldots, q(h, t), p(1, t), \ldots, p(h, t), t) = -Dp(i, t)$;
  (ii) $D_{h+1} H(q(1, t), \ldots, q(h, t), p(I, t), \ldots, p(h, t), t) = Dq(i, t)$.

In the case of Hamiltonian mechanics, we have no mapping like F which maps *every* PM into some SM. Rather what we have is this. For those PM's in which the resultant force on each particle is the gradient of some scalar potential, we can define a mapping, analogous to F, which carries any one of these systems into some SM. PM's having this property are sometimes called 'conservative systems'. Thus we can define a mapping that carries every conservative PM into some SM. Moreover, we can show that, if the conservative PM satisfies Newton's second law, then SM's into which it is mapped satisfy Hamilton's equations. Going the other way, we also have a result like (T24). Thus, we should expect to find that Hamiltonian mechanics is related to the fragment of classical particle mechanics dealing with conservative systems in the same way that Lagrangian mechanics is related to the whole of particle mechanics. We shall not go further into the details here, but simply summarize in the following way. The Hamiltonian formulation of particle mechanics, when formulated as a distinct theory of mathematical physics, is formally equivalent (at least in the weak sense) to the fragment of classical particle mechanics dealing with conservative systems.

Having examined the relations of identity and equivalence, we want now to examine a third important relation that may hold between theories of mathematical physics – that of reduction. It is commonly said that thermodynamics reduces to statistical mechanics, that rigid body mechanics and hydrodynamics reduce to particle mechanics, that geometrical optics reduce to electromagnetic theory, and that classical particle mechanics reduces to relativistic mechanics (in the limiting case of velocities which are small in comparison to the velocity of light). We want now to use the characterization of theories of mathematical physics that has been developed to give a precise account of what it means to say of two theories of mathematical physics, T and T', that T' reduces to T.

In developing an account of the reduction relation, I shall follow very closely the work of Adams ([1], pp. 16–36). Adams develops an account of the reduction based on what I have called earlier 'the weak view of formal identity' (see above pp. 185–186). This view makes no distinction between theoretical and non-theoretical functions and takes the formal part of the theory to be identified with what we have called 'the basic mathematical structure'. The empirical claim of theories, according to this account, is simply that the range of intended applications is a sub-set of the basic mathematical structure. Since our account of the formal identity of theories, as well as our account of the way the formal part of a theory is "used" to make the empirical claims of the theory, is somewhat more complex than the corresponding parts of Adams' views, it is natural to expect that our account of the reduction relation will also be more complex than his. It is, however, in the "limiting case" where the theories involved have no theoretical functions and no constraints on any of the functions, that our account will be substantially the same as that of Adams.

Adams distinguishes two intuitive features of the reduction relation between theories that must be reflected in any successful account of the relation. They are roughly the following. First, a reduction relation is a correspondence between the fundamental concepts of the reduced theory T' and the fundamental concepts of the reducing theory T. Second, from this correspondence and the fundamental laws of T, it is possible to deduce the fundamental laws of T'. Let us examine both these conditions in turn.

To begin, it appears that this correspondence between the fundamental concepts of the theories T' and T has the set effect of establishing a correspondence between, at least, some of the possible applications of T. Thus, for example, in a putative reduction of rigid body mechanics to particle mechanics, the claim is made that every rigid body "consists of" a set of particles, and further, that the mechanical properties of this rigid body "correspond to" certain mechanical properties of this set of particles. In effect this is a claim that every rigid body system – rigid body together with mechanical functions defined on it – corresponds to some system of particle mechanics. This is to say that the members of some set of possible applications of rigid body mechanics – those that actually *are* rigid bodies or those that are intended applications of the theory – each correspond to at least one possible application of particle mechanics. Indeed, each one is claimed to correspond to a system of particle mechanics whose domain *is* actually a set of "physical" particles, i.e. one of the intended applications of particle mechanics. Likewise, the reduction of thermodynamics to statistical mechanics is commonly thought of in this way. Every thermodynamic object "consists of" a set of particles and the thermodynamic functions defined on this object "correspond to" the expected values of certain mechanical functions defined on this set of particles. In effect, this is to claim that every thermodynamic system – thermodynamic object together with functions defined on it – corresponds to some statistical mechanical system – a set of particles together with mechanical functions defined on them and a probability measure defined on the possible values of these functions.

This suggests that the very least that we should want to require of a reduction holding between T' and T is that it determine a mapping from some sub-set of the possible applications of T', N'_0, into the set of possible applications of T, N_0. Typically, this mapping will be determined in a "piecemeal" fashion by describing a correspondence between the domains, e.g. rigid bodies correspond to the set of particles composing them, telling how values of the functions on one domain are related to values of the functions on the corresponding domain. But, for the time being, we can ignore the exact way this mapping is determined and simply consider the mapping itself.

The examples we have just mentioned suggest some features that this mapping should have. First, they suggest roughly that the intended applications of T' and T must be such that every intended application of T' is mapped into *some* intended application of T.

One of the intuitive claims made for the reduction relation is this. If T' is reducible to T then *everything* that can be explained, or accounted for, by T' can be as well explained by T. That is, by showing that T' can be reduced to T we have shown that, in some sense, we can get on without T'. Roughly, the way reduction works is this. We show that *every* intended application of T' corresponds to *some* intended application of T and further that what T' claims about the behavior of this application "follows from" what T claims about this corresponding application. Clearly, for reduction to work this way, every intended application of T' must correspond to at least one intended application of T. This might be called 'the requirement of completeness for the reduction relation'. It should be noted, however, that this is not simply a requirement on the mapping from N_0' to N_0. It is a claim about how the sets of intended applications of the theories T' and T must be related by this mapping. We shall be more precise about this requirement below. However, it is expedient now to examine some other properties that this mapping must have.

Another strong intuitive notion about the correspondence between N_0' and N_0 is this. If $x' \in N_0'$ corresponds to $x \in N_0$, then it should be possible to regard x' and x as essentially the "same" thing – the same physical system, in some ultimate sense – albeit "described in different ways". This is to say, the domains of x' and x, if not in fact identical, should be related in some way that makes an "intuitive" identification possible. For example, it is possible to regard a rigid body and the particles that compose it as being, in some sense, the same thing. The same is true of thermodynamic objects and the particles that compose them. In both these cases, the domain of the possible application of the reduced theory may be thought of as a set of one member. This *member* is identical with the *set* of objects that form the domain of the possible application of the reducing theory. In some reductions, like that of classical particle mechanics to relativistic mechanics the correspondence between the domains is obviously going to be identity.

It is the functions defined on these "intuitively" identical domains in the different theories that provide what we intuitively regard as alternative descriptions of essentially the same thing. However, there is an asymmetry between these modes of description. Typically, we regard the reducing theory as, in some sense, more fundamental or more basic than the reduced theory. What this seems to mean is that the way of describing the world provided by the reduced theory is less complete, less precise, or provides the means for making fewer distinctions than the way provided by the reducing theory. This is to say, *one* description of a given physical system in the "vocabulary" of the reduced theory will typically be expected to correspond to *several* descriptions of the *same* system in the "vocabulary" of the reducing theory. All of these corresponding descriptions in the reducing theory are "compatible" with the description in the reduced theory, but each "fills out" this description in a different way. Even if this is not the case, we should at least expect that the reduced theory would not provide more complete descriptions of the same physical system than the reducing theory. That is, we should not expect that several descriptions of the same system in the "vocabulary" of the reduced theory correspond to one-and-the-same description of this system in the reducing theory. Roughly speaking, this simply says that the "vocabulary" of the reducing theory must be *at least* as rich as the "vocabulary" of the reduced theory.

In our formal account of the reduction relation, we shall not attempt to reproduce the intuitive requirement that it be possible to regard the domains of corresponding possible applications as, in some sense, identical. But we will reproduce the requirement that the "vocabulary" of the reducing theory be at least as rich as the "vocabulary" of reduced theory. We shall do this by requiring that the mapping from N_0' to N_0 be one–many. This explicitly rules out the possibility that several members of N_0' may correspond to one member of N_0 and leaves open the *possibility* that several members of N_0 correspond to one member of N_0'. We will see a realization of this possibility in the reduction of rigid body mechanics to particle mechanics where one rigid body system corresponds to a number of particle *systems* – each system, however, having the *same* set of particles.

Next, let us consider Adams' second intuitive criterion on the reduction relation: that the fundamental laws of T' be deducible from the fundamental laws of T, together with the correspondence relation. We have already suggested that this means roughly the following. If S is a statement of T about a certain physical system and S' is a statement of T' about a *corresponding* physical system, then S is true only if S' is true. On Adams' account of the statements of a theory, this means that the correspondence relation must be such that, for any x', if there is an x which corresponds to x' and "has" the basic mathematical structure of T ('has the characteristic property of T' in Adams' terminology) then x' "has" the basic mathematical structure of T'. Note, first, that this account of the second intuitive requirement says nothing about the intended applications of the theories involved. It speaks only of the mathematical structures that are a part of these theories. For this reason, it might be called 'an account of the formal aspect of the reduction relation'. What we want to do now is to provide a parallel account of this formal aspect of the reduction relation which is adequate to our more complex account of the relation between the mathematical structure of the theory and its statements.

In the account of the statements associated with a theory of mathematical physics that has been developed here, it is the expanded cores of the theory that play a role analogous to that of the basic mathematical structure (characteristic property) in Adams' account. In a case where there are no theoretical functions and no constraints on any of the functions, the expanded core is effectively the same as the basic mathematical structure and our account of statements reduces to that of Adams. This suggests that the appropriate way to begin is by considering how one might define a notion of "reduction" for expanded cores which is a generalization of Adams' account of the formal aspect of the reduction relation. From this, we shall work up to an explication of the reduction relation for theories.

(D51) If $\mathscr{E}' = \langle M'_0, N'_0, r', M', C', L', C_{L'}, \alpha' \rangle$ and $\mathscr{E} = \langle M_0, N_0, r, M, C, L, C_L, \alpha \rangle$ are expanded cores for theories of mathematical physics then **R** *reduces* $\mathscr{E}'$ to $\mathscr{E}$ if and only if:
(1) $\mathbf{R} \subseteq N'_0 \times N_0$;
(2) For all $y \in N_0$, if there exist x and $x' \in N'_0$ such that $\langle x, y \rangle$ and $\langle x', y \rangle \in \mathbf{R}$ then $x = x'$;
(3) If $\bar{\mathbf{R}}$ is such that
   (i) $\bar{\mathbf{R}} \subseteq \mathscr{P}(N'_0) \times \mathscr{P}(N_0)$;
   (ii) $\langle X', X \rangle \in \bar{\mathbf{R}}$ if and only if for all $x' \in X'$ there is an $x \in X$ such that $\langle x', x \rangle \in \mathbf{R}$
   then, for all $X' \in \mathscr{P}(N'_0)$, if there exists an $X \in \mathscr{P}(N_0)$ such that $X \in N_{\mathscr{E}}$ and $\langle X', X \rangle \in \mathbf{R}$ then $X' \in N_{\mathscr{E}'}$.

In (D51), (1) simply says that the reduction relation **R** is a mapping from some sub-set of N'_0 into N_0. Since our purpose at this point is only to characterize the formal aspect of the reduction relation, we do not say anything about how the intended applications of the theories must be related by the **R**-relation. However, we do take account of our previous discussion in requiring, by (2), that the **R**-relation be one–many.

(D51-3) is an attempt to characterize the way in which the reduction relation allows us to deduce what T' says about some physical system from what T says about a corresponding system. The generalization is that we must now talk about *sets* of physical systems, rather than single systems. This is because our account of the statements of theories takes them to be, in general, about sets of physical systems. What (D51-3) requires is roughly this. For all $X' \subseteq N'_0$, if there is an $X \subseteq N_0$ which "corresponds" to X' and $X \in N_{\mathscr{E}}$, then $X' \in N_{\mathscr{E}'}$. That is if X' "corresponds" to some X which is in the set of sets of possible applications determined by $\mathscr{E}$, then X' is in the set of sets of possible applications determined by $\mathscr{E}'$. This condition is sufficient to allow us to conclude, from the fact that $S = \langle \mathscr{E}, I \rangle$ is a successfully applied core and I' "corresponds" to I, that $S' = \langle \mathscr{E}', I' \rangle$

is a successfully applied core. That is, it reproduces the intuitive requirement that the truth of a statement about I made with T, together with the fact that I' corresponds to I, allows us to deduce that a statement about I', made with T' is true. Note the "entailment relation" associated with the reduction relation, on the account given here, holds between two sweeping claims about *all* intended applications of the respective theories and not between two statements about *single* applications of these theories. The notion of "correspondence" between *sets* of possible applications that is appropriate here is the $\bar{\mathbf{R}}$-relation, defined by (D51-3-i) and (ii) in terms of the R-relation. The requirement is that every member of X' must stand in the **R**-relation with at least one member of X. Note that this allows that there may be members of X which do not stand in the **R**-relation to any member of X' and that several members of X stand in the R-relation

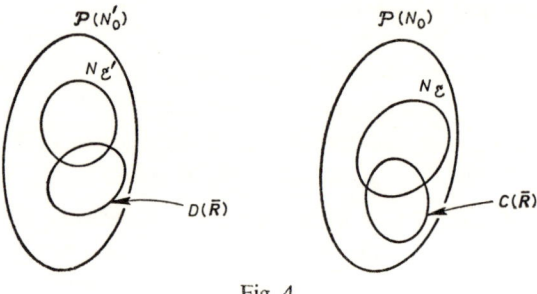

Fig. 4

with the same member of X'. To see why this sort of correspondence between sets is appropriate, think of X as the set of intended applications of particle mechanics and X' as the set of intended applications of rigid body mechanics. It seems clear that there are going to be some intended applications of particle mechanics that do not correspond to applications of rigid body mechanics, and, as mentioned before, some applications of rigid body mechanics may correspond to more than one application of particle mechanics.

Note that both **R** and $\bar{\mathbf{R}}$ are functions from their counter domains onto their domains. Thus we may speak of the image of members of sub-sets of these counter domains under the **R** and $\bar{\mathbf{R}}$ functions. If we let $D(\bar{\mathbf{R}})$ denote the domain of **R** and $C(\bar{\mathbf{R}})$ denote the counter domain of $\bar{\mathbf{R}}$, then we can express the content of (D51-3) in the following way. Figure 4 illustrates this situation.

Thinking of the "entailment condition" in this way leads one to note the following facts about reduction relations as defined by (D51). First, note that any relation satisfying (D51-1) and (D51-2) for which $C(\bar{\mathbf{R}}) \cap N_{\mathscr{E}}$ is void satisfies (D51-3). In particular, if $N_{\mathscr{E}'}$ is void, i.e. if any statement made with the expanded core $\mathscr{E}'$ is false, then any relation satisfying (D51-1) and (D51-2) will serve as a reduction relation. Further, even if $C(\bar{\mathbf{R}}) \in N_{\mathscr{E}'}$ is not void, it still might be the case that there are members of $D(\bar{\mathbf{R}}) \in N_{\mathscr{E}}$ which correspond, via the $\bar{\mathbf{R}}$-relation to no member of $N_{\mathscr{E}'}$. Roughly speaking, these situations admit the possibility that a core of a theory T' which is successfully applied to a set of intended applications of T' may be reduced by $\mathbf{R}$ to an expanded core of T which is not successfully applied to any set of intended applications of T, and perhaps even an expanded core that is not capable of being successfully applied to anything. These facts are analogous to well known properties of material implication. Ultimately they result in our account of reduction for theories admitting certain cases of "vacuous" reduction which seem intuitively unacceptable. We shall discuss these in more detail below when the full account of reduction for theories has been given.

In at least one important way the requirements placed on a reduction relation by (D51) are much too weak to serve as a basis for an intuitively satisfactory account of the reduction relation for theories. To see this, remember how $N_{\mathscr{E}}$ is determined. Its members are just those sets of possible applications that can be "filled out", by adding theoretical functions which satisfy the constraints, to produce models for the basic mathematical structure which also satisfy the special laws that have been added. Now, intuitively, if $\mathbf{R}$ reduces $\mathscr{E}'$ to $\mathscr{E}$, we should expect that the theoretical functions which assure that X is a member of $N_{\mathscr{E}}$ would correspond in some way to the theoretical functions that assure that X' is a member of $N_{\mathscr{E}'}$. For example, it is very unlikely that we would regard any putative reduction of rigid body mechanics to particle mechanics as successful if it did not exhibit some correspondence between moments of force and forces and between moments of inertia and masses. If we simply produce a correspondence between the kinematical descriptions of rigid bodies and kinematical descriptions of sets of particles, even if it did have the property (D51-3), it is doubtful that we would regard this correspondence as a successful reduction. It is more likely that we would regard it as a happy accident, but otherwise, not very significant.

To remedy this defect, we shall first characterize a notion of *close reduction* for expanded cores. Roughly, $\overline{\mathbf{R}}$ closely reduces $\mathscr{E}'$ to $\mathscr{E}$ if and only if $\mathbf{R}$ is a one–many mapping from M'_0 to M_0 having the following property. If $X' \subseteq M'_0$ and $X \subseteq M_0$ "correspond" via $\mathbf{R}$, and X satisfies the constraints and laws of $\mathscr{E}$ then X' satisfies the constraints and laws of $\mathscr{E}'$.

(D52) If $\mathscr{E}' = \langle M'_0, N'_0, r', M', C', L', C_{L'}, \alpha' \rangle$ and $\mathscr{E} = \langle M_0, N_0, r, M, C, L, C_L, \alpha \rangle$ are expanded cores for theories of mathematical physics then $\mathbf{R}$ *closely reduces* $\mathscr{E}'$ *to* $\mathscr{E}$ if and only if:

(1) $\mathbf{R} \subseteq M'_0 \times M_0$;

(2) For all $x \in M_0$, if there exist x' and $y' \in M'_0$ such that $\langle x', x \rangle$ and $\langle y', x \rangle \in \mathbf{R}$, then $x' = y'$;

(3) If $\overline{\mathbf{R}}$ is such that
   (i) $\overline{\mathbf{R}} \subseteq \mathscr{P}(M'_0) \times \mathscr{P}(M_0)$;
   (ii) $\langle X', X \rangle \in \overline{\mathbf{R}}$ if and only if for all $x' \in X'$ there is an $x \in X$ such that $\langle x', x \rangle \in \mathbf{R}$,
then for all $X' \in \mathscr{P}(M'_0)$, if there exists an $X \in \mathscr{P}(M_0)$ such that:
   (iii) $X \in (\mathscr{P}(M) \cap C \cap C_L)$;
   (iv) for all $x \in X$, $x \in \alpha(r(x))$;
   (v) $\langle X', X \rangle \in \overline{\mathbf{R}}$;
then
   (vi) $X' \in (\mathscr{P}(M') \cap C' \cap C_{L'})$;
   (vii) for all $x' \in X'$, $x' \in \alpha'(r'(x'))$.

Now, what (D52–3) is getting at is the "entailment condition". But, it does not quite make it – simply because the class reduction relation is not a correspondence between possible applications of the theories, but between extensions of these. Crudely speaking, it is not a relation between the things that the theories are really talking about. However, a close reduction relation does "impose", in a quite natural way, a correspondence on possible applications of the theories. If $x' \in N'_0$ and $x \in N_0$ are such that they have expansions that stand in the close reduction relation, we shall say that $\langle x', x \rangle \in \mathbf{R}^*$. Intuitively, what one would like to find is that the $\mathbf{R}^*$-relation reduces $\mathscr{E}'$ to $\mathscr{E}$. It obviously satisfies (D51–1) and (D51–2) and one might hope that (D52–3) would assure that it satisfies (D51–3). But such is not the case. (D52–3) does assure us that whenever X' and X have extensions related by $\overline{\mathbf{R}}$, and *this* extension of X assures that $X \in N_\mathscr{E}$, then the corresponding extension of X' assures that $X' \in N_{\mathscr{E}'}$. But, suppose that the *only* extension of X that assures it a place in $N_\mathscr{E}$ fails to correspond,

IDENTITY, EQUIVALENCE AND REDUCTION 225

via $\bar{\mathbf{R}}$, to any extension of X', though some *other* extension of X does correspond to an extension of X'. Then, even though $\langle X', X \rangle \in \mathbf{R}^*$, and $X \in N_{\mathscr{E}}$, we have no guarantee that $X' \in N_{\mathscr{E}'}$. Of course, it may be that quite "by accident" $\mathbf{R}^*$ satisfies (D51–3), but (D52–3) does not assure this.

We can remedy this defect in the notion of close reduction by adding the following requirement. If x' and x stand in the $\mathbf{R}^*$-relation and x has an extension in M, then this extension of x corresponds, via $\mathbf{R}$, to at least one extension of x'. Relations having this additional property we shall call 'strong reduction relations'. These are precisely defined below.

(D53)   If $\mathscr{E}' = \langle M_0', N_0', r', M', C', L', C_{L'}, \alpha' \rangle$ and $\mathscr{E} = \langle M_0, N_0, r, M, C, L, C_L, \alpha \rangle$ are expanded cores for theories of mathematical physics then $\mathbf{R}$ *strongly reduces* $\mathscr{E}'$ to $\mathscr{E}$ if and only if:
(1) $\mathbf{R}$ closely reduces $\mathscr{E}'$ to $\mathscr{E}$;
(2) For all $x' \in N_0'$ and $x \in N_0$, if there exist $y' \in M_0'$ and $y \in M_0$ such that $x' = r'(y')$, $x = r(y)$ and $\langle y', y \rangle \in \mathbf{R}$, then, for all $z \in M$ such that $x = r(z)$, there exists some $z' \in M'$ such that $x' = r'(z')$ and $\langle z', z \rangle \in \mathbf{R}$.

The condition (D53–2), together with (D52–3), assures us that the $\mathbf{R}^*$-relation will be a reduction relation, i.e. that it will satisfy the "entailment condition". For, if $\langle X', X \rangle \in \bar{\mathbf{R}}^*$ and X has some extension that satisfies (D52–3–iii) and (iv), then (D53–2) assures us that this extension of X stands in the $\bar{\mathbf{R}}$-relation with *some* extension of X' and (D52–3) assures us that this extension of X' satisfies the conditions (D52–3–vi) and (D52–3–vii). This means that, whenever $\langle X', X \rangle \in \bar{\mathbf{R}}^*$ and $X \in N_{\mathscr{E}}$, then $X' \in N_{\mathscr{E}'}$. This fact is stated precisely in the following theorem.

(T30)   If $\mathscr{E}' = \langle M', N_0', r', M', C', L', C_{L'}, \alpha' \rangle$ and $\mathscr{E} = \langle M_0, N_0, r, M, C, L, C_L, \alpha \rangle$ are expanded cores for theories of mathematical physics, $\mathbf{R}$ strongly reduces $\mathscr{E}'$ to $\mathscr{E}$, and $\mathbf{R}^* = \{\langle x, y \rangle \mid \text{there is a } \langle u, v \rangle \in \mathbf{R} \text{ and } x = r(u), y = r(v)\}$ then $\mathbf{R}^*$ reduces $\mathscr{E}'$ to $\mathscr{E}$.

In addition to providing us with the "neat" relationship between strong reduction and reduction summarized in (T30), the condition (D53–2) is recommended by some strong intuitive considerations. These, however, are most expediently considered in the light of a complete account of reduction relations for theories. It should only be noted here that a condition parallel to (D53–2) can be formulated by interchanging the role of x' and x. That is, we might require that, if $\langle x', x \rangle \in \mathbf{R}^*$, then every extension of x' that is in M' corresponds to some extension of x. We do not use this requirement to establish (T30). Moreover, we shall subsequently see that there are strong intuitive grounds for *not* adopting this parallel requirement.

The following two theorems make explicit the claim that, for expanded cores with no theoretical functions and no constraints on any function, our account of reducibility for expanded cores is the same as Adams' account of the formal aspect of reducibility for theories. (T31) deals with the case where there are no non-trivial constraints, though there still may be theoretical functions. The absence of constraints simply means that it is no longer *essential* that we consider the statements of the theory to be about sets of intended applications. The (weak) reduction relation then simply has the property that, for any $x' \in N_0'$ if there is an $x \in N_0$ which stands in the **R**-relation to x' and x is extendible to a model for M then x' is extendible to a model for M'. Note that there is no guarantee here that the theoretical functions used to produce these extensions are "related". The strong reduction, in this case, has the property that, for any $x' \in M_0'$, if there is an $x \in M_0$, which stands in the **R**-relation to x' and is in M, then x' is in M'. This is essentially the requirement that Adams offers to account for the formal aspect of the reduction of theories. However, since theoretical functions still may be present, we still have the requirement of (D52–3) for strong reduction. (T32) states that, in the absence of theoretical functions, strong reduction and (weak) reduction are coextensive. In this case the requirement (D52–3) is vacuously satisfied and the only relevant requirement is (T31–2–iv). Thus, in the absence of constraints and theoretical functions the account given here reduces to that of Adams.

(T31) If $\mathscr{E}' = \langle M_0', N_0', r', M', \mathscr{P}(M_0'), \{M'\}, \mathscr{P}(M_0'), N_0' \times \{M'\} \rangle$ and $\mathscr{E} = \langle M_0, N_0, r, M, \mathscr{P}(M_0), \{M\}, \mathscr{P}(M_0), N_0 \times \{M\} \rangle$ are expanded cores for theories of mathematical physics then:

(1) **R** reduces $\mathscr{E}'$ to $\mathscr{E}$ if and only if
   (i) (D51–1);
   (ii) (D51–2);
   (iii) For all $x' \in N_0'$, if there exists an $x \in N_0$ such that $\langle x', x \rangle \in \mathbf{R}$ and $x \in \mathbf{R}(M)$, then $x' \in \mathbf{R}'(M')$.

(2) **R** strongly reduces $\mathscr{E}'$ to $\mathscr{E}$ if and only if
   (i) (D52–1);
   (ii) (D52–2);
   (iii) (D52–3);

(iv) For all $x' \in M'_0$, if there exists an $x \in M_0$ such that
   (a) $x \in M$
   (b) $\langle x', x \rangle \in \mathbf{R}$, then $x' \in M'$.

(T32) If $\mathscr{E}' = \langle N'_0, N'_0, r', N', \mathscr{P}(N'_0), \{N'\}, \mathscr{P}(N'_0), N'_0 \times \{N'\} \rangle$ and $\mathscr{E} = \langle N_0, N_0, r, N, \mathscr{P}(N_0), \{N\}, \mathscr{P}(N_0), N_0 \times \{N\} \rangle$ are expanded cores for theories of mathematical physics then $\mathbf{R}$ strongly reduces $\mathscr{E}'$ to $\mathscr{E}$ if and only if $\mathbf{R}$ reduces $\mathscr{E}'$ to $\mathscr{E}$.

Now, how are we to use this notion of reduction for expanded cores to get a notion of reduction for theories. Every theory has associated with it a vast number of expanded cores, each corresponding to some way that special laws and constraints associated with them may be "postulated" to hold. All these expanded cores are, however, "expansions" of the same core – the one that is characteristic of the theory. This suggests that the appropriate way to get at the formal aspect of the reduction of theories is to use the notion of reduction for expanded cores to define a notion of reduction for the cores of which they are "expansions".

One obvious way this might be done is this. One could say that the core H' reduces to the core H if and only if every expansion of H is reducible to some expansion of H. That is H' reduces to H if and only if, for every expansion of H', $\mathscr{E}'$ there exists some expansion of H, $\mathscr{E}$, and some $\mathbf{R}$-relation that reduces $\mathscr{E}'$ to $\mathscr{E}$. Intuitively, this means that, for every way that we postulate special laws and constraints in an attempt to make T' "fit the facts", we can find some way to postulating special laws and constraints in T which reduces to it by *some* $\mathbf{R}$-relation. This allows that different expansions of H' *may* require different $\mathbf{R}$-relations to reduce them to some expansion of H, and yet we would still say that H' was reduced to H. That is, different ways of postulating laws in T' might require different reduction relations to reduce them to ways of postulating laws in T. This would be the case, for example, if the postulating of different special laws about torques on rigid bodies required us to employ different reduction relations to reduce them to ways of postulating special force laws for sets of particles.

Intuitively, it seems that we want to regard the reduction relation as a more or less enduring property that theories possess. If thermodynamics is "really" reducible to thermodynamics, then we expect it to remain so throughout its historical development.

We even expect it to remain reducible when the range of intended applications of thermodyanmics is expanded, provided that the new applications still correspond via the reduction relation to statistical mechanical systems. Thus, when the range of applications of thermodynamics is expanded to include systems with electromagnetic energies, it is expected that the same reduction relation that had been used previously will continue to work, since these systems can still be regarded as being composed of particles with forces acting on them. We have only to "postulate" new, special force laws – electromagnetic forces – in some of the statistical mechanical systems. Were we forced to contemplate, even the *possibility*, that this same reduction relation might not work, and further, that we *might not* be able to find another intuitively acceptable one, it seems that we would come to doubt the acceptability of the initial reduction. It would begin to appear *ad hoc*. All this is to say that we expect that whether or not two theories satisfy the *formal* conditions of reduction will remain unchanged as the range of intended applications of the reduced theory expands. The only way that we expect a reduced theory to cease being reduced is by adding new intended applications that fall outside the domain of the reduction relation – as, for example, if we come to apply thermodynamics to systems that can not be regarded as sets of particles.

This discussion suggests that the appropriate account of the reduction relation between cores should be roughly this. We should require that there be *one* reduction relation that will serve to reduce *all* the expansions of H' to *some* expansion of H. Intuitively, we should require that there be some correspondence between the concepts of the two cores H' and H which serves to reduce any application of H' to some application of H. This is stated precisely in (D54) where we define a strong, close and weak sense of reduction for cores corresponding to the strong, close and weak senses of reduction for expanded cores.

(D54) If H' and H are cores for theories of mathematical physics then **R** (strongly, closely) reduces H' to H if and only if, for all L', $C_{L'}$, and α' such that $\mathscr{E}' = \langle H', L', C_{L'}, \alpha' \rangle$ is an expanded core for a theory of mathematical physics, there exists some L, C_L, and α such that $\mathscr{E} = \langle H, L, C_L, \alpha \rangle$ is an expanded core for a theory of mathematical physics and **R** (strongly, closely) reduces $\mathscr{E}'$ to $\mathscr{E}$.

In the case of close reduction, a very useful sufficient condition can be given. Roughly it is this. If the "null expansions" of two cores are closely

IDENTITY, EQUIVALENCE AND REDUCTION

reduced by **R**, then the cores themselves are closely reduced by **R**. This is because in the case of close reduction, the **R**-relation itself provides a natural way to construct the expansion of the core H to which any given expansion of H' may be reduced by **R**. A bit more precisely, if the "null expansions" are reduced by **R**, then given any expansion of H', it is reducible by **R** to its "image" under the **R**-relation. The "image" of an expansion of H' is obtained by adding to H laws and constraints which are "images" under the **R**-relation of the laws and constraints that have been added to H' to produce this expansion. These ideas are made precise in the statement and proof of the following theorem.

(T33) If $H' = \langle M_0', N_0', r', M', C' \rangle$ and $H = \langle M_0, N_0, r, M, C \rangle$ are cores for theories of mathematical physics and **R** strongly reduces $\mathscr{E}_0' = \langle H', \{M'\}, \mathscr{P}(M_0'), N_0' \times \{M'\} \rangle$ to $\mathscr{E}_0 = \langle H, \{M\}, \mathscr{P}(M_0), N_0 \times \{M\} \rangle$ then **R** strongly reduces H' to H.

We are now in a position to define reduction for theories of mathematical physics. Parallel to the preceding development, we shall define a strong, close and weak sense of reduction for theories. Let us consider the weak sense first. Given the two theories $T' = \langle H', I' \rangle$ and $T = \langle H, I \rangle$, we shall say first that **R** reduces T' to T only if **R** reduces H' to H. This constitutes the formal aspect of the reduction relation – that part of it that is independent of any facts about the intended applications of the two theories. In addition we want to require, that every intended application of T' be **R**-related to some intended application of T. In the cases of close and strong reduction, the situation is much the same. If **R** is to closely reduce T' to T, we require that x closely reduce H' to H. Further we require that every $x' \in I'$ "correspond" via **R** to some $x \in I$. The only difference is that 'corresponds' is now defined in the following way. Here $x' \in I'$ "corresponds" to $x \in I$ if and only if there are extensions of x' and x that stand in the **R**-relation.

(D54) If $T' = \langle H', I' \rangle$ and $T = \langle H, I \rangle$ are theories of mathematical physics then:
(A) **R** *reduces* T' *to* T if and only if:
(1) **R** reduces H' to H,
(2) for all $x' \in I'$, there is an $x \in I$ such that $\langle x', x \rangle \in \mathbf{R}$,
(B) **R** *strongly* (*closely*) *reduces* T' *to* T if and only if:

(T34) If $H'=\langle M'_0, N'_0, r', M', C'\rangle$ and $H=\langle M_0, N_0, r, M, C\rangle$ are cores for theories of mathematical physics and **R** closely reduces $\mathscr{E}'_0=\langle H', \{M'\}, \mathscr{P}(M_0), N'_0\times\{M'\}\rangle$ to $\mathscr{E}_0=\langle H, \{M\}, \mathscr{P}(M_0), N_0\times\{M\}\rangle$ then **R** closely reduces H' to H.

To see this, let $\mathscr{E}'=\langle H', L', C_{L'}, \alpha'\rangle$ be any expansion of H'. Construct $\mathscr{E}=\langle H, L, C_L, \alpha\rangle$ in the following way. For any $1'\in L'$, let $1(1')=\{x\,|\,x\in M$, and there is an $x'\in 1'$ such that $\langle x'_0, x\rangle=\mathbf{R}\}$. Let $L=\{y\,|\,$there is an $1'\in L'$ and $y=1(1')\}$. Let α be such that, for all $x\in N_0$, if there is an $x'\in N'_0$ such that $\langle x', x\rangle\in\mathbf{R}^*$ (see (T30)) then $\alpha(x)=1(\alpha'(x'))$. (Note that the fact that **R**, and thus **R***, is one–many assures that this can be done.)

Let $C_L=\{X\,|\,X\in\mathscr{P}(M_0)$ and there is an $X'\in C_{L'}$ such that $\langle X', X\rangle\in\bar{\mathbf{R}}$. If **R** closely reduces $\mathscr{E}'_0$ to $\mathscr{E}_0$ then **R** satisfies (D52-1), and (D52-2). Thus, all that remains to be shown is that **R** satisfies (D52-3) for $\mathscr{E}'$ and $\mathscr{E}$. Suppose $\langle X', X\rangle\in\bar{\mathbf{R}}$. Now, if $X\in\mathscr{P}(M)\cap C$, then $X'\in\mathscr{P}(M')\cap C'$, since **R** closely reduces $\mathscr{E}'_0$ to $\mathscr{E}_0$. Further, if $X\in C_L$, then $X'\in C_{L'}$ since to be in C_L at all, X must correspond to some $Y\in C_{L'}$, and since $\bar{\mathbf{R}}$ is one–many $Y=X'$. Thus, if $X\in(\mathscr{P}(M)\cap C\cap C_L)$, then $X'\in(\mathscr{P}(M')\cap C'\cap C_{L'})$. Now suppose, for all $x\in X$, $x\in\alpha(r(x))$. Since $\langle X', X\rangle\in\bar{\mathbf{R}}$, for all $x'\in X'$ there is some $x\in X$ such that $\langle x', x\rangle\in\mathbf{R}$ and $x\in\alpha(r(x))$. Further, since $\langle x', x\rangle\in\mathbf{R}$, $\langle(r(x))=1(\alpha'(r'(x')))$ and if $x\in 1(\alpha'(r'(x')))$ there is some y such that $\langle y, x\rangle\in\mathbf{R}$ and $y\in\alpha'(r'(x'))$. But since **R** is one–many $y=x'$. Thus, $x'\in\alpha'(r'(x'))$ and hence, for all $x'\in X'$, $x'\in\alpha'(r'(x'))$. Thus **R** reduces $\mathscr{E}'$ to $\mathscr{E}$.

As a trivial corollary of (T34) and (D54) we have:

(1) **R** strongly (closely) reduces H' to H,
(2) for all $x'\in I'$ there is an $x\in I$, a $y'\in M'_0$, and a $y\in M_0$ such that $x'=r'(y')$, $x=r(y)$ and $\langle y', y\rangle\in\mathbf{R}$.

The following theorem makes explicit the claim that our account of the reduction for theories of mathematical physics has the two intuitive properties of reduction relations mentioned by Adams. The first: that if **R** reduces T' to T then every intended application of T' "corresponds" to some intended application of T via the reduction relation, is a trivial consequence of (D54-A-2). The second: that for every application of T' there is a "corresponding" application of T, such that, if this application is successful, then the application of T' is also successful, follows from the formal properties of the reduction relation (D54-A-1). In the case that **R** strongly reduces T' to T then the **R***-relation reduces T' to T (T31) and thus has these properties.

(T35) If $T' = \langle H', I' \rangle$ and $T = \langle H, I \rangle$ are theories of mathematical physics and **R** reduces T' to T then:
(1) For all $x' \in I'$ there is an $x \in I$ such that $\langle x', x \rangle \in \mathbf{R}$
(2) If $S' = \langle H', L', C_{L'}, \alpha', I'_t \rangle$ is an application of T' then there is an application of T, $S = \langle H, L, C_L, \alpha, I_t \rangle$ such that:
   (i) For all $x' \in I'_t$ there is an $x \in I_t$ such that $\langle x', x \rangle \in \mathbf{R}$,
   (ii) If S is a successfully applied core for a theory of mathematical physics then S' is a successfully applied core for a theory of mathematical physics.

It should be noted that (D54) separates the requirements for a reduction into a formal and an "applied" requirement. It is a purely formal question, one that can be answered by looking at the mathematical structures of the theories involved, as to whether some given **R**-relation reduces, closely reduces or strongly reduces two cores. On the other hand, it will generally be an "empirical" question as to whether (D54–A–2) and (D54–B–2) are satisfied. That is whether every member of I' "corresponds" in the appropriate way to a member of I. Thus, whether every physical system – rigid body – that we would count as an intended application for rigid body mechanics "corresponds" to some particle system that we would count as an intended application of classical particle mechanics, is a question that can only be answered by examining these systems of determining whether they do, or do not, stand in the **R**-relation to one another. In this case, this amounts, roughly, to examining the rigid body systems to see if all of them are "composed" of particles. Of course, if our conceptions, both of what it is to be a rigid body and what it is to be a particle are a bit vague, then this question will also be vague. However, it is clear that is a different sort of question from the question of whether **R** reduces two cores.

Further, the possibility of "vacuous" reductions should be noted. Properties of the reduction relations (both strong and weak) for expanded cores already mentioned (see above p. 223) allow for the following possibilities. It might be the case that **R** reduces T' to T, even though one or more successful applications of T' corresponds via **R** to unsuccessful applications of T. Indeed, it could even be the case that all applications of T' were successful and yet correspond, via **R** only to unsuccessful applications of T. Roughly, what this amounts to is the following. Given any "real, live" theory of mathematical physics which has at least some successful applications, it is always going to be possible to construct wierd and fanciful "possible" theories of mathematical physics to which this theory reduces, via some reduction relation. The major difficulty with these constructed theories, aside from their novelty, will be that all, or most, of their applications will be unsuccessful. In particular, successful applications of the given theory will correspond to unsuccessful applications of the constructed theory. I do not believe that the possibility of such vacuous reductions constitutes a serious objection to the account of reduction relations offered here. After all, we are really interested only in the question of whether two "real, live" theories stand in some reduction relation with one another. So long as our account is adequate to these cases, there seems to be no reason not to allow that our intuitive notion of reducibility be stretched a bit to admit these "vacuous" examples.

We are now in a position to appreciate more fully the intuitive significance of (D53–2). Suppose we have two theories, $T' = \langle H', I' \rangle$ and $T = \langle H, I \rangle$ and that **R** closely reduces T' to T. Then $I' \subseteq D(\mathbf{R}^*)$ and $\langle I', I \rangle \in \overline{\mathbf{R}}^*$. Now consider the following situation. Suppose there is a member of I', x', that can be extended to a model for M' in such a way that the extension does not stand in the **R**-relation with any extension of any member of I that stands in the $\mathbf{R}^*$-relation with x'. That is, there is a $y' \in M'$ such that $x' = r(y')$ but there is no $x \in I$ such that $\langle x', x \rangle \in \mathbf{R}^*$ and there is a y such that $x = r(y)$ and $\langle y', y \rangle \in \mathbf{R}$. Note that there will be *other* extensions of x' in M', besides y', that stand in the **R**-relation with extensions of members of L which stand in the relation with x'. Also note that *this* situation is not ruled out by (D53–2). (D53–2) rules out there being extensions of x which correspond to no extensions of x' when $\langle x', x \rangle \in \mathbf{R}^*$.

Essentially, what happens in the situation just described is this. The reduction relation **R**, together with the set of intended applications of the reducing theory, operate to besmirch the character of some, otherwise perfectly respectable models for M'. If one takes the $\mathbf{R}^*$-relation holding between x' and x as intuitively indicative of something like identity of the individuals involved and if T is such as to endow its theoretical functions with an aura of "concreteness", then extensions of x' which do not correspond to any extension of x are likely to be regarded as arbitrary or *ad hoc*, even though they are models for M'. For example, if a rigid body p corresponds to a set of particles P, via some relation that reduces rigid body mechanics to particle mechanics, we are likely to look with suspicion upon an assignment of total mass and moment of inertia to *p* which can not be "realized" by *some* assignment masses to members of *P*, even if this assignment does produce a model for the laws of rigid body mechanics. (We shall consider this example in more detail below.) Roughly, we feel that, if we were *compelled* to use one of these "bastard" extensions of *x'* to successfully apply *H'* to *I'*, the intuitive appeal of the reduction would be considerably diminished. Fortunately, we are not so compelled, except perhaps in certain "vacuous" cases of reduction where *H* can not be successfully applied to *I*, even though *H'* can be successfully applied to *I'*. Intuitively, it is just this situation that is ruled out by (D53–2).

An appealing way of looking at this situation is the following. Roughly speaking, the successful reduction of *T'* to *T* has brought to light "empirical" relations among the *theoretical* functions of *T'* that do not appear in the basic mathematical structure of *T'*. This reduction has shown that, as a matter of fact, some possible models for *M'* do not occur "in nature". These are the models that are produced by bastard extensions of members of *I'*. That **R** reduces *T'* to *T* shows us that, so long as we make claims with *H'* about sets of intended applications which stand in the **R***-relation to some set of intended applications to which *H* can be successfully applied, we never need these bastard models. Since we regard the fact that **R** reduces *T'* to *T* as an empirical fact, it is natural to call the fact that we do not need certain models for *M'* 'an empirical fact'. We might even be clever enough to restrict the definition of 'is an *M''* in such a way as to rule out just these bastard models. If we did this, then this new predicate would describe the empirical result of the reduction of *T'* to *T*.

Described in this way, the situation we have been considering appears to be perfectly natural and in accord with our intuitions that the reducing theory ought to be more fundamental or more complete than the reduced theory. When the appropriate relation is seen between a "coarse grained" way of looking at certain phenomena and a "fine grained" way of looking at the same phenomena, it is not surprising that new facts come to light that can not be expressed in the "coarse grained" vocabulary. It is not surprising that the reduction of thermodynamics to statistical mechanics occasions the discovery of new relations among the thermodynamic functions. On the other hand – to go the other way – to let the reduced theory impose new laws in the reducing theory runs counter to those intuitive ideas about reduction. Indeed, to allow this would be to depart from what is perhaps *the* central motivation for attempts at reduction – showing that the reduced theory can "in principle" be dispensed with. If the possibility existed that new special laws added to the reduced theory could impose new special laws in the reducing theory, a case could hardly be made for dispensing with the reduced theory.

As an example of a reduction relation holding between two theories of mathematical physics, let us consider classical rigid body mechanics and classical particle mechanics. The claim is frequently made that the former is reducible to the latter, or that it is reducible to a "special case" of the latter. If this is so, and if our account of the reduction relation is correct, then we should be able to do the following. Given a description of the cores and admissible sets of intended applications for both theories, we should be able to describe a relation, call it '$\mathscr{L}$', and show, first that $\mathscr{L}$ strongly reduces the core of classical rigid body mechanics to the core of classical particle mechanics, and, second, that the admissible sets of intended applications have the property (D54–B–2) with respect to the relation.

In the case of classical particle mechanics, as has already been noted, we have a much more adequate characterization of the core than we have of the intended applications. Thus, it seems unlikely that we will be able to carry out, with the same degree of precision, both parts of the task of showing that rigid body mechanics reduces to classical particle mechanics. It is very likely that we shall achieve precision, if at all, only with respect to the formal aspect of the reduction relation. For this reason, most of our attention will be focused on this formal aspect. That is we shall want to show that the core of classical rigid body mechanics is strongly reduced by some relation $\mathscr{L}$ to the core of classical particle mechanics.

To carry this out we shall need a description of the core of classical rigid body mechanics. The central part of this description is given by a set-theoretic axiomatization of this theory. This axiomatization is essentially that due to Adams and Rubin ([1], Chapter 6, [2]), with two modifications. First, their axiomatization is modified in such a way that a distinction between theoretical and non-theoretical functions can be drawn. This is done by introducing the possibility of describing the motion of the rigid body with respect to an arbitrary point in the body, rather than the center of mass. This allows us to regard the position (both translational and rotational) of the body with respect to the center of mass as a theoretical function. The need for this will be explained below. Second, the axiomatization we shall employ requires that systems of rigid body mechanics contain only *one* body. The reason for restricting systems of rigid body

IDENTITY, EQUIVALENCE AND REDUCTION

mechanics in this way is mainly that the axioms become somewhat simpler and the essential ideas of the reduction relation thereby more perspicuous. Further, many significant applications of rigid body mechanics do appear to involve only one body, e.g. the motion of a top. There may however be others which involve more than one rigid body. If this is so, then the axiomatization given here would not be sufficient to provide a complete logical reconstruction of the theory. But since our purpose here is to illustrate the ideas involved in the reduction of theories of mathematical physics, rather than to provide a logical reconstruction of classical rigid body mechanics, this simplification seems justified. In any event, the discussion that follows generalizes in a natural and obvious way to apply to the more general situation in which there is more than one rigid body in a system of rigid body mechanics.

To begin we define the predicate 'is an RBM' (is a rigid body mechanics). This predicate plays a role in classical rigid body mechanics analogous to the role of 'is a PM' in classical particle mechanics. It determines the set of possible models for the basic mathematical structure of the theory, i.e. the set M_0 in the core of the theory. Roughly, it tells us what sort of entities and functions the theory is going to deal with, but does not say anything explicit about the relations between these functions.

(D55)   *x is an* RBM if and only if there exists a $\rho, T, \bar{u}, \varphi, \bar{r}, \theta, g, \mu, \bar{h}, \bar{p}$ such that:

(1) $x = \langle \{\rho\}, T, \bar{u}, \varphi, \bar{r}, \theta, g, \mu, \bar{h}, \bar{p} \rangle$;
(2) $\langle \{\rho\}, T, \bar{u}, g, \bar{h} \rangle$ and $\langle \{\rho\}, T, \bar{r}, g, \bar{h} \rangle$ are PM's;
(3) For all $t, t' \in T$, $|\bar{u}(\rho, t) - \bar{r}(\rho, t)| = |\bar{u}(\rho, t') - \bar{r}(\rho, t')|$;
(4) φ and θ are functions whose domain is $\{\rho\} \times T$ and whose range is the set of all 3×3, real, orthogonal matrices such that, for all $t, t' \in T, i \in \{1, 2, 3\}$,
   (i) $D_{2,2}[\varphi(\rho, t)]$ and $D_{2,2}[\theta(\rho, t)]$ exist,
   (ii) $|(\bar{u}(\rho, t) + \bar{\varphi}_i(\rho, t)) - (\bar{r}(\rho, t) + \theta_i(\rho, t))|$
        $= |(\bar{u}(\rho, t') + \bar{\varphi}_i(\rho, t')) - (\bar{r}(\rho, t') + \theta_i(\rho, t'))|$;
(5) $\bar{p}$ is a function whose domain is $\{\rho\} \times T \times I_+$ and whose range is the set of ordered triples of real numbers such that, for all $t \in T$ and $i \in I_+$, $\sum_{i \in I+} \bar{h}(p, t, i) \otimes \bar{p}(\rho, t, i)$ is absolutely convergent;
(6) μ is a function whose domain is $\{\rho\}$ and whose range is the set of 3×3, symmetric positive definite matrices.

Some of the notation in (D55) is not standard and must be clarified. First, in (D55-4), the i-th row of a 3×3 matrix φ is considered to be a 3-dimensional vector and denoted by φ_i. Second, the symbol $\otimes$ in (D55-5) denotes a binary operation on the set of all real matrices defined by the equation:

$$A \otimes B = A^*B - B^*A,$$

where A^* denotes the transpose of A, i.e. $(A^*)_{ij} = A_{ji}$. A 3-dimensional vector $\bar{a} = \langle a_1, a_2, a_3 \rangle$ is regarded as a 1×3 matrix. The transpose of $\bar{a}$ is a 3×1 matrix or a "column vector", i.e.

$$(\bar{a}^*) = \begin{pmatrix} a_1 \\ a_2 \\ a_3 \end{pmatrix}.$$

Thus, for two 3-dimensional vectors $\bar{a}$ and $\bar{b}$

$$\bar{a} \otimes \bar{b} = \begin{pmatrix} 0 & a_1b_2 - b_2a_2 & a_1b_3 - b_1a_3 \\ a_2b_1 - b_2a_1 & 0 & a_2b_3 - b_2a_3 \\ a_3b_1 - b_3a_1 & a_3b_2 - b_3a_2 & 0 \end{pmatrix}.$$

Note that $\bar{a} \otimes \bar{b}$ is skew symmetric, i.e. $\bar{a} \otimes \bar{b} = -(\bar{a} \otimes \bar{b})^*$. This is true in general. Also, note that, in this case the elements of $\bar{a} \otimes \bar{b}$ correspond to the components of the ordinary vector cross product (denoted by x) in the following way:

$$\bar{a} \times \bar{b} = \langle (\bar{a} \otimes \bar{b})_{2,3}, (\bar{a} \otimes \bar{b})_{3,1}, (\bar{a} \otimes \bar{b})_{1,2} \rangle.$$

In the intended interpretations of classical rigid body mechanics the primitive symbols in (D55) have roughly the following intuitive significance. ρ is *the* rigid body and T is the time interval during which the motion of ρ is considered. $\bar{u}$ is the position vector of some arbitrary point "in" ρ relative to some coordinate system, call it 'C'. $\bar{r}$ is the position vector of the center of mass of ρ relative to the coordinae system C. (D55-3) assures us that it is possible to regard both $\bar{u}$ and $\bar{r}$ as the positions of points "in" the rigid body ρ. g is the total mass of ρ and $\bar{h}$ is a function from $\{\rho\} \times T \times I_+$ such that $\bar{h}(\rho, t, i)$ is the i-th force acting on ρ at time t, relative to the coordinate system C. Essentially, (D55-2) says that the physical system just described are systems of particle mechanics. It does not say that they constitute a classical particle mechanics, i.e. that they obey the second law. Indeed, we shall subsequently only claim that the "center of mass system" $\langle \{\rho\}, T, \bar{r}, g, \bar{h} \rangle$ is a classical particle mechanics. Roughly speaking (D55-2) characterizes the mathematical apparatus to be used in

IDENTITY, EQUIVALENCE AND REDUCTION   237

talking about the translational motion of the rigid body. The rest of (D55) characterizes the mathematical apparatus to be used in talking about the rotational motion of the rigid body.

To understand the intuitive significance of the remainder of (D55) it is necessary to consider two coordinate systems, in addition to C: first, C' whose origin is at the center of mass of the body ρ, and whose axes always remain parallel to the corresponding axes of C; and second, C'' whose origin is likewise at the center of mass of ρ, but whose axes are "imbedded" in the rigid body and move with it as it rotates. The following 2-dimensional picture illustrates this (Figure 5).

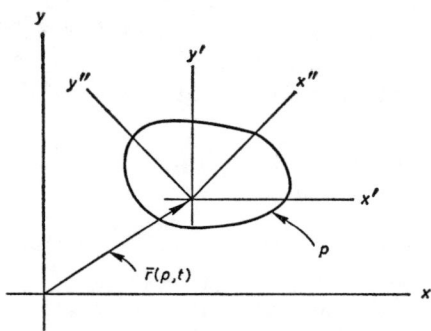

Fig. 5

For any $t \in T$, the 3×3 real, orthogonal matrix $\theta(\rho, t)$ may be interpreted in the following way. The vectors

$$\theta_i(\rho, t) = \langle \theta(\rho, t)_{i,1}, \theta(\rho, t)_{i,2}, \theta(\rho, t)_{i,3} \rangle$$

for $i \in \{1, 2, 3\}$ are mutually orthogonal unit vectors. They may be thought of as "being" (or "lying along") the i-th coordinate axis of the coordinate system C'' at time t. Intuitively, the C'' system is envisioned as being imbedded in the body ρ and moving with it as it rotates. Thus, the orientation of the coordinate axis of C'', relative to C', at t is essentially the orientation of ρ, relative to C', at t. Hence, the values of the function θ may be interpreted as specifying the "angular position", or orientation of ρ, relative to a coordinate system with fixed orientation whose origin is the center of mass of ρ.

Likewise, the 3×3 real, orthogonal matrix $\varphi(\rho, t)$ may be interpreted as a coordinate system whose origin is at $\bar{u}(\rho, t)$ and whose axis are fixed in the body. It describes the orientation of ρ, relative to a coordinate system with fixed orientation whose origin is at $\bar{u}(\rho, t)$. That we may regard the values of both φ and θ as being fixed in the body is assured by (D55–4–ii). We may think of $\bar{u}$ and φ as providing us with a means of describing the motion of ρ (both translational and rotational) that is independent of our being able to locate the center of mass of ρ.

The 3×3, symmetric, positive definite matrix $\mu(\rho)$ is interpreted to be the moment of inertia of ρ, relative to the coordinate system C'' which rotates with ρ. It is independent of time because, intuitively, the mass distribution of ρ, relative to C'', remains fixed. Note that $\mu(\rho)$ is taken as primitive in this axiomatization and not defined in terms of a mass distribution. The relation between $\mu(\rho)$ and the mass distribution of ρ will be illuminated by the relation which reduces rigid body mechanics to particle mechanics. The usual notions of principal moments of inertia and products of inertia may be defined in terms of $\mu(\rho)$ in the following way:

i-th principal moment $= Tr(\mu(\rho)) - (\mu(\rho))_{ii}$
ij-th product $= (\mu(\rho))_{ij}$.

For all $t \in T$ and $i \in I_+$ the vector $\bar{p}(\rho, t, i)$ is interpreted to be the position of ρ at t where the i-th force is applied to ρ – the point of application of $\bar{h}(\rho, t, i)$ – relative to the coordinate system C''. The skew-symmetric matrix

$$\tau(\rho, t, i) = \bar{h}(\rho, t, i) \otimes \bar{p}(\rho, t, i)$$

is the moment about the center of mass of ρ at time t due to the i-th force applied to ρ. For example, $\tau(\rho, t, i)_{2,3}$ is the moment due to $\bar{h}(\rho, t, i)$ about the I-axis of the C'' coordinate system. (D55–4) requires that the sum of these moments, due to all the forces, be absolutely convergent. This requirement is motivated by the same considerations that motivated the analogous requirement for forces in our definition of 'is a PM' (D7).

Having determined the set M_0 in the core for classical rigid body mechanics, we can now draw a distinction between theoretical and non-theoretical functions in this theory by defining the predicate 'is an RBK' (is a rigid body kinematics). Essentially, we maintain that only the position of some arbitrary point in the rigid body and its orientation relative to a fixed axis moving with this point are non-theoretical functions in this theory. It should not be surprising that we take the position of the center of mass of the body and the angular position function, relative to the center of mass to be theoretical functions, since they can apparently only be described by reference to a theoretical function – mass – in another theory – classical particle mechanics. However, this reason for regarding them as theoretical functions is misleading. Ultimately, the only justification for regarding them as theoretical functions is that they play the role of theoretical functions in the empirical claims of classical rigid body mechanics. This is a substantive claim about the existing expositions of this theory which will not be defended in detail here. However, it will be noted below that only by taking these as theoretical functions can an intuitively plausible candidate for a reduction relation between classical rigid body mechanics and particle mechanics be shown to have the properties we demand of a reduction relation.

(D56) x is an RBK if and only if there exists a $\rho, T, \bar{u},$ and φ such that:

(1) $x = \langle \{\rho\}, T, \bar{u}, \varphi \rangle$;

(2) $\langle \{\rho\}, T, \bar{u} \rangle$ is a PK;

(3) φ is a function whose domain is $\{\rho\} \times T$ and whose range is the set of all 3×3 real, orthogonal matrices such that, for all $t \in T$, $D_{2,2}[\varphi(\rho, i)]$ exists.

We now define the predicate 'is a CRBM' (is a classical rigid body mechanics). This predicate plays a role in classical rigid body mechanics analogous to the role of 'is a PM' in classical particle mechanics. It determines the basic mathematical structure of the theory – the set M in the core of the theory. Essentially (D57) requires that the motion of the center of mass of a rigid body be a system of classical particle mechanics (D57-2) and that the rotational motion satisfy an analogous law (D57-3).

(D57) x is a CRBM if and only if:

(1) $x = \langle \{\rho\}, T, \bar{u}, \varphi, r, \theta, g, \mu, \bar{h}, \bar{p} \rangle$ is an RBM;

(2) $\langle \{\rho\}, T, \bar{r}, g, \bar{h} \rangle$ is a CPM;

(3) For all $t \in T$, $\theta(p, t) \otimes \mu(\rho) D_{2,2}[\theta(\rho, t)] = \sum_{i \in I_+} \bar{h}(\rho, t, i) \otimes \bar{p}(\rho, t, i)$.

The intuitive significance of (D57-3) can perhaps best be seen by noting how it reduces to more familiar formulations of the "second law", for rotational motion. Consider a situation in which the mass of a rigid body is confined to the $1'$, $2'$-plane, its rotation is only in this plane, and all the forces acting as it lie in this plane. In this case

$$\mu(p) = \begin{pmatrix} \mu_{11} & \mu_{12} & 0 \\ \mu_{21} & \mu_{22} & 0 \\ 0 & 0 & 0 \end{pmatrix},$$

$$\tau(p, t) = \begin{pmatrix} 0 & \tau_{12}(t) & 0 \\ -\tau_{12}(t) & 0 & 0 \\ 0 & 0 & 0 \end{pmatrix}$$

and

$$\theta(p, t) = \begin{pmatrix} \theta_{11}(t) & \theta_{12}(t) & 0 \\ \theta_{21}(t) & \theta_{22}(t) & 0 \\ 0 & 0 & 1 \end{pmatrix}.$$

Here, we can express the elements of $\theta(p, t)$ in the terms of α, the angle between the $1'$ and $1''$ axis and obtain

$$\theta(p, t) = \begin{pmatrix} \cos \alpha & \sin \alpha & 0 \\ -\sin \alpha & \cos \alpha & 0 \\ 0 & 0 & 1 \end{pmatrix}.$$

When these values are substituted into the equation (D57–3) it reduces to the single equation:

$$(\mu_{11} + \mu_{22}) D_2 \alpha(t) = \tau_{12}(t).$$

We recognize $\mu_{11} + \mu_{22}$ as the moment of inertia of the rigid body relative to the $3''$-axis, $\tau_{12}(t)$ as the torque about this axis, and $D_2\alpha(t)$ as the angular acceleration about this axis.

In the case of 3-dimensional motion, it can be shown that (see [1], pp. 107–111) by identifying

$$W(p, t) = \theta^*(p, t) D_{2,2} \theta(p, t)$$

as the angular velocity matrix and

$$\bar{\omega}(p, t) = \langle W_{2,3}(p, t), W_{3,1}(p, t), W_{1,2}(p, t) \rangle$$

as the usual angular velocity vector, axiom (D57–3) leads directly to Euler's equation for the motion of a rigid body.

Leaving aside the matter of constraints and special laws, we can say very roughly, that the empirical claim of classical rigid body mechanics is this. Given any description of the motion of a rigid body as the translational motion of an arbitrary point in the body and its orientation relative to a coordinate system of fixed orientation moving with the point, we can find some point in the body – its center of mass – such that if θ describes

IDENTITY, EQUIVALENCE AND REDUCTION   241

the orientation of the body relative to a coordinate system of fixed orientation moving with this point then: (1) there exist functions μ, $\bar{h}$, and $\bar{p}$ which satisfy (D57–3); and (2) there exists a g such that $\langle\{\rho\}, T, \bar{r}, g, h\rangle$ is a CPM. Still more roughly, we can say the claim is that there is some point in the body – the center of mass – whose translational motion can be made a model for CPM by the same force function that can be used to make the body's rotational motion, relative to this point, a model for (D57–3). This rough condition of the empirical claim of classical rigid body mechanics at least serves to give some idea of how the position functions relative to the center of mass could serve as theoretical functions. It should be noted that the "connection" between the non-theoretical description of the bodies' motion and the theoretical description is provided by (D55–3) and (D55–4–ii). These requirements, in effect, assure us that these are "alternative" descriptions of the same motion. It just happens that the theoretical description allows the motion to be neatly "separated" into translational and rotational components both of which obey relatively simple laws.

Let us leave aside, for the moment, the question of the constraints that belong in the core of classical rigid body mechanics. We can define a relation which appears likely to strongly reduce this theory to classical particle mechanics without explicitly considering these constraints. First, if the relation we define is to strongly reduce classical rigid body mechanics to classical particle mechanics, it must be a one–many mapping from some sub-set of M'_0 – in this case, the set of all models for 'is an RBM' – into M_0 – in this case, the set of all models for 'is a PM'. We begin to describe this relation by defining its counter domain. That is we single out a sub-set of the set of models for 'is a PM' *onto* which the relation we shall define maps its domain – some sub-set of the set of all models for 'is an RBM'. The counter domain of the relation we shall define is simply the set of all systems of particle mechanics in which the particles move in such a way that the distances between them remain constant in time. We shall call such a model for 'is a PM', 'a rigid particle mechanics' and determine the class of such models by defining the predicate 'is an RPM'.

(D58)   *x is an* RPM if and only if
      (1) $x = \langle P, T, \bar{s}, m, \bar{f}\rangle$ is a PM;
      (2) For all $p_1, p_2 \in P$; $t_1, t_2 \in T$,
          $|\bar{s}(p_1, t_1) - \bar{s}(p_2, t_1)| = |\bar{s}(p_1, t_2) - \bar{s}(p_2, t_2)|$.

1

The putative reduction relation $\mathscr{L}$ is defined by (D59). The intuitive content of this definition is essentially this. An RBM and an RPM stand in the $\mathscr{L}$-relation if and only if the following conditions are satisfied. First, the rigid body is identical with the set of particles in the particle mechanics system and the time intervals during which their motion is considered are identical (D59–3) and (D59–4). Second, the position in the rigid body whose motion we describe by $\bar{u}$ is coincident with one to the particles in P (D59–5). Actually, we do not need a requirement this strong. Merely making the distance between the position described by $\bar{u}$ and one of the particles a constant in time would be sufficient. However, (D59–5) is intuitively satisfying in that it lets us envision the vector $\bar{u}$ as actually being "anchored" to one of the particles comprising the rigid body.

The mass of the rigid body is equal to the sum of the masses of the particles (D59–6) and the position vector of the center of mass of the rigid body is identical with the center of mass of the system of particles (D59–7). There is some correspondence v between the forces on the particles and the forces on the rigid body such that the $v(p, i)$-th force on p at t is equal to the i-th force on the particle p at t (D59–8–i) and the point of application of this force on ρ is identical with the position vector of the particle p relative to the center of mass of the system of particles (D59–8–ii). There is a coordinate system, represented by θ_0, and a time t_0 such that the following is true. The moment of inertia $\mu(\rho)$ is equal to the sum of the "moments" of the particles about the center of mass of the particle system measured in the coordinate system represented by θ_0 (D59–9–i). The coordinate system fixed in the rigid body ρ which describes its angular position coincides with the coordinate system represented by θ_0 at time t_0 (D59–9–ii). And finally, the position vectors of all the particles in the particle system relative to the center of mass of the system, in the coordinate system $\theta(\rho, t)$ remain constant in time (D59–9–iii). Intuitively, this means that $\theta(\rho, t)$ may be thought of as a coordinate system with its origin at the center of mass of the particle system, "tied to" the particles and rotating with them.

(D59) $\mathscr{L}$ is a binary relation such that $\langle x, y \rangle \in \mathscr{L}$ if and only if:
(1) $x = \langle \{\rho\}, T, \bar{u}, \varphi, \bar{r}, \theta, g, \mu, \bar{h}, \bar{p} \rangle$ is an RBM;
(2) $y = \langle P, T', \bar{s}, m, f \rangle$ is an RPM;
(3) $\rho = P$;
(4) $T = T'$;

IDENTITY, EQUIVALENCE AND REDUCTION

(5) There is some $p \in P$ such that, for all $t \in T$, $\bar{u}(\rho, t) = \bar{s}(p, t)$;

(6) $g(\rho) = \sum_{p \in P} m(p)$;

(7) For all $t \in T$, $\bar{r}(\rho, t) = \frac{1}{g(\rho)} \sum_{p \in P} m(p)\bar{s}(p, t)$;

(8) There is a one–one mapping ν from $P \times I_+$ onto I_+ such that, for all $p \in P$ and $i \in I_+$, $t \in T$:
   (i) $\bar{h}(\rho, t, \nu(p, i)) = \bar{f}(p, t, i)$;
   (ii) $\bar{p}(\rho, t, \nu(p, i)) = \bar{s}(p, t) - \bar{r}(\rho, t)$;

(9) There is a 3×3, orthogonal matrix θ_0 and a $t_0 \in T$ such that:
   (i) $\mu(\rho) = \sum_{p \in P} m(p) L^*(p) L(p)$ where $L(p) = [(\bar{s}(p, t_0) - \bar{r}(\rho, t_0))]\theta_0^*$;
   (ii) $\theta(\rho, t_0) = \theta_0$;
   (iii) For all $p \in P$ and $t \in T$,
   $(\bar{s}(p, t) - \bar{r}(\rho, t))\theta^*(\rho, t) = (\bar{s}(p, t_0) - r(\rho, t_0))\theta^*(\rho, t_0)$

Does the $\mathscr{L}$-relation we have just defined strongly reduce classical rigid body mechanics to classical particle mechanics? Let us consider first the formal aspect of the reduction relation. What we want to show is that the $\mathscr{L}$-relation strongly reduces the core of classical rigid body mechanics to the core of classical particle mechanics. That is, we want to prove the following theorem:

(T36) If $H' = \langle M'_0, N'_0, r, M', C' \rangle$ is the core of classical rigid body mechanics and $H = \langle M_0, N_0, r, M, C \rangle$ is the core of classical particle mechanics, then $\mathscr{L}$ strongly reduces H' to H.

Now, (T36) would follow as an immediate consequence of (T33), provided we could show that $\mathscr{L}$ strongly reduced the "null expansions" of the cores H' and H. It is trivial to see that $\mathscr{L}$ satisfies conditions (D52–1) and (D52–2). It should be noted, however, that though $\mathscr{L}$ is one–many, it is not one–one. It is possible that more than one assignment of position masses and forces to the particles in P have the properties demanded by (D59–3) through (D59–9).

To see that $\mathscr{L}$ satisfies (D53–2) suppose $x \in \langle \{p\}, T, \bar{u}, \varphi, \bar{r}, \theta, g, \mu, \bar{h}, \bar{p} \rangle$, $y \in \langle P, T', \bar{s}, m, \bar{f} \rangle$ and $\langle x, y \rangle \in \mathscr{L}$. Then $r(x) = \langle \{p\}, T, \bar{u}, \varphi \rangle$ and $r(y) = \langle \{p\}, T, \bar{s} \rangle$. Let $y' = \langle \{p\}, T, \bar{S}, x', \bar{f}' \rangle$ be any extension of $r(y)$. First, note that y' is an RPM since y is an RPM and $\bar{s}$ is the same in both y and y'. Next let:

$$g'(\rho) = \sum_{p \in P} m'(p)$$

$$\bar{r}'(\rho, t) = \frac{1}{g'(\rho)} \sum_{p \in P} m'(p)\bar{s}(p, t),$$

and let

$$\bar{h}(\rho, t, \nu(p, i)) = \bar{f}'(p, t, i)$$
$$\bar{p}'(\rho, t, \nu(p, i)) = \bar{s}(p, t) - \bar{r}'(\rho, t)$$

where ν is any one–one mapping from $P \times I_+$ onto I_+. Also let θ' be any function from $\{\rho\} \times T$ such that, for all $t, t' \in T$, $\theta'(\rho, t)$ and $\theta'(\rho, t')$ are 3×3, real orthogonal matrices such that, for all $p \in P$,

$$(\bar{s}(p, t) - \bar{r}'(\rho, t))\theta'^*(\rho, t) = (\bar{s}(p, t) - \bar{r}'(\rho, t'))\theta'^*(\rho, t'),$$

and let

$$\mu'(\rho) = \sum_{p \in P} m'(p) L'^*(\rho) L'(\rho)$$

where

$$L'(\rho) = (\bar{s}(p, t) - \bar{r}'(\rho, t))\theta'^*(\rho, t).$$

It is almost obvious that $x' = \langle \{\rho\}, T, \bar{u}, \varphi, \bar{r}', \theta', g', \mu', \bar{h}', \bar{p}' \rangle$ is an RBM. The only thing to note is that the fact that (D59–5) assures us that the position described by $\bar{u}$ coincides with a particle in P, and the fact that all the particles in P keep a fixed distance from one another, assure us that x' satisfies (D55–3) and (D55–4–ii). Further, the method of constructing x' assure that (D55–6) through (D55–8) are satisfied. Thus x' is an extension of $r(x)$ and $\langle x', y' \rangle \in \mathcal{L}$.

The essential idea in the above construction is this. The non-theoretical functions in RBM's, $\bar{u}$ and φ are not related, via $\mathcal{L}$, to any theoretical function in the corresponding RPM's. Indeed, they are only related at all by (D59–5). Thus, we can arbitrarily change the theoretical functions which are associated with $r(y)$, use the relations (D59–6) through (D59–9) to crank out corresponding theoretical functions for RBM's, and add these to $r(x)$ to obtain an extension of $r(x)$ that corresponds to an arbitrary extension of $r(y)$. Had the non-theoretical functions in the RBM's been correlated with theoretical functions in the RPM's this might not have been possible.

The fact just noted provides with some justification for augmenting the Adams-Rubin axiomatization of rigid body mechanics with the functions $\bar{u}$ and φ, and taking them to be non-theoretical functions. The models for the Adams-Rubin axiomatization have essentially the form:

$$\langle \{\rho\}, T, \bar{r}, \theta, g, \mu, \bar{h}, \bar{p} \rangle$$

when restricted to one-body systems. The primitive symbols have essentially the same meaning as is given to them by our (D55). In this axiomatization virtually the only, even plausible, choice for non-theoretical functions is $\bar{r}$ and θ. If we make this choice, *and* if we believe that $\mathscr{L}$ (with (D59–5) deleted) is intuitively the obvious reduction relation, then we are in trouble. For, $\mathscr{L}$ does not, in this case, satisfy (D53–2). To see this, suppose that $x = \langle \{\rho\}, T, \bar{r}, \theta, g, \mu, \bar{h}, \bar{p} \rangle$ and $y = \langle P, T', \bar{s}, m, \bar{f} \rangle$ correspond via $\mathscr{L}$. Let $y' = \langle P, T, \bar{s}, m', \bar{f}' \rangle$ be an arbitrary extension of $r(y)$. Now *any* x' corresponding to y' must have an $\bar{r}'$ function such that

$$\bar{r}'(\rho, t) = \frac{\sum_{p \in P} m'(p) \bar{s}(p, t)}{\sum_{p \in P} m'(p)}.$$

But, in general, this $\bar{r}'$ will not be identical with $\bar{r}$ and thus x' will not be extension of $r(x)$. Thus there may be extensions of $r(y)$ which do not correspond, via $\mathscr{L}$, to any extensions of $r(x)$.

At this point, it should be noted that $\mathscr{L}$ fails to satisfy the parallel of (D53–2) for the reduced theory. That is, even if $\langle x, y \rangle \in \mathscr{L}$, there may be an extension of $r(x)$ which fails to correspond to some extension of $r(y)$. To see this, suppose we add arbitrarily chosen μ and g-functions to $r(x)$. There is no guarantee that we can find *any* m-function to add to $r(y)$ so that both (D59–6) and (D59–9–i) are satisfied. Remember that we can not change the position functions in $r(y)$; we can only add theoretical functions. That we can not find such an m-function is a reflection of the fact that μ and g are independent, and apparently unrelated, in our axiomatization of classical rigid body mechanics. Indeed, one is initially inclined to feel somewhat uneasy about this feature of the axiomatization. We do feel that there ought to be *some* relationship between these two concepts. Intuitively, it seems odd that all arbitrary choices of μ and g-functions can be added to $r(x)$ and produce a model for RBM. But, on our account of the matter, the relation between μ and g is an empirical one, and not one that should be reflected in the structure of RBM's. Rather, it appears in the fact that all *intended applications* of rigid body mechanics *in fact* correspond, via a reduction relation having properties (D59–6) and (D59–9–i), to an intended application of particle mechanics. If we assume that particle mechanics can be successfully applied to its range (or all admissible ranges) of intended applications, then "bastard" extensions of rigid body systems *need* never be used to satisfy the empirical claims of that theory and we can say, roughly, that they "do not occur in nature". In this sense, the reduction shows us that only certain combinations of total mass and moment of inertia functions will occur.

It should be noted too that we could attempt to restrict our definition of 'is a RBM' to rule out bastard configurations of μ and g-functions. Doing this would exhibit directly in the basic mathematical structure of the theory the relation between these functions that is imposed by the reduction relation.

However, it is doubtful that the syntactical apparatus needed for such a restriction would ever appear to be intuitively independent of the reduction relation. That is, there do not seem to be any way of expressing the relation between μ and g except to say roughly that there is some configuration of "particles" such that (D59-6) and (D59-9-i) are true.

To complete the proof that $\mathscr{L}$ reduces the null expansion of the core of classical rigid body mechanics to the null expansion of the core of classical particle mechanics, we need to show the following. If $\langle x, y \rangle \in \mathscr{L}$ and $y \in \mathscr{P}(M) \cap C$, then $x \in \mathscr{P}(M') \cap C'$. In his discussion the reduction of rigid body mechanics to particle mechanics, Adams proves essentially the following theorem ([1], pp. 148–166):

(T37)   If $\langle x, y \rangle \in \mathscr{L}$ and y is a CPM, then x is CRPM.

The proof of this theorem is not difficult; however, it is rather long and tedious. For these reasons, it will not be repeated here. Using Adams' result, we now have the following. If $\langle x, y \rangle \in \mathscr{L}$ and $y \in \mathscr{P}(M)$, then $x \in \mathscr{P}(M')$. It remains only to be shown that, if $\langle x, y \rangle \in \mathscr{L}$ and $y \in C$, then $x \in C'$.

Now, at this point in the discussion, we can no longer afford to ignore the question of just exactly what constraints belong in the cores of the theories under consideration. Unfortunately, this is a question that cannot be answered in an entirely satisfactory way without a more complete logical reconstruction of these theories than has been provided by our discussion. However, it does seem at least plausible to claim that the only constraints appearing in the core of classical particle mechanics are roughly the following: (1) particles have the same mass values when they

IDENTITY, EQUIVALENCE AND REDUCTION 247

appear in different applications; (2) mass is extensive with respect to particle concatenation; and (3) forces acting on the same particle, at the same time, in different applications have the same values. Further, it seems plausible to think that the total mass function g in classical rigid body mechanics is constrained in essentially the same way as the mass function in particle mechanics. Also, it appears likely that the $\bar{h}$ and $\bar{p}$ functions are constrained in a way analogous to (3). That is, roughly, forces acting on the same body, at the same time, in different applications, have the same values *and* act at the same place in the body. Now, *if* we take this to be a correct account of the core constraints of the theories – if we suppose that all the remaining constraints on the force functions are associated with special force laws, then it is not difficult to see that (D59–6) and (D59–8) assure us that, if $\langle x, y \rangle \in \mathcal{L}$ and $y \in 0$, then $x \in C'$. Thus, provided we are willing to accept certain plausible assumptions about the constraints in the cores, we can show that the $\mathcal{L}$-relation we have defined strongly reduces the core of classical rigid body mechanics to the core of classical particle mechanics.

What about the "applied" aspect of the reduction of theories? (D54–B–2) roughly requires that, for every set of rigid body systems to which we might try to apply classical rigid body mechanics, there must be some "corresponding" set of particle systems to which we might apply classical particle mechanics. 'Corresponding' here means that every rigid body in the set of rigid bodies must correspond, via $\mathcal{L}$, to some particle system in the set of particle systems. In view of the obscurity that surrounds the question of exactly what is to count as a "particle" (see Chapter VI), it seems very unlikely that a satisfactory answer can be given to the question of whether this requirement is satisfied. Intuitively the requirement is that every rigid body *is* composed of particles. Whether this is true surely depends, to some extent, on what one is willing to count as "a particle". Indeed, one might conceivably feel that our conception of what a particle is could stand quite radical revisions in order to preserve the reducibility of rigid body mechanics to particle mechanics. Roughly speaking, almost *any* way of "regarding" a rigid body as a set of particles which satisfied the reduction relation *and* which did not falsify the empirical claim of particle mechanics would probably be acceptable. Still more roughly, we think that both classical rigid body mechanics and classical particle mechanics can be used to make tenable empirical statements, *and* we think the former theory is reducible to the latter by some *specific* reduction relation. We are willing to make repeated revisions in the way we carve up the world into rigid bodies and particles to maintain these claims.

NOTES

[1] Note that we might also be inclined to say that the theoretical functions had been Ramsey eliminated in the strong sense from the theory whose core is H since H' can be used to make any statement that H is used to make, regardless of what I is.

[2] Here, and in the remainder of this section, I rely heavily on the work of Jamison [19].

[3] In (D44) in addition to the notation explained in connection with (D7), we use the following notation. If f is a real or vector valued function of n real variables and $i \in I_n$, we denote the partial derivative of f with respect to its i-th argument at a point $\langle x_1, \ldots, x_n \rangle$ in its domain by

$$D_i f(x_1, \ldots, x_n).$$

We denote the function whose value at every $\langle t_1, \ldots, x_n \rangle$ where $D_i f(x_1, \ldots, x_n)$ exists is $D_i f(x_1, \ldots, x_n)$ by '$D_i f$'. In this notation

$$D_j D_i f(x_1, \ldots, x_n)$$

denotes the value of the j-th partial derivative of $D_i f$ at $\langle x_1, \ldots, x_n \rangle$. If $h_1, \ldots, h_n$ are real valued function of one real variable, then we have

$$D_i f(h_1(t), \ldots, h_n(t))$$

to denote the value at t of the derivative of the function g defined by the equation

$$g(t) = f(h_1(t), \ldots, h_n(t)).$$

We denote the set of all real numbers by R and the set of ordered n-tuples of real numbers by R^n.

[4] It should be noted that our previous usage of the word 'constraint' is in no way related to the usage mentioned here.

[5] Actually, it is somewhat more common to refer to the system of equations that result from (D45-2) in the special case where the generalized forces can be derived from a scalar potential function as 'Lagrange's equations'.

CHAPTER VIII

THE DYNAMICS OF THEORIES

In this final chapter, I shall attempt to bring together and elaborate upon several points that have been "mentioned in passing" in the previous discussion. Most of these points have to do with what I have called 'the dynamic aspect of physics theorizing' – that is, with the way theories of mathematical physics grow and develop in time. Some of these questions were discussed near the end of Chapter IV and were mentioned again in discussing the identity conditions for theories of mathematical physics in the preceding chapter. We shall begin the discussion by attempting to say more about the nature of the characteristic range of intended applications in a theory of mathematical physics. This will provide something closer to necessary and sufficient conditions for being a theory of mathematical physics than was provided in the previous chapter. Then we shall consider in some detail the question of what it is for a person to have a theory of mathematical physics. This will provide an account of the characteristic ways that a person's beliefs may change while he still has the same theory. In doing this we will examine some objections to our account of the identity criteria for theories of mathematical physics. In particular, we shall examine a certain view about the genesis of such theories which suggests that this view requires modification. Finally, we shall have something to say about how people come to have theories of mathematical physics, how they cease to have them, and how the conceptual apparatus in theories once held is related to the conceptual apparatus in theories held subsequent to them.

To begin, let us see if we can say more about the properties of the set of intended applications I in the theory of mathematical physics $\langle H, I \rangle$. In the preceding chapter we stated necessary conditions for $\langle H, I \rangle$'s being a theory of mathematical physics. These conditions dealt primarily with the formal, mathematical structure of the theory H. Of I, we required only that it be a set of partial possible models for the basic mathematical structure of the theory, i.e. we required that $I \subseteq N_0$. We have indicated previously that the set I will probably not be very precisely described in the existing expositions of the theory $\langle H, I \rangle$ and have suggested that this may be something more than an "accidental" fact about the way theories of mathematical physics are expounded. But, before examining this aspect of mathematical physics in detail, let us suppose that we have succeeded in obtaining a precise characterization of I and consider the properties it must have if $\langle H, I \rangle$ is to be a theory of mathematical physics. Our aim will be to add more necessary conditions to (D38) – necessary conditions involving I – that will all together provide necessary and sufficient conditions for a theory of mathematical physics. It is important to understand that in giving these conditions we are no longer restricting ourselves to the purely semantic aspects of mathematical physics – what sorts of statements mathematical physicists make in their professional capacity. We are now explicitly concerned with what statements of this sort might be regarded as "interesting", "fruitful" or in some other way worthy of serious attention. Similar "pragmatic" considerations appeared in our discussion of the special status of constraints on theoretical functions and theoretically expanded cores in the preceding chapter.

The most obvious insufficiency in (D38) is the failure to require that the members of I be sets of physical systems, in some appropriate sense of this term. I have previously confessed to having almost nothing to say about what counts as a physical system. At this point, I should like to make clearer the sense in which I have nothing to say. In doing this, I shall go as far as I can in removing *this* insufficiency in (D38).

First, it should be noted that I have been using the term 'physical system' in a way that entails that some physical systems, at least, are the sorts of things that could be models for N_0. That is, my usage entails that some physical systems are sets of individuals *together with* numerical functions on these individuals. One might contend that it would be closer to ordinary usage to regard just the domains of these models as physical systems. Whatever customary usage might dictate, I think this is a slightly misleading way to employ the term 'physical system'. It is misleading in that it tends to obscure a rather important fact about the way theories of mathematical physics work. The fact I have in mind is very roughly this. Theories of mathematical physics deal with a very precisely circumscribed, and perhaps very limited, range of properties of the individuals that appear in the applications of the theory. For example, of all the properties of the individuals that appear in applications of classical particle mechanics, this theory deals with the property that is measured by the position function and ignores the rest. To explicitly point out this exclusiveness, it seems appropriate to say that the entities that this theory is "about" are individuals-cum-position-functions rather than simply individuals. I do not want to maintain that entities of this sort are the *only* things that it is appropriate to call 'physical systems'. For example, it might be appropriate to call models for certain relational structures with no numerical functions 'physical systems'. But, so far as theories of mathematical physics go, I believe that individuals-cum-numerical-functions are the sort of entities that these theories are 'about'.

This view about the nature of physical systems that may be applications of theories of mathematical physics is not *completely* neutral about the nature of the individuals that can be in the domains that are parts of these systems. Roughly, they must be the sort of individuals that have properties that admit of quantitative measurement. There are well known accounts of what these properties must be like (see Chapter II, pp. 17–23). To my knowledge, the individuals that may have such properties do not define nor happen to be coextensive with, any "interesting" ontological category. Some philosophers (e.g. N. R. Campbell) have apparently believed the possession of relational properties that admit of quantification to be characteristic of "physical objects". However, the widespread introduction of quantitative theories into the behavioral sciences has rendered this view untenable. Thus, the precise sense in which I have nothing to say about what may be a physical system is this. I have nothing to say about the sorts of entities that may appear in the domains that are in physical systems.

Despite the rather minimal nature of my claim about the sorts of things that are physical systems, it is not entirely without interesting implications. For example, it illuminates at least one sense in which it may be said that the facts accounted for by theories of mathematical physics always presuppose the acceptance of some theory. Before we can even begin to theorize using the core $H = \langle M_0, N_0, r, M, C \rangle$, we must have some reason to believe that we have some physical models for N_0 lying around. There are essentially only two ways we could be convinced of this. First, we could be convinced that some individuals are models for some relational structure that yields functions having the properties required to produce models for N_0. Roughly, one might say that these functions were "theoretical" with respect to certain relational properties (see Chapter IV, pp. 86–89).

This appears to be a natural way to regard the partial possible models for classical particle mechanics. Second, we could be convinced that a claim (of form (5)) of some *other* theory of mathematical physics was true, and this claim could entail that, in certain applications, some theoretical functions of this theory had the properties required to produce models for N_0. Thus, it appears that we may regard the individuals in some applications of rigid body mechanics to be partial possible models for classical thermodynamics because a claim of rigid body mechanics entails that the pressure function in this application has the requisite properties. It should be evident that both these ways of convincing ourselves that we have models for N_0 require that we believe what amounts to a claim of some theory. In the first case, the theory *may be* an intuitive and not explicitly formulated belief that some relation has certain properties. In the second, it is a claim of the sort characteristically associated with theories of mathematical physics.

This rough account of how the non-theoretical functions in a particular theory of mathematical physics are related to theoretical functions in other theories might raise the following question. How can we be certain that this is the same function appearing in both theories? This question can be taken in two ways. It might mean: 'How is someone providing a logical reconstruction of a particular theory certain that a non-theoretical function in some intended application of this theory is the same as a theoretical function in some intended application of another theory?' The answer here is simply that he cannot be "certain" of this. This is an empirical claim about the theory he is investigating – a claim about the statements believed by people who propound this theory. At best, one can hope to find evidence for the claim in existing expositions of the theory. On the other hand, the question might be: 'How is one who propounds a particular theory certain that a non-theoretical function in some intended application of this theory is the same as a theoretical function in some intended application of another theory?' The answer here is that this is just the way he identifies, or describes, this intended application. There might, of course, be some difficulty for him in deciding whether some list of ordered pairs of individuals and numbers *is* the function he described in this way. But this is in principal no different from my claiming that Jones' oldest son is red-headed and then encountering some difficulty in deciding whether some particular man is, in fact, Jones' oldest son.

What has just been said can be regarded as just a restatement of the truism that the enterprise of mathematical physics cannot even begin until the raw data has been put in quantitative form. It is at this point that we must begin to look at the phenomena "as if" they were models for abstract mathematical structures. Typically, the assumptions that one must make about "pre-theoretical" relational properties such as 'is-longer-than' in order to guarantee the existence of numerical functions needed for mathematical physics already entail a significant amount of idealization. For example, the development of the concept of instantaneous velocity to be employed in describing the motion of bodies represents, in itself, a tremendous intellectual advance. To the completely uninitiated, the gap between the everyday concepts employed to describe motion and a twice differentiable position function probably appears much greater than the gap between this concept and the concepts of mass and force. Indeed, the relational properties that underlie the position function must be assumed to have rather subtle properties in order to assure the existence of this function. It is not even clear that our everyday concepts of 'is-longer-than' and 'is-later-than' have these properties. In the strict sense of 'theoretical' defined in Chapter II, functions like mass and force are theoretical with respect to functions like position which may be "theoretical", in a closely related sense (see Chapter IV, pp. 86–89), with respect to certain relational properties. Yet, even these properties will typically be abstractions, and in this sense "theoretical" with respect to the everyday concepts from which they arise.

Besides requiring that the members of I be physical systems, in the sense just explained, what else can we say about I? Remember that theories of mathematical physics, on our account, typically function by permitting us to make claims that "tie together", via constraints on the theoretical functions, values of the non-theoretical functions in various different intended applications. If there were not some reasons to think that the functions in different members of I "ought" to be tied together in this way, then there would be very little reason to seek the sort of explanation of I's behavior that a theory of mathematical physics can provide. This suggests, at least, that the members of I must be such that their domains overlap in ways that would permit a significant use of the device of constraining the values of theoretical functions.

For example, consider an I that could be partitioned into two sub-sets such that no domain in a member of one sub-set shared members with the domain of any member of the other sub-set. Clearly there would simply be no point in including *all* of *I* within the scope of *one* claim of form (5). Essentially what constraints acting across intended applications do is this. They make it possible to use information obtained in one intended application to make predictions about another intended application. If there is no possibility of doing this throughout the whole of *I*, then there is very little reason to make a claim of form (5) for *I*. In this example there would be no way that the constraints could justify predictions across the gap between the two classes of domains. Of course it might still be of some interest to make separately a claim of form (5) for each of the "connected" sub-sets of *I* – be sub-sets that have suitably overlapping domains.

It must be emphasized that these considerations do not show that one *could not* employ the mathematical structure of a theory of mathematical physics to make a significant empirical claim about some *I* of the sort just described. There are no *semantic* reasons why this could not be done. What has been claimed is that there do appear to be convincing *pragmatic* reasons why this *is not*, in fact, done.

Having agreed that *some* overlap is essential among the domains of the numbers of *I*'s that are amenable to "explanation" by theories of mathematical physics a natural question is: "How much overlap?". Surely we want to rule out the situation just described in which *I* can be divided into two sub-sets whose domain intersect among themselves, but never intersect with those of members of the other set. On the other hand, requiring too much overlap would appear to rule out *I*'s that are intuitively unobjectionable.

Let us consider some possible proposals for specifying the degree of overlapping that ought to be present among the domains of the members of *I*. Let *J* be the set consisting of just the domains of members of *I* and $\bar{D}$ be the union of all the members of *J*. Suppose we require that *I* be such that, if $D \subseteq \bar{D}$ then $D \in J$. That is *I* must be such that *any* set of individuals chosen from the domains of members of *I* produces the domain of another member of *I*. This clearly rules out the objectionable example. But it also rules out much more. Surely not *all* sets of particle–time pairs chosen from the range of applications of classical particle mechanics form the domain of another such application. Indeed, most of them will not even be partial possible models for this theory since the times will not be an interval. Even if we succeeded in taking the particles themselves as the individuals, this principle does not appear to fare much better when we recognize the wide variety of things that may be taken as particles. It does not appear that we consider applications in which both a "composite" particle (with its position taken as the position of its center of mass) and one of its component particles appear in the same domain. One might consider weakening this requirement by stipulating only that every pair of individuals chosen from the domains of members of I appear together in some domain in a member of I. This too rules out the objectionable situation, but it falls victim to the same counterexamples as the initial proposal. One possibility for ruling out objectionable situations of the sort mentioned which survives these counterexamples is the following. I must be such that, for all $A, B \in J$, there exists $D_0, D_1, \ldots, D_n$ such that $D_i \in J$, $D_i \cap D_j \neq \Lambda$ and $A = D_0, B = D_n$. That is there is some sequence of overlapping domains "connecting" any pair of domains of members of I.

It is worthwhile formulating the last proposal a bit more precisely. To this end, consider the following.

(D60) If J is a set of sets then, for all $A, B \in J$, A is *connected* with B if and only if there exist $D, D_1, \ldots, D_n$ such that, for all $i, j \in I_n$, $D_i \in J$, $D_i \cap D_j \neq \Lambda$ and $A = D_0, B = D_n$.

Now note that the 'is connected with' relation is an equivalence relation on J. Thus this relation partitions J into sub-sets all members of which are connected with each other, but not connected with any member of any other of these sub-sets. The requirement we are considering simply says that the partitioning that this relation produces on the set of domains of members on I must be trivial – that is, the partition of J consists only of J. Viewed in this way, it is clear that this requirement rules out our objectionable example and other situations similar to it. Intuitively, what this requirement rules out is the possibility that the domains in J can be partitioned into sub-sets across the boundaries of which constraints can have no influence. Moreover, there is a sense in which this is the weakest requirement that could effect this. It requires the minimal amount of sharing of elements, for a given set *could be* connected to every other set in J and yet share only one individual with another set. For this reason it appears plausible to *require* at least the amount of overlap specified by this condition. Though it may be that more is actually present in most I's.

Thus far we have been considering situations in which constraints *cannot* operate significantly. Are there also situations in which constraints *could* operate, but where they *should not* operate? That is, are there situations in which we would feel some justifiable reluctance in seeking an explanation of the sort typically provided by theories of mathematical physics, even though there was no reason to expect that constraints could not operate significantly.

Consider, for example, an I that contains only partial possible models for the mathematical formalism of hydrodynamics – some of these partial possible models being systems of fluids flowing in pipes and others being electric circuits. Intuitively, it appears that we would not expect the *whole* of I to be a candidate for a hydrodynamical explanation. Why is this so?

A possible answer might run like this. Whatever the individuals in these models might be, it seems reasonable to expect that there are going to be individuals that are "parts" of these models in the sense that parts of one model might also be parts of another model. That is, electric circuits are the sort of thing that one can build up out of component parts (the individuals in the model): resistors, capacitors, etc. . . . It makes sense to take components out of one circuit and put them into another. The same is true of systems of fluids flowing in pipes. This is to say that the domains of the partial possible models that are electric circuits might be expected to overlap each other, as might the domains of these that are systems of fluids. But, it might be argued, what we cannot do is take a part of an electrical circuit and combine it with a part of a system of fluid flow to produce a new physical system of either of these kinds. That is, a domain in one of these classes of partial possible models for hydrodynamics could not be expected to share individuals with a domain in the other class. Thus we have an example of the type of situation we have just described in which the device of constraints on theoretical functions *could not* function significantly.

But is this really an example of this sort? Is it really plausible to expect that the domains in the partial possible models that are electric circuits will not overlap, in the appropriate way, with the domains in those that are systems of fluid flow. Closer examination reveals that this claim is not so plausible. *Some* pipes are, after all, the sorts of things that may conduct electricity. There is no reason to expect that some components of systems of fluid flow might not also turn up as parts of electric circuits. The real force of this example appears to be somewhat more subtle. It appears that we would *resist* the imposing of constraints across these two classes of partial possible models, not simply that it would be impossible to impose constraints because the domains do not overlap. Why do we resist imposing such constraints? Intuitively, the answer seems to be simply that there is no reason to expect that the properties that an individual has when it is a component of an electrical circuit have anything to do with the properties that it has when it is a component of a system of fluid flow.

Another way of describing the situation illustrated by this example is the following. We have two sets we believe to be partial possible models for the basic mathematical structure of the theory – two sets we believe to be models for N_0. The individuals in the domains of these models may overlap in a way that makes possible a significant use of constraints on theoretical functions. What differentiates these two sets of models is the underlying theory we appeal to in justifying our claim that they are, in fact, models for N_0. We use different, albeit perhaps not precisely formulated, theories in arguing that electrical circuits are models for the appropriate mathematical structure and in arguing that systems of fluid flow are models for the same structure. Is this fact alone sufficient to explain why we, intuitively, do not expect the values of functions in one of these classes of models to be related to the values in the other? More precisely, is a necessary condition for I's demanding an explanation provided by using the core $H = \langle M_0, N_0, r, M, C \rangle$ that our belief that $I \subseteq N_0$ be supported by appealing to some single underlying theory?

We can get some insight into the answer to this question by considering the theory of length discussed briefly in Chapter IV (pp. 86–89). Let us suppose that a set of relational structures J all members of which are models for (ES1)–(ES3) is the largest successful application of this theory known. That is, we know of no way to add more models of (ES1)–(ES3) to J and still have a single extensive length function on all the individuals in the domains in J. Intuitively, this means that we know of no additional models for (ES1)–(ES3) in which it is possible to regard R and O as determining the same property that they determine in the other members of J. Were we initially, on intuitive grounds, to think that they did, a failure to produce an h-function satisfying $\langle =, = \rangle$ for this enlargement of J would convince us that we were in error.

Now suppose that H is such that N_0 is characterized by the predicate 'is an extensive scale system' defined by (ES). That is, suppose the partial possible models for the theory which the theory of length underlies are just extensive scale systems. Now let I be some set of extensive scale systems produced by adding a set of h-functions to the members of J which satisfies the constraint $\langle =, = \rangle$. Clearly no set larger than I could be claimed to be a sub-set of N_0 by appealing to the theory of length. Now suppose we had good reason to believe that there were some other models for N_0 whose domains overlapped in the appropriate way with the domains in members of I. What would be wrong with adding these to I? One obvious difficulty is this. In the domains in I that overlapped with these additional domains there would be some ambiguity as to the values the h-function ought to have for the shared members since, by hypothesis, there is no way to make the h-functions in the enlarged set satisfy $\langle =, = \rangle$. One could resolve this by stipulating that h-function values in I were to be determined by the theory of length and those in the additional models were to be determined by whatever theory it is that gives us reason to believe they are models for N_0. The same individual might have a different value of the h-function depending on the model in which it appeared. But, having done this, it is difficult to see what grounds we might have for expecting some claim with constraints on the theoretical functions operating across the entire enlarged range of applications to be true. For the "cash value" of this claim is going to be that there are certain (typically rather complicated) relations holding among the h-function values in different applications. But it appears that we have already implicitly denied that $h(a)$ in an application in I can be dependent upon $h(a)$ in an application outside I.

It is clear that these considerations do not conclusively establish, even for this special case, the claim that a necessary condition for I's demanding explanation by H is that $I \subseteq N_0$ be supported by appealing to a single underlying theory. At best, this discussion merely makes this claim plausible. However, without some refinement, even its plausibility might be questioned.

One objection to the claim runs as follows. What about a situation in which the models for N_0 contain two functions and we appeal to a different underlying theory in claiming that each of these functions, in the members of I, have the properties required of models for N_0? Thermodynamics might be an example of this situation, depending upon whether or not temperature is treated as a theoretical function. I suspect that it should be treated as a theoretical function, but *if* it were not, the situation would be roughly this. The mechanical functions (e.g. pressure, volume) would be claimed to have the appropriate properties by appealing to some branch of mechanics, while the claim for the temperature function would rest on some theory of empirical temperature analogous to the theory of length.

Here there is no *single* underlying theory that shows $I \subseteq N_0$, but rather a single theory *for each function* in the members of N_0.

One might also object that this necessary condition rules out the possibility that two apparently unrelated underlying theories might be "unified" by incorporating them in a single theory of mathematical physics. One way this might happen was suggested in the previous paragraph, and apparently this necessary condition must be modified to allow for such a possibility. But, one might contend that there are still other possibilities for unifying underlying theories that are ruled out by this condition. For example, might it not be that two theories, analogous perhaps to the theory of length, were thought to be different *until* it was shown that they could be successfully incorporated into a single range of intended applications for some theory of mathematical physics? Only then was it discovered that these were "really" theories about the same property – in the sense that there was some single function that could be regarded as measuring this property of individuals in all situations where these individuals appeared. Perhaps no one had even bothered to consider the possibility that there might be such a function before these underlying theories were incorporated into a single theory of mathematical physics. Surely there are no *a priori* grounds for ruling out such a possibility. Yet, I am not aware of a "real life" example of this situation.

In summary, it appears plausible to say that the members of I must process some kind of "coherence" that goes beyond the overlapping of their domains. Intuitively, I must consist of the same kind of physical systems. That is, it must consist of physical systems that can be "naturally" grouped together – whose behavior "ought" to be accounted for within a single conceptual framework. It is not too surprising that the full content of the "principle of collection" that justifies our grouping together the members of I cannot be explained solely by the way individuals are shared among the domains of members of I. It is also pretty clear that what makes us take two physical systems to be systems of the same kind has something to do with the fact that we believe the same underlying theory, or theories, to be true of them. But, it is difficult to be more precise than this. Again, this should not be surprising. We have already noted that, in the case of a specific theory, e.g. classical particle mechanics, it is very difficult to say with exactness what is in the range of intended applications. This may be so, even though we are quite certain that some particular physical systems are in this range. Given that this difficulty arises for all specific theories, it is not surprising that it appears difficult to provide a general account of how members of the range of intended applications for theories of mathematical physics are collected together. There is a bit more to be said on this point, but it appears expedient to defer this until we discuss, in more detail, what it is for a person to have a theory of mathematical physics.

We can now summarize the discussion of I by stating necessary and sufficient conditions for a theory of mathematical physics. To do this it is necessary to use some admittedly vague and questionable notions like 'is a physical system' and 'is a physical system of the same kind as'. This means that the claim that the stated conditions are sufficient is not very informative. The notions employed in stating the conditions are not obviously less problematic than the notion being analyzed. But the claim is not entirely vacuous. It could be wrong in claiming that the notion of a theory of mathematical physics can be analysed in *just* these terms and no others. At any rate, the following expansion of the necessary conditions of (D38) serves to bring the results of our preceding discussion more clearly into focus.

(D61) x is a *theory of mathematical physics* if and only if there exists an H and I such that:
(1) $x = \langle H, I \rangle$;
(2) $H = \langle M_0, N_0, r, M, C \rangle$ is a core for a theory of mathematical physics;
(3) $I \subseteq N_0$;
(4) If $y \in I$, then y is a physical system;
(5) If J is the set consisting of all and only domains of members of I and $y, z \in J$, then y is connected with z;
(6) If $y, z \in I$, then y is the same kind of physical system as z.

It should be noted that (D61-5), in mentioning the domains of models for N_0, implicitly assumes that we have a characterization of a frame for a theory of mathematical physics like that given by (D25) rather than the more non-committal one given by (D26). I believe that it would be possible to carry through the present discussion without this assumption, but to do so would needlessly complicate matters.

We are now in a position to consider the question of what it is for a person to "have" or "accept" a certain theory of mathematical physics. In everyday discourse there are a number of things one might mean by saying that a person has a certain theory. To become clearer about just which one of these meanings we wish to single out for closer examination, it is expedient to consider a view of theories that we have already rejected as inadequate for theories of mathematical physics. If one regards a theory as simply a set of statements, then there are some very natural distinctions to be made about what it is to have such a theory. These distinctions will help to single out the notion of 'having a theory' that we want to illuminate for our, somewhat different, conception of what a theory is.

According to the "statement view", to have a particular theory is, at least, to *believe* that the statements comprising the theory are true. But, one might maintain that 'having a theory' means something more than this. For example, it seems clear that a person who has a theory must believe the statements of the theory for *good reasons*. That is, one might claim that completely groundless belief in the statements of the theory is simply not what we mean to attribute of a person when we say he has a theory. One might go even further and insist that the person believe the statements of the theory *for special sorts of good reasons*. For example, simply believing the statements of the theory because one believes them to be affirmed by a reputable authority is not to count as having a theory. To really have the theory, one must believe its claims for the same reasons that the reputable authority believes them. That is, one must see, in detail, how the available empirical evidence supports the claims of the theory; what the theory entails about yet to be examined situations; what modifications might be made in the theory were these predictions to be faulted; and perhaps many other things about the theory. It is this strong sense of 'having a theory of mathematical physics' that we want to single out for examination. Roughly, this is the sense in which competent, mature, professional mathematical physicists may be said to have the theories they promote in their professional capacity. It is to bring students to have in this sense certain theories of mathematical physics that a large part of graduate education in the physical sciences is directed.

We have already suggested that the statement view of theories is not adequate to account for this strong sense of 'having a theory of mathematical physics'. The major objection we noted to this view was that having a theory of mathematical physics seemed to be much more loosely tied to beliefs in particular statements than this view appears to require. We have suggested roughly that the empirical content of a theory of mathematical physics – the claim believed by those who have the theory – may be rendered as a single claim of form (5). However, the particular claim of form (5) that these people believe will typically change during the time that we would intuitively want to say that they had the same theory. Two ways were noted in which this might happen. First, additional applications may be brought within the range of intended applications in this claim and others, perhaps, deleted from this range. Second, further investigation of the range of intended applications might suggest the possibility of successfully postulating new special laws (perhaps with new constraints) or demonstrate the need to modify or reject special laws previously postulated. Considerations of this sort led us to reject the statement view of theories of mathematical physics and adopt a view in which the theory *is* a formal mathematical structure – the core – together with a characteristic range of intended applications. We can begin to become more precise about what is to have a theory of this sort by explaining how the pair $\langle H, I \rangle$ is related to the various different statements that might be believed, at different times, by those who have the theory.

Having identified a theory of mathematical physics as the ordered pair $\langle H, I \rangle$, the most obvious way to associate various statements of form (5) with the theory is this. We simply consider the various ways that the core H might be applied to the characteristic range of intended applications I. Each of these produced a statement of form (5). However, doing this accounts for only the second way in which we have intuitively noted that statements associated with the same theory may differ. It takes no account of the fact that the scope of the theory's claim is typically being enlarged by adding new applications. On the face of it, this appears to be an objection to *identifying* the theory with $\langle H, I \rangle$. One might think that, since the ranges of applications that appear at different times in *the* central empirical claim of the theory are different, it is impossible to find a single set of intended applications that is characteristic of the theory. On the other hand, the various set of intended applications that appear in the different claims of the theory surely share some common property that is "essential" to regarding them as statements of the same theory of mathematical physics. Something about the range of applications must appear in the identity conditions for the theory. Our solution to the problem is roughly this. We maintain that there is single characteristic range of intended applications that remains fixed throughout the history of the theory, while people's beliefs about what is actually in this characteristic range change. *The* empirical claim of the theory at a given time is, very roughly, some application of H to the largest explicitly listed set of physical systems that are believed to be in I. This view will be elaborated in detail below. For the present, however, let us consider only the fixed characteristic range of intended applications and concentrate our attention on the second of the ways that the statements of the theory might change.

The most natural way that the theory $\langle H, I \rangle$ could generate a number of statements is this. One could employ different expansions of H, say $\mathscr{E}_1, \mathscr{E}_2, \ldots, \mathscr{E}_i, \ldots$ and use each of them to make a statement about I of the form $I \in N_{\mathscr{E}_i}$. Doing this might make some sense in the case that the $N_{\mathscr{E}_1} \supseteq N_{\mathscr{E}_2} \supseteq \cdots \supseteq N_{\mathscr{E}_i} \ldots$. One could envision theorizing about I commencing with the claim that $I \in N_{\mathscr{E}_1}$ and progressing successively through the sequence of claims of this form generated by the sequence $\mathscr{E}_1, \mathscr{E}_2, \ldots$ $\mathscr{E}_i, \ldots$. We would begin by making a claim that placed the set of intended applications I in the set of sub-sets of N_0, $N_{\mathscr{E}_1}$. We would then proceed to "refine" this claim by placing I in successively smaller sets of sub-sets of N_0. Intuitively, we would begin by postulating special laws that are compatible only with I's being in $N_{\mathscr{E}_1}$. But this claim allows some latitude in the observable states of affairs that are compatible with it. The values of the non-theoretical functions in members of I could have any configuration that appears in $N_{\mathscr{E}_1}$. We might like to say something sharper about I. To do this we attempt to postulate special laws that further restrict the configurations of non-theoretical functions that would satisfy our claim. That is we attempt to postulate special laws that come closer and closer to being satisfied by just exactly the observed configurations of non-theoretical functions and no others. If we were doing this by postulating *theoretical* laws, we might expect each successive, stronger claim to afford more possibilities to determine values of theoretical functions and thus offer more opportunities to obtain a crucial check of the claim.

The situation just described might be regarded as an "ideal" picture of the development in time of the beliefs of a person who has the theory $\langle H, I \rangle$. It is a picture of steady progress toward a unique characterization of the behavior of the members of I. But, in other than the best of all possible worlds, one might envision something less than a monotone convergence on the truth. Mistakes might be made in the postulating of special laws. An attempt to produce an expansion of $H_{\mathscr{E}_{i+1}}$ such that $N_{\mathscr{E}_{i+1}} \subseteq N_{\mathscr{E}_i}$ may be unsuccessful simply because it turns out that I is not a member of $N_{\mathscr{E}_{i+1}}$. This is to say that at the time one considers $\mathscr{E}_{i+1}$ as a possible refinement of $\mathscr{E}_i$ the claim that $I \in N_{\mathscr{E}_{i+1}}$ is consistent with the available information about I. But subsequent investigation, turning up new information about values of non-theoretical functions in members of I, might reveal conclusively that one cannot fill in theoretical functions in the way that $I \in N_{\mathscr{E}_{i+1}}$ requires. Were this to happen, one *might* still be able to retreat to $I \in N_{\mathscr{E}_i}$, i.e. this claim might still be consistent with all the available data about I. Intuitively, we would be inclined to say there that the special laws postulated with $\mathscr{E}_{i+1}$ were revealed to be "false" by further investigation. This failure, however, need not shake our confidence that we can find *some* way of postulating special laws that improve upon $\mathscr{E}_i$. We can always try again until we come up with an expansion that improves on $\mathscr{E}_i$ and remains tenable in the light of further investigation about I.

Now, this suggests that a person who has the theory $\langle H, I \rangle$ must at least believe some statement of the form $I \in N_{\mathscr{E}}$ where $\mathscr{E}$ is an expansion of H. Indeed, he will typically believe several statements of this form. For, if someone believes that $I \in N_{\mathscr{E}_i}$ for some $\mathscr{E}_i$ in a sequence like the one we have described, then it is natural to expect that he will also believe the weaker claims made with earlier members of this sequence. Thus, if a person has the theory $\langle H, I \rangle$ at a time t, we might single out the strongest claim of the form $I \in N_{\mathscr{E}}$ that he believes at this time and call the expansion of H used to make this statement $\mathscr{E}_t$. Intuitively, the statement that $I \in N_{\mathscr{E}_t}$ may be thought of as *the* empirical claim that a person who has this theory $\langle H, I \rangle$ at time t makes with this theory at this time.

If we are after an account of the strong sense of a person having the theory $\langle H, I \rangle$ characterized earlier, it is clear that much more is involved in addition to simply his believing that $I \in N_{\mathscr{E}_t}$. The most apparent additional requirement is that the person in question must also have good reasons, indeed good reasons of a special sort, for believing that $I \in N_{\mathscr{E}_t}$. Roughly, what we have in mind is that this person believe that $I \in N_{\mathscr{E}_t}$ because this statement is consistent with the entirety of some non-empty collection of data about the values of non-theoretical functions in I that is available to him at t. That is, we want to require that the person actually have at t *some* beliefs about the values of non-theoretical functions in members of I; that $I \in N_{\mathscr{E}_t}$ be consistent with *all* such beliefs: and that his belief in $I \in N_{\mathscr{E}_t}$ actually be "occasioned" by its relation to this data. Clearly, this does not exhaust the requirements connected with what one might call 'the epistomological petigree' of our person's belief in $I \in N_{\mathscr{E}_t}$. For example, we would surely want to say something about how he got his putative data about I. Nor is this even near to an adequate treatment of the requirements that have been mentioned. I do not propose to venture further into this area and attempt to remedy these deficiencies. My reasons for stopping short are two. First, I do not believe that statements of the form $I \in N_\mathscr{E}$ raise any essentially new epistomological issues – issues beyond those traditionally discussed by philosophers. Second, I have nothing new to say about these traditional epistemological issues. Consequently, I propose to lump all epistemological requirements on a person's having the theory $\langle H, I \rangle$ into the requirement that he have "observational evidence" for his belief that $I \in N_{\mathscr{E}_t}$. In doing this, I admit the notion of 'observational evidence' into my account as an unanalysed primitive notion. I do this, not because I believe the notion to be unproblematic, but only because I have nothing to add to the solution of the problems connected with it.

Another aspect of theories of mathematical physics that are still "going concerns" – theories that some people still have in the strong sense – is this. There are still "problems" to be solved within the framework of the theory. There are still "new" applications of the theory to be worked out, and perhaps ways open to improve upon old applications. Frequently the latter problems simply amount to tidying up the mathematics or refining the experimental techniques used in an application. Beyond this, however, one who has the theory $\langle H, I \rangle$ at t and thus believes that $I \in N_{\mathscr{E}_t}$ would surely regard the discovery of some expansion of H, $\mathscr{E}$, such that $N_\mathscr{E} \subset N_{\mathscr{E}_t}$ and $I \in N_\mathscr{E}$ was consistent with his data about I as an improvement in his situation. Intuitively, this simply amounts to his discovering new special laws that accord with the data he presently has and further restrict the possible states of affairs that are consistent with the claim he makes with the theory thus allowing him to make sharper predictions about future observations.

Is it plausible to maintain that anyone who has a theory of mathematical physics, in the strong sense of 'have', must *always* expect that "progress" of this sort is possible? That is, can we regard the expectation that the conceptual apparatus of the theory is adequate to producing ever more specific predictions, and, perhaps, the "commitment" to attempting to use the theory in this way as necessary conditions for having the theory? Of course, in degenerate cases where $N_{\mathscr{E}_t}$ contains only one member, this would obviously not be a plausible requirement, but I know of no "real life" examples of such a degenerate case. In most instances it does appear that practitioners of a theory believe that, in at least some intended applications, the final word has not been said. In some cases they seek to discover new special laws that are directly relevant to accounting for the behavior of the application in which they are claimed to hold – for example, the force laws determining the motion of charged particles. In other cases they seek to discover special laws in one application that are relevant – via constraints that sometimes appear as "boundary conditions" in certain problems – to accounting for the behavior of some other application, for example, when dielectric properties of solids – special force laws holding in the solids – are postulated with the aim of determining the electromagnetic field in regions near them. It does appear that a significant part of the scientific activity carried on within the tradition of a given theory of mathematical physics can be viewed in this way. Those working within the confines of the theory seek to discover more restrictive special laws, usually theoretical laws, that can be regarded as holding in some already familiar applications, thus more uniquely specifying the possible configurations of non-theoretical functions in these, and perhaps other, intended applications. The way this activity is connected with the other way that theories "develop" – the apparent adding of new intended applications will be considered below.

The results of the discussion thus far can be summarized in a putative definition of what it is for a person to have a particular theory of mathematical physics at a given time. We shall shortly have occasion to reconsider parts of this definition, so it is best regarded as simply a summary of the results of our inquiry up to this point.

(D62) If p is a person and $\langle H, I \rangle$ is a theory of mathematical physics then p has $\langle H, I \rangle$ at time t if and only if:
(1) There is an expansion of H, $\mathscr{E}$, such that p believes at t that $I \in N_{\mathscr{E}}$;
(2) If $\mathscr{E}_t$ is an expansion of H such that, for all expansions $\mathscr{E}$ of H such that p believes at t that $I \in N_{\mathscr{E}}$, $N_{\mathscr{E}_t} \subseteq N_{\mathscr{E}}$ and p believes at t that $I \in N_{\mathscr{E}_t}$ then:
  (a) p has observational evidence at t that $I \in N_{\mathscr{E}}$;
  (b) p believes at t that there exists an $\mathscr{E}$ such that $I \in N_{\mathscr{E}}$ and $N_{\mathscr{E}} \subset N_{\mathscr{E}_t}$.

(D62–1) just requires a person who has the theory $\langle H, I \rangle$ to believe at least one of the statements associated with this theory. (D62–2) characterizes the strongest statement associated with $\langle H, I \rangle$ that the person believes at time t – the statement that $I \in N_{\mathscr{E}_t}$. It further requires that he have observational evidence for this statement and that he believe that he can, in some sense, make an even stronger statement of the theory than this. (D62-2-a) is an attempt to lump together all the epistemological aspects of having a theory into one requirement and then forget about them. It should be noted that it is only plausible if we suppose that the evidence that p has for $I \in N_{\mathscr{E}_t}$ "carries over" to the weaker claims of the form $I \in N$ that are entailed by $I \in N_{\mathscr{E}_t}$. (D62-2-b) is an attempt to state precisely p's belief that he can make progress working within the framework of the theory toward a narrower specification of the possible observable states of affairs. A possible objection to (D62-2-b) might be that 'belief' is too weak here. There is much more involved in a person's "commitment" to employ the theory in this way than simply a belief that it can be done. One might plausibly contend that those who *really* have the theory $\langle H, I \rangle$, in the strong sense, at least *want* to use the theory in this way. Another, and apparently much more serious, objection to (D62) is this. There is no guarantee that all persons who have $\langle H, I \rangle$ at the same time will believe something close to the same statements of the theory. Intuitively, it appears that something like this is, in fact, the case and it should be reflected in our account of what it is for a person to have a theory. We shall return to this point below.

It should be fairly clear, at this point, that what a person's having a theory of mathematical physics $\langle H, I \rangle$ actually amounts to – in terms of the phenomena he is able to explain and the predictions he is able to make – will depend quite heavily on what this person knows or believes about the contents of the characteristic range of intended applications I. It depends on this in much the same way that the "cash value" (in terms of nutrition, perhaps) to someone of knowing that artichokes are edible depends on his ability to pick out the things that are, in fact, artichokes. We can be somewhat more precise about this and, in doing so, come to see more clearly the way that those who have a theory of mathematical physics might be said to increase the scope of the theory by enlarging the range of intended applications in some applications of the theory.

First, let us recall a distinction mentioned in Chapter II of two ways that a set might be described. One might describe a set by simply producing an exhaustive list of its members, or one might describe it by naming some property possessed by all and only members of the set. For example, I might describe the set of books in my office by providing a list of their titles or I might describe this set by naming the property 'is a book in Sneed's office at 7:45 a.m., May 20, 1969'. Let us call a description of the set S given by providing a list of the members of S 'an extensional description of S' and call a description of S given by naming a property possessed by all and only members of S 'an intensional description of S'. For our purposes, the most important difference between these two ways of describing sets is this. If S is described extensionally then the question of whether or not some person has "examined" all the members of S can always be easily answered. Intuitively, one simply runs down the (finite) list checking to see if any member has been overlooked. On the other hand, if S is described intentionally then it may be, and in typical cases is, impossible to be certain that someone has examined all the members of S. If we describe the set of bodies in the solar system by naming the property 'moves in a closed path around the sun' we can never be certain that we have discovered all the bodies having this property. (Note that this is true even if we are perfectly clear about our identity criteria for bodies.) It is also worth noting that only finite sets admit of extensional description while finite or infinite sets may be described intensionally.

Let us now consider some aspects of intensional descriptions that will turn out to be particularly relevant to situations in which a person who has the theory $\langle H, I \rangle$ describes the range of intended applications I, or members of it intensionally. Suppose that a person describes the set S intensionally. First note that it is quite possible in this situation that there are sets S' that this person is able to describe extensionally and such that he believes that $S' \subseteq S$. We may only be able to provide an intensional description of the set of bodies in the solar system, yet we are quite confident that the extensionally describable set consisting of the nine planets is a sub-set of this set. Further, we may think of the largest S', call it S_m, for which a person has an extensional description and believes that $S' \subseteq S$ as characterizing that person's "explicit" beliefs about the contents of S. Roughly, his extensional description of S_m is just the longest list he can make of individuals that he believes to be in S. But he makes no claim that this is an exhaustive list of individuals in S. He is willing to add names to the list as he discovers further individuals that have the characteristic property of members of S. That is, the set S_m might grow as he discovers more individuals having this property. The S_m associated with the set of bodies in the solar system probably does not grow for most people, but it surely grows for those concerned with searching for new comets, asteroids, etc. On the other hand, the set S_m might also shrink. One could discover that he had been mistaken. Some body, initially thought to describe a closed path about the sun, might, upon closer observation, be found to describe a hyperbolic path.

Now suppose the following situation arises regarding some intensionally described S. As a person's S_m grows and shrinks in response to his discovering new facts about the properties that individuals do and do not have, suppose that there is some set S_0 that always remains in S_m. To take an example that has had some currency among philosophers, consider the set determined by the predicate 'is a game'. Presumably, I could produce a not-very-long list of all those activities like golf, bridge, football, etc., that I now believe to be games. As my experience with different cultures widens, I might conceivably add to this list. I might move to an Italian neighborhood, observe my neighbors playing bocce and add this to my list. I might delete some things from my list after learning more about them. I first might think that an activity, bearing some superficial similarity to basketball, engaged in by pre-Columbian Indians in Mexico was a game. After learning of the activity's religious significance, I might change my mind. But there are some things on my list like baseball, basketball, poker about which I never, in fact, change my mind. Indeed, it might even be that there were some things in my list of games that I *could not* delete. These entries on my list might be called my "paradigm examples" of games. I could not delete them from my list in the sense that, in doing so, I would recognize that I was changing the meaning of the predicate 'is a game'. In the situation envisioned here, at least some of the things that always, in fact, remain on my list of games *must* remain there unless I am ready to recognize that I am changing my notion of what is to count as a game.

The situation envisioned here can be described a bit more precisely in the following way. There is some extensionally described set S_0 that I believe to be a sub-set of S and, no matter what I discover about the properties of individuals in S_0, my belief that $S_0 \subseteq S$ will not be changed. In this situation the properties that I may use to provide an intensional description of S are limited. Only properties that members of S_0 could not, in some sense, be discovered not to have may be used. In the example, this suggests that I am not actually in a position to state "interesting" sufficient conditions for something's being a game. Were my conditions interesting enough that my paradigm examples of games could be discovered not to satisfy them, then they would not do. The best I can do in giving an intensional description of the set of games is to name a property like being a game that my paradigm examples could not fail to have. But one might say that what I am "really" doing is not best understood as naming a property that characterizes this set. Rather, what I am doing is listing my paradigm examples of games and saying that the set of games consists of these individuals, together with other individuals that are like these in some significant number of "relevant respects".

One way of understanding what is happening here is the following. In effect, I describe the set S by listing the members of S_0. The description works in this way. There will be a large number (indeed an infinite number) of properties shared by all, or almost all, members of S_0. Some of the properties shared by all members of S_0 will typically be necessary conditions for membership in S. For example, all games are human activities. But the situation with sufficient conditions is somewhat different. An individual will be counted as a member of S as it is found to have a "significant number" of the properties that all, or almost all, the members of S_0 have. But there is no single list of properties that assures membership in S. Nor is there any finite list of lists of properties such that having all the properties on any one of these lists assures membership in S. To qualify as a member of S an individual must be found to have a "significant" overlapping of properties with members of S_0. This means that one must be able to find some list of properties of the individual that are shared by all, or almost all, members of S_0. But, in general, there will be an infinite number of possible ways this requirement could be satisfied. Indeed, this would still be the case even if one specified how many properties of the individual were "a significant number" and how many members of S_0 counted as "almost all". Roughly, what I do here is describe S by giving an example of the sort of things that I want to include in S, but I am a bit vague on just exactly how much "like" my examples things have to be in order to get into S.

Previously, it has been suggested that something like the situation just described is typical of the descriptions of the range of intended applications I that are given by people who have the theory of mathematical physics $\langle H, I \rangle$. This is to say that the range of intended applications is described by listing the members of a set of paradigm applications of the theory. This view will be considered in some detail below, but one question about this view should be raised now. The question is roughly this. Can describing a set in this way actually single out one and only one set? Is it not the case that descriptions of this sort are inherently so vague as to what is to be included in the described set that it is misleading to speak of there being a single set that is described? In particular, if the I in $\langle H, I \rangle$ is actually described in this way, is it not misleading to speak of *the* theory as if there were a single, well defined range of intended applications? Would it not be more realistic, at least in cases where such descriptions were employed, to speak of a collection of "admissible" ranges of intended applications spanning the range of vagueness that is apparently permitted by such a description?

What might lead one to think that listing the members of S_0 – the paradigm examples – does not really describe a single set S? Roughly, this. Though one can not give a single set of necessary and sufficient conditions for membership in the S that we putatively describe by listing the members of S_0, it is typically possible to give several sets of conditions that "almost" work. That is, each of these sets of conditions fail because it lets in, or rules out, "just a few" things that run counter to our intuitions about what is to count as being enough like the members of S_0. It might appear natural to regard each of these sets of conditions as an intensional description of a set and to say that, in listing the members of S_0, we describe this entire class of sets. That is, in listing the members of S_0, we do not succeed in picking out a single set S but rather we pick out an entire class of sets. The members of this class are all equally plausible candidates for being "the" set that we "intend" to pick out when we list the members of S_0. If we really want to pick out a unique set, we must, in addition to listing the members of S_0, "arbitrarily" decide which one of these equally acceptable sets of necessary and sufficient conditions we really intend to specify the content of S. We must settle on *one* of the various possible ways of eliminating the vagueness in our description of S. Viewed in this way, a putative description of S by listing paradigm examples must *always* be supplemented by some essentially arbitrary decision that removes the vagueness inherent in such descriptions.

Is the possibility of producing different sets of necessary and sufficient conditions that appear to be equally adequate (or inadequate) in circumscribing the class of individuals that are "enough like" the paradigm examples really incompatible with thinking that the paradigm examples determine a single set? Consider how we might decide whether some particular individual is in the set S putatively determined by listing the members of S_0. It might turn out that, for every individual we encounter there is a clear cut answer to the question of whether it is a member of S. But suppose we do encounter a difficult case – an individual that shares some properties with the paradigm individuals, but there is doubt as to whether it shares enough. In this situation we would, typically, try to find out more about the individual, and perhaps also to find out more about the members of S_0. We would try to find additional ways in which this problematic individual was similar or dissimilar to the paradigm individuals. The expectation would be that this investigation would ultimately yield enough information to give a clear-cut answer to the question of whether the individual belongs to S. Of course, there is no guarantee that this would be the result. It might be that we are not, in fact, able to improve on our initial uncertainty about this individual's membership in S. Indeed, it might even be that we become convinced that further investigation would be fruitless. We might become convinced that the question of whether this individual belongs in S was one that could only be decided arbitrarily.

Were this last situation to occur, there would appear to be good reason to think that, in listing the members of S_0, we had not described a single set S. Our views about what is and is not in the set S are subject to *arbitrary* modification when we encounter new individuals. Our putative description of S simply does not allow us to settle all problematic questions of membership in S – not even if we are willing to carry out extensive further investigation about the individual in question and the paradigm individuals. But, there is no reason to expect that this last situation will *always* arise in *every* attempt to describe a set by listing paradigm examples. It might be that, for some of these attempts, the problematic cases present the possibility of being resolved by further empirical investigation. Even though we can not give satisfactory necessary and sufficient conditions for membership in S, it may be that every case we are actually called upon to decide is such that we never lose hope that further empirical investigation will ultimately result in a clear-cut resolution of the question that has no appearance of being arbitrary or *ad hoc*. We may not be able to say precisely what a game is. But we might feel rather confident that all questions as to whether a particular activity is a game could be resolved, in a non-arbitrary way, by finding out more about this activity and our paradigm examples of games. Were this the case, it would appear natural to regard our list of paradigm examples as determining a single set S, rather than a class of sets consisting of all the "almost" successful attempts to capture the set by stating necessary and sufficient conditions.

The two situations just described may be summarily characterized by noting a distinction between being *vague* about what is to count as a member of S and being *arbitrary* about what is to count as a member of S. One is vague about this question when he is unable to give precise necessary and sufficient conditions for membership in S. Vagueness is an essential feature of descriptions of S given by listing paradigm individuals. One is arbitrary about this question when he is unable to provide reasons of a certain kind for some of his answers to instances of this question – namely, reasons having to do with properties of the individuals in question. Arbitrariness is not an *essential* feature of descriptions of S given by listing paradigm individuals, though it may be a feature of some of these descriptions.

Is there any reason to think that, when the characteristic range of intended applications I of the theory of mathematical physics $\langle H, I \rangle$ is described by listing some paradigm applications, it is always going to appear to be an empirical question as to whether some newly discovered physical system is a member of this set? That is, decisions about what counts as an intended application of the theory will not be made by fiat, but will hinge on the outcome of further investigation of the properties of the physical system in question. I know of no knock-down argument that demonstrates that this *must* always be the case. However, it does appear to be in fact the case for the theories of mathematical physics that are familiar to me. At least, it appears that, where the possibility of an arbitrary decision arises, there is a systematic and natural way of making these decisions. Roughly, it is possible to let the theory itself resolve the cases where further empirical investigation is not decisive. That we can do this makes it plausible to regard the listing of paradigm examples as uniquely determining the characteristic range of intended applications. This question will be examined in more detail shortly.

Keeping in mind these general remarks about descriptions of sets, let us draw some distinctions among the ways that the range of intended applications I might be described by a person who has the theory of mathematical physics $\langle H, I \rangle$. That is, how might such a person describe I in making a statement that $I \in N_\mathscr{E}$ for some expansion of $H, \mathscr{E}$? First, suppose he uses an extensional description of I and consider how he might describe each member of I. That is, what sort of descriptions might the entries in his list of members of I be? The members of I are ordered n-tuples of sets – a domain and functions on that domain (here again we employ (D25)). Thus, one way of describing a member of I would be to provide a list of the individuals in its domain and lists of ordered pairs corresponding to each function on that domain. This would be a fully extensional description of a member of I. One would, however, describe the domain of the member of I extensionally and provide intensional descriptions of the functions of this domain. (Recall the discussion of intensional descriptions of numerical functions in Chapter II, p. 24.) Finally, one could use a fully intensional description of a member of I – describing both the domain and the functions on it intensionally. In the case that I itself is described intensionally these same possibilities of description arise for the particular individuals claimed to be in I. For future reference, it is useful to have an outline of these various possibilities.

(A) Extensional description of I:
  (1) Extensional description of each member of I.
  (2) Intensional description of some members of I.
    (a) Functions only,
    (b) Both domains and functions.
(B) Intensional description of I.
  (1) Extensional description of each member of I.
  (2) Intensional description of some members of I.
    (a) Functions only,
    (b) Both domains and functions.

Intuitively, in each of the situations just described a person who has the theory $\langle H, I \rangle$ has different amounts of explicit information about the contents of I. In each of these situations, what does having $\langle H, I \rangle$, in the sense of (D62), actually amount to? First, consider the situation in which the person who has $\langle H, I \rangle$ has an extensional description of I and extensional descriptions of each member of I (case (A–1)). In this case the person has exhaustive knowledge about the values of the non-theoretical functions in members of I, and it is doubtful if anything like the employment of successively more restrictive expansions of some H to make claims about I would occur. The most likely way to deal with this situation would be simply to employ a core with no theoretical functions – $\langle N_0, N_0, r, N, C \rangle$ – where N and C describe the completely known values of the non-theoretical functions in members of I. In this case, it appears that the only justification for employing theoretical functions would be syntactical simplicity. But if this syntactical simplicity had to be achieved at the cost of sacrificing a maximally sharp description of the observed facts, it would appear to be a dubious value. That is, if we employed theoretical functions and a claim of form (5) which was satisfied by configurations of non-theoretical functions other than the known configuration, it is hard to see that we have gained anything by this claim.

On the other hand, it is possible to cite examples of situations in which persons who have theories of mathematical physics appear to be doing the sort of thing just described. Suppose we regard the solar system as the *single* intended application of classical particle mechanics and that we have an exhaustive description of the bodies in this system and their paths. Does it make any sense to make a claim of form (2) for this system with the predicate 'is a GNCPM' (D23))? The paths which satisfy this claim are not *uniquely* determined. All that is determined is that they will be conic sections. Which particular conic section is determined by the "initial conditions". Yet, someone might maintain that this claim does make sense – that it is explanatory. I know of no way to show conclusively that this is not so. However, it appears to me that this claim – taken in isolation – is not explanatory. It is only when viewed as a consequence of the more general claim of classical particle mechanics – a claim of form (5) involving other intended applications about which we do not have complete information and some of which may be unknown to us – that this claim is explanatory. Consequently, it is misleading to regard this as a real example of a situation in which we employ theoretical functions in a way that fails to "explain" *all* that we know about a closed collection of data.

Thus it appears that, in a situation in which we have complete information, both about the membership of I and about the values of the non-theoretical functions in the members of I, it is unlikely that anything like a sequential development of empirical claims of the theory will occur. If theoretical functions are used at all, they will be employed simply to reproduce the description of the values of the non-theoretical functions given by $\langle N_0, N_0, r, N, C \rangle$ in a syntactically more tractable manner. Were one to do this, however, there would be no particular advantage to be gained from using a core in which only the theoretical functions were constrained, nor from using only theoretical expansions of the cores. Here the vulnerability of the claim to refutation in the light of new data and the usefulness of new means for determining the values of theoretical functions for making predictions are not relevant considerations. The class of data is closed. No new data is going to be added and *a fortiori* no interesting predictions can be made about what the new data will be like. It is worth noting that in a trivial sense, theoretical functions that were used in this situation would be Ramsey eliminable. One could always replace the theory with theoretical functions with the non-theoretical description of the data.

In either case, whether theoretical functions are used or not, a person's having $\langle H, I \rangle$ amounts simply to his believing that $I \in N_\mathscr{E}$ for some $\mathscr{E}$ such that $N_\mathscr{E} = \{I\}$. A little reflection about claims of this sort will quickly convince one that there is little likelihood that they could account for the beliefs characteristic of people who have significant, "real life" theories of mathematical physics. The fact that the claim that $I \in N_\mathscr{E}$ could be verified by purely formal calculations appears to make it even dubious that we would want to call it an empirical claim at all. This suggests that an essential feature of the physical theories that are actually interesting to people is that they deal with open classes of data. This means roughly, that people who have these theories must have only intensional descriptions for at least some "parts" of the range of intended applications.

Let us now consider the situation in which the person who has the theory $\langle H, I \rangle$ has an extensional description of I, extensional descriptions of the domains in members of I, but only intensional descriptions of some of the functions on these domains. The first point to note about this situation is that the enterprise of seeking successively more restrictive expansions of H to make claims about I is significant. Intuitively, one seeks expansions $\mathscr{E}$ such that $I \in N_\mathscr{E}$ is consistent with what one knows at a particular time about the values of the non-theoretical functions in members of I. One seeks to find the smallest $N_\mathscr{E}$ that is consistent with this data and whereby to make the "sharpest" predictions about the yet-to-be-observed values of these functions. If we choose to use a theory with theoretical functions to single out the set $N_\mathscr{E}$ and if there is an infinite number of configurations of non-theoretical functions – models for N_0 – in this class, there is a possibility that these theoretical functions will not be Ramsey eliminable (see Chapter III, p. 49).

THE DYNAMICS OF THEORIES

In this situation, there is a clear sense in which constraints on theoretical functions are more tenable in the face of recalcitrant data than are constraints on non-theoretical functions, provided that we restrict ourselves to theoretical expansions of H (see Chapter VII, p. 180). To see this, suppose we have $H = \langle M_0, r, N_0, M, C \rangle$ and C constrains only non-theoretical functions. Let $\mathscr{E} = \langle H, L, C_L, \alpha \rangle$ be any expansion of H such that C_L is either vacuous or contains only constraints on non-theoretical function, and let $\mathscr{E}' = \langle H, L', C_L, \alpha' \rangle$ be such that $L \subseteq L'$ and α' restricted to L is α. $\mathscr{E}'$ is obtained from $\mathscr{E}$ by adding additional special laws (either theoretical or non-theoretical) to $\mathscr{E}$, but no additional constraints. Now $N_{\mathscr{E}'} \subseteq N_{\mathscr{E}}$, but note that nothing in $N_{\mathscr{E}}$ can be ruled out of $N_{\mathscr{E}'}$ because of failure to satisfy the constraint $C \cap C_L$. All the laws in $L' - L$ do is rule out configurations of non-theoretical functions in separate members of sub-sets of N_0. Alternatively, any array of non-theoretical functions that satisfies the laws in $\mathscr{E}'$ but fails to satisfy the constraints in $\mathscr{E}'$ will also fail to satisfy the constraints in $\mathscr{E}$. Confronted with such an array of non-theoretical functions, we could not "save" the constraints in $\mathscr{E}'$ by dropping some of the laws in $\mathscr{E}'$. But now, suppose the situation is the same except that C and C_L constrain only theoretical functions. In this case, were the laws in $L' - L$ theoretical laws, it might be that configurations of non-theoretical functions in $N_{\mathscr{E}}$ are ruled out of $N_{\mathscr{E}'}$ solely because no theoretical function can be added to this configuration that satisfies both the new laws in $L' - L$ and the old constraints. We may have added theoretical laws in such a way that some values of theoretical functions satisfying some claims of form (2) become more uniquely determined in some members of I than they were before adding these laws. This may make it impossible to satisfy a theoretical constraint that we could satisfy before adding this new law. Thus, confronted with an array of non-theoretical functions which could not be supplemented with a theoretical function satisfying the new law and old constraint, we might be able to "save" the old constraint by giving up the new law.

Intuitively, adding additional theoretical laws to $\mathscr{E}$ may give us more ways of determining theoretical function values and these new values *may* fail to satisfy our original constraints on theoretical functions. In contrast, adding new laws (theoretical or non-theoretical) can provide no new ways of determining non-theoretical function values. These values are determined in ways that are independent of any laws of $\langle H, I \rangle$ that might be postulated. The upshot of this is that whether or not a constraint on non-theoretical functions in H is satisfied by a given $I \subseteq N_0$ is independent of any special laws that might be added to H. Modifying the laws that have been added in trying to refine the account of I provided by $\langle H, I \rangle$ to save the constraints in H which are satisfied by I will depend on the nature of the special laws that are postulated. If they are theoretical laws, it may be possible to retain the constraints in H and modify the postulated laws. Saving the constraints in this way amounts to admitting that some physical systems that we initially thought might provide means of determining theoretical function values, in fact, do not because certain laws thought to hold in these systems do not hold. For example, one might initially believe that the spring in a certain spring balance "produced" forces obeying a certain force law. On the basis of assuming this force law, one might calculate the mass ratios of some individuals and discover that these mass ratios differed from those calculated in (perhaps several) other ways. This is merely to say that the $\langle =, = \rangle$-constraint on mass can not be satisfied. In this situation, we would very likely simply admit that our initial belief about the forces produced by the spring were in error. In attempting to deal more precisely with systems containing this spring we had postulated a special force law that we were unable to maintain in the face of further investigation.

Another way to illustrate the difference between constraints on theoretical and non-theoretical functions is this. Note that, if the constraints in H are on theoretical functions, the null expansion of H, $\mathscr{E}_n$, may be such that $I \in N_{\mathscr{E}_n}$ is "vacuously" true for all $I \subseteq N_0$. Such appeared to be the case in classical particle mechanics. Nevertheless, in non-null expansions of H the constraints may be doing real work in that $I \in N_{\mathscr{E}}$ may fail to be true for some I because these constraints can not be satisfied in the presence of the theoretical laws in $\mathscr{E}$. On the other hand, if the constraints in H are only on non-theoretical functions and the null expansion produces a vacuously true claim, these constraints can never do any work in ruling out I's. Anything which satisfies them in the null expansion (which in this case is everything) will continue to satisfy them no matter what laws we add in non-null expansions.

It is important to understand precisely how the postulating of theoretical laws in the situation we are now considering opens up the possibility of discovering new methods of determining values of theoretical functions. In this situation we have an exhaustive list of all the members of I. There is no possibility of coming onto, heretofore unnoticed, physical systems that turn out to be in I. There is no way that we could discover *new* physical systems that happen to have properties allowing us to calculate what the values of theoretical functions in these systems must be if our current claim of form (5) for I is to be satisfied. But, what we can do is "discover" new physical laws – theoretical laws – that hold in the physical system we already know about. Since we do not have complete information about the values of some non-theoretical functions, it makes sense to attempt to predict what these unknown values might be like. If we attempt to do this using theoretical expansions of a core H which contains theoretical functions, some of the special laws we postulate to hold in some of the known physical systems may allow us to calculate values for these theoretical functions.

In this situation a person's having the theory $\langle H, I \rangle$ at a particular time can be understood in the following way. At that time t, he believes several statements of the form $I \in N_{\mathscr{E}}$ where $\mathscr{E}$ is an expansion of H. The strongest such statement is that $I \in N_{\mathscr{E}_t}$. Typically $N_{\mathscr{E}_t}$ will contain more than one member so that it "makes sense" for this person to seek an expansion $\mathscr{E}_{t'}$ such that $N_{\mathscr{E}_{t'}} \subseteq N_{\mathscr{E}_t}$ and $I \in N_{\mathscr{E}_{t'}}$ remains consistent with all the data about I that is available to him at time t. Roughly, this means the person is trying to make successively more restrictive predictions about what the values of non-theoretical functions in members of I that are unknown to him at t will be like. He is further committed to using restrictions of the same core H in making these successively more restricted claims. That is, he is committed to pushing further in accounting for the behavior of I using the conceptual apparatus in H, in the sense that he believes that it is always possible to do a little bit better – to postulate more special laws that more nearly uniquely determine the "observable" behavior of members of I. These attempts to make more restrictive predictions may initially appear to be successful in that the claim is consistent with currently available data, but later turn out to be unsuccessful in the light of new data about I. This new data *could be* such that *no* claim of form (5) made with H would be consistent with all available data about I. In this case, its discovery should compel the person who has $\langle H, I \rangle$ to give it up, i.e. (D62–2), should cease to be satisfied. On the other hand, the new data might just show that $I \in N_{\mathscr{E}_{t'}}$ is false and still be consistent with $I \in N_{\mathscr{E}_t}$. In this case, the data should convince the person who has $\langle H, I \rangle$ that this *particular* attempt at using H to make a further claim about I is unsuccessful, yet it might plausibly leave unshaken this belief that *some* attempt to "extend" the theory beyond the statement that $I \in N_{\mathscr{E}_t}$ will be successful. That is, a reasonable person might continue to have the belief described in (D62-2-b) even in the face of, perhaps repeated, failure to actually find an $\mathscr{E}$ that improves upon $\mathscr{E}_t$. That is he might continue to have $\langle H, I \rangle$.

In this case, the account of a person's having the theory $\langle H, I \rangle$ given by (D62) shows how situations might naturally arise in which it was appropriate to say that a person was maintaining the theory – continuing to have it – even in the face of newly discovered data that conclusively refuted what one might call '*the* empirical claim of the theory'. But what the person is continuing to maintain is *some* claim of the theory, plus the belief that he can better this claim in some way. This latter belief amounts roughly to confidence in the "usefulness" of the conceptual apparatus in the core H in dealing with the range of intended applications in I. It is important to understand the intimate relation between this confidence and the presence of theoretical functions and constraints on only these functions in H. In the case where we have only constraints on theoretical functions and the null expansion of H is vacuously true, there is a sense in which new information about I can never *conclusively* compel a person to give up $\langle H, I \rangle$. The possibility of finding some non-null expansion that is consistent with this data is always open. (What does compel a person to give up $\langle H, I \rangle$ in these situations will be discussed below.) But even in the case where the null expansion is not vacuously true, we have seen that constraints on theoretical functions admit of possibilities of being maintained that are not present for constraints on non-theoretical functions. This strongly suggests that confidence in the continuing usefulness of a core with constrains on only theoretical functions may be occasioned by the plausibility of regarding the core constraints as "essential" features of the functions in the core. This plausibility is, in turn, occasioned by the difficulty in conceiving of empirical evidence that would compel one to give up these constraints. Thus there is some reason to believe that people will only have theories of mathematical physics which contain only theoretically constrained cores.

Consider now the situation in which the domain in a member of I, as well as the functions on that domain, is described intensionally (Case (A-2-b)). Here again the enterprise of seeking expansions of H to use in making successively more restrictive claims about I is a significant one. Again too, if there are an infinite number of configurations of non-theoretical functions in an $N_\mathscr{E}$, it is possible that the theoretical functions in the claim that $I \in N_\mathscr{E}$ are not Ramsey eliminable. In this case too, constraints on theoretical functions are more tenable than constraints on non-theoretical functions. But there is here an additional feature of the tenability of constraints that merits attention. If additional information about I appears to make it impossible to satisfy a constraint (on either theoretical or non-theoretical functions) it may be possible to question whether the individuals causing the difficulty actually "should" be in the domains in question. If these domains are described intensionally by naming a property, it is always possible to question whether a particular individual has this property. Perhaps we made a mistake in thinking that it did. In the case where the domain is described intensionally in the somewhat looser way of listing some paradigm examples of individuals in this domain, there is even more room for casting doubt as to whether some non-paradigm individual "really" belongs in the domain.

Another new possibility introduced by intensional descriptions of the domains of some members of I is this. It now becomes possible to significantly speak of "searching" for new ways of measuring the values of theoretical functions. Since we do not know, prior to empirical investigation, how many individuals are in some of the domains in members of I, we may search for domains that have appropriate (usually small) numbers of members allowing for calculation of theoretical function values required to satisfy some claim of form (2). This may, but need not necessarily, be carried on in conjunction with postulating additional theoretical laws which more nearly uniquely determine the values of theoretical functions in some domains. For example, we know that mass ratios are uniquely determined in models for classical particle mechanics plus third-law central forces when the domains of these models contain small numbers of members. We can then go out and look for members of the range of intended applications of this theory having these properties (double star systems and the Fletcher trolley "approximately" have these properties) and employ them to determine mass ratios.

One final feature introduced by intensional descriptions of the domains needs to be mentioned. A person who has the theory $\langle H, I \rangle$ is now in a position to make predictions, not only about the values of non-theoretical functions in some domains, but also about the number of individuals in some domains. That is, it becomes possible to use the theory to predict the presence of heretofore undetected individuals. The discovery of new planets has been mentioned before as an example of this use of a theory of mathematical physics. The presence of intensional descriptions of some domains does not significantly change the ways in which a person who has the theory $\langle H, I \rangle$ can retain it in the face of unsuccessful efforts to sharpen the claim made about I. It only provides another way that these efforts could be revealed to be unsuccessful by further investigation. One could postulate theoretical laws in such a way that $I \in N_\mathscr{E}$ entailed that the domain of some member of I contained members in addition to those presently known. Failure to actually turn up these individuals would be cause for giving up the claim that $I \in N_\mathscr{E}$.

Let us now turn to consider the situation in which the person who has the theory $\langle H, I \rangle$ has only an intensional description of I (Case (B)). There is good reason to believe that this is the typical, if not indeed the only, case for significant, "real life" theories of mathematical physics. However, this does not mean that our consideration of the case of an extensionally described I has been a superfluous exercise. It will be seen shortly that when a person has $\langle H, I \rangle$ and an intensional description of I, he typically also has a theory of mathematical physics $\langle H, I_t \rangle$ and an extensional description of I_t. Thus what we have discovered about the case of extensionally described I's is also relevant to the case of intensionally described I's. It will not prove necessary to consider the sub-cases of this case individually, but it should be noted that the case of an intensional description of I and an extensional description of each member of I (Case (B-1)) is not vacuous. For one may have an intensional description of I and nevertheless just "happen" to have a list of everything that, in fact, satisfies this description. This list could of course consist of extensional descriptions.

The first thing to note about this situation is this. If the range of intended applications I is in the set $N_\mathscr{E}$ then any sub-set of I is also in $N_\mathscr{E}$. This is a trivial consequence of (D27-2) and (D35). Intuitively, this means that if $\mathscr{E}$ is a successful application of the core H to I then $\mathscr{E}$ is also a successful application of H to any sub-set of I. For future reference, we state this fact as:

(T38) If $\langle H, I \rangle$ is a theory of mathematical physics, $\mathscr{E}$ is an expansion of H, and $I \in N_\mathscr{E}$, then if $I' \subseteq I$ then $I' \in N_\mathscr{E}$.

Now suppose that a person p has $\langle H, I \rangle$ at time t and an intensional description of I. Let I_t be the union of all the I' such that:

(i) p believes at t that $I' \subseteq I$;

(ii) p has an extensional description of I'.

Roughly, I_t consists of all those individuals that p could list as being members of I at time t. Clearly, p believes at t that $I_t \subseteq I$ and he has an extensional description of I_t at t. Since p has $\langle H, I \rangle$ at t, there is some expansion of H, $\mathscr{E}_t$, that satisfies (D62–2). Intuitively, the statement that $I \in N_{\mathscr{E}_t}$ is the strongest statement of the theory that p believes at t, but he believes also that it is possible to find an $\mathscr{E}$ that can be used to make a stronger true statement of this form. In view of (T38), it is natural to expect that p believes at t that $I_t \in N_{\mathscr{E}_t}$, and further that he believes it is possible to find an $\mathscr{E}$ that can be used to make a stronger true statement of this form about I_t. That is, if we assume that p believes the logical consequences of statements that he believes (D62–2), for I entails a similar statement for I_t. Roughly, what this amounts to is the following. Were $\langle H, I_t \rangle$ a theory of mathematical physics, then p would have it at t. This is to say that, whenever p has the theory of mathematical physics $\langle H, I \rangle$ at t and has an intensional description of I, he also has at t the theory of mathematical physics obtained from $\langle H, I \rangle$ by taking just those things p is able to list as being in I, *provided that* this turns out to be a theory of mathematical physics.

Of course $\langle H, I_t \rangle$ might fail to be a theory of mathematical physics even though $\langle H, I \rangle$ is a theory of mathematical physics. For we have, in effect, required that p *believe* that $I_t \subseteq I$, not that, in fact, $I_t \subseteq I$. It could be that I_t fails to have any one, or even all, the properties (D61–3) through (D61–6) necessary for it to be the range of intended applications in a theory of mathematical physics with the core H. Nevertheless, in considering how the person p who has $\langle H, I \rangle$ and believes that $I_t \subseteq I$ regards the statement that $I_t \in N_{\mathscr{E}_t}$ it is useful to think of p as having a theory of mathematical physics $\langle H, I_t \rangle$ and an extensional description of I_t. Then all our previous remarks about having theories of mathematical physics and an extensional description of the range of intended applications can be brought to bear on this situation.

Intuitively, it is possible to consider the statement that $I_t \in N_{\mathscr{E}_t}$ as the actual "cash value" for p of this having the theory $\langle H, I \rangle$ at t. It is the strongest consequence of the beliefs associated with his having $\langle H, I \rangle$ at t that is "about" only members of I known to him at t. It is what his having the theory $\langle H, I \rangle$ tells him about that part of the world that he is actually acquainted with at t. All the checking of the theory that he can do at t, without looking for additional physical systems in I, amounts to checking the consequences of $I_t \in N_{\mathscr{E}_t}$. All the predictions he can make that do not extend to yet undiscovered physical systems are predictions based on the claim that $I_t \in N_{\mathscr{E}_t}$. It should be noted that both $\mathscr{E}_t$ and I_t may be different for different persons who have the theory $\langle H, I \rangle$ at t. There is no guarantee on our account thus far that persons who have the same theory of mathematical physics at a given time believe essentially the same things about the world as a result of having this theory. They may differ both as to how far they have pushed the theory in specifying the behavior of its range of applications and as to what parts of its range of applications they believe they have discovered.

In the situation we are considering now, there are two distinct ways that a person who has the theory $\langle H, I \rangle$ can be regarded as making progress in applying the theory to I – two ways the "theory" may be regarded as growing over time. First, he might succeed in finding an $\mathscr{E}'_t$ such that $N_{\mathscr{E}'_t} \subset N_{\mathscr{E}_t}$. This way of progressing is essentially no different than in the case of an extensionally described I. Indeed, if one envisions I_t as remaining fixed while successively more restrictive expansions of H are discovered, the situation is exactly the same. On the other hand, progress is also made in applying the theory when additional members of I are discovered, that is when $I_t \subset I_{t'}$ for $t < t'$. There are at least two kinds of motivation for this second activity, each of which provides a reason for viewing it as progress. In seeking to enlarge I_t, one may be attempting to find new physical systems which can be used to determine the values of theoretical functions (see the discussion in connection with (E3–b), Chapter IV, p. 79). Here the motivation for extending the scope of the theory is to obtain sharper predictions about the applications that we already know about without postulating any more laws about them. In contrast, one might be seeking to enlarge I_t, with no aim other than acquiring new knowledge. One can envision expanding the scope of the theory in this way by simply "happening onto" new physical systems that are found to belong to I – as, for example, discovering another planetary system. But it is perhaps more typical to see a conscious effort being made to find some way of regarding an already familiar physical phenomenon as a partial possible model for the non-theoretical formalism of the theory. Thus, one attempts to regard systems of fluids as systems of particles in order to bring them within the scope of classical particle mechanics.

It is important to understand that, on the view presented here, the theory – the ordered pair $\langle H, I \rangle$ – remains the same throughout all the time that a person who has it uses it to make claims about I. The intuitive fact that the scope of theories – the things to which they are claimed to apply – appears to change over time is accounted for by noting that I_t typically changes over time. Roughly, this is saying that the scope claimed for the theory – the characteristic range of intended applications I – never changes. What does change is the list of things people who have the theory are willing "explicitly" to claim as being within the scope of the theory.

Now, there is an obvious alternative to this view. One could regard the second member of the pair $\langle H, I \rangle$ as changing over time – putting I_t in the place of I. That is, we do not regard the range of intended applications of the "theory" as specified initially, once-and-for-all, but rather as something that can be expanded or contracted as the work with the "theory" progresses. On our account of the identity conditions for theories of mathematical physics, this would mean that person had different theories at different times depending upon what the present range of intended applications was. Intuitively, this appears to be an implausible thing to say. We do regard all this activity as working with the same theory. But, one might contend, what is at fault here are our identity criteria for theories. What is really characteristic of the theory is the core H. The range of intended applications is irrelevant. However, this too is an implausible view. $\langle H, I \rangle$ and $\langle H, I' \rangle$ will simply not always be regarded as two applications of the same "theory", no matter what I and I' are like. There must be some relation between them. It is the fact that one can not avoid saying *something* about the range of intended applications in the identity criteria that leads to the view expounded here. Another possible alternative will be considered below.

Let us consider now what might happen when a person who has the theory $\langle H, I \rangle$ discovers that $I_t \in N_{\mathscr{E}_t}$ is false. As before, there is the possibility that he might give up his belief that $I \in N_{\mathscr{E}_t}$ and attempt to find another expansion of H such that $I_t \in N_{\mathscr{E}_t}$ is consistent with his data about I_t. Roughly, this amounts to admitting that the special laws postulated with $\mathscr{E}_t$ do not hold in I. In this case, however, it is possible to maintain $\mathscr{E}_t$ and simply give up the belief that $I_t \subseteq I$. That is, if $I_t \in N_{\mathscr{E}_t}$ is discovered to be false one can continue to maintain that $I \in N_{\mathscr{E}_t}$ by admitting that a mistake was made initially in thinking that $I_t \subseteq I$. Roughly, this amounts to admitting a mistake about what is, in fact, in an intensionally described range of intended applications. Both of these ways of reacting to the discovery that $I_t \in N_{\mathscr{E}_t}$ is false can be taken as ways of continuing to maintain the theory $\langle H, I \rangle$ in the face of recalcitrant data. The second way of maintaining the theory appears to give a good account of what happens when unsuccessful attempts are made to "carve up" some already familiar phenomenon in a way that it becomes an intended application of the theory. When we fail in an attempt to apply classical particle mechanics to optical phenomena, we conclude that light does not consist of "particles" after all.

What we have said thus far applies to any situation in which a person has the theory $\langle H, I \rangle$ and an intensional description of I. What can be said, in addition, about the case in which the intensional description is provided by listing paradigm examples of individuals that are in I? If person p has $\langle H, I \rangle$ and p describes I by listing paradigm examples of members of I, let us call the set of paradigm examples I_0. According to our earlier account of this method of describing sets, it is clear that I_0 will be a sub-set of any I_t that p believes to be a sub-set of I. Intuitively, this is just to say that p's paradigm examples of things that are in I will always be in his most exhaustive list of the things he believes to be in I. This means that, for any $\mathscr{E}_t$ that arises, p will believe that $I_0 \in N_{\mathscr{E}_t}$. That is, there is some, one extensionally described set of intended applications, namely I_0, to which p always applies H in some way or another.

In this case, there is some question about one situation that might arise when the person who has the theory $\langle H, I \rangle$ discovers that $I_t \in N_{\mathscr{E}_t}$ is false and yet retains the theory. Suppose I_t contains some "borderline" individuals – individuals which share some properties with some members of I_0, but not enough so that they clearly belong to I. What are we to say, in this case, about giving up the belief that $I_t \subseteq I$? If the description of I given by listing paradigm examples is simply vague, then whether or not this belief can be given up will depend on further empirical investigation of the properties of the borderline individuals and the members of I_0. On the other hand, if there were some arbitrariness surrounding decisions about membership in I, it might be that one could arbitrarily decide that I_t was not a sub-set of I and thereby "save" the theory. Indeed, it might be that I was described in such a way that *every* decision about what to do in the face of evidence that some $I_t \in N_{\mathscr{E}_t}$ is false was resolved in this arbitrary fashion. Were this the case, one might be inclined to say that it was "almost" impossible that someone who has $\langle H, I \rangle$ would give it up. For, having the theory only amounts to believing that H can be successfully applied to I_0 in some way. The further claim that it can be made to apply successfully to physical systems "like" the members of I_0 turns out to be vacuous – 'vacuous' in the sense that nothing could convince us that it was not true.

The significance of the possibility just mentioned can be seen more clearly by considering the circumstances in which one might actually describe the characteristic range of intended applications by listing paradigm examples. We have already mentioned that there is some reason to believe that the *typical* way of describing the characteristic range of intended applications of a theory of mathematical physics is to list paradigm examples of individuals that are in it. The principal reason for thinking that this is true is a certain view about how theories of mathematical physics come into being, or about how people come to have theories of mathematical physics.

To see how this is so, consider the following highly schematic account of how someone might become the first person to have the theory of mathematical physics $\langle H, I \rangle$. Leaving aside the question of how this person might happen to have at hand the core $H = \langle M_0, N_0, r, M, C \rangle$, consider how he might come to apply this core successfully for the first time. He might first discover how to regard some physical phenomena as an extensionally describable set of models for the mathematical structure N_0. His discovering how to do this may be closely connected with his discovering how to construct the core. But let us ignore this aspect of the situation for the time being. This extensionally described set I_0 can be expected to have all the properties that (D61) requires of the range of intended applications in a theory of mathematical physics. It will serve as the range of his first successful application of the core H. That is, he discovers some expansion of H, $\mathscr{E}_0$, such that $I_0 \in N_{\mathscr{E}_0}$ in an "interesting" (i.e. explanatory) empirical claim that is compatible with all his information about I_0. In the case of classical particle mechanics, one might regard the person in question as Newton and the set I_0 as consisting of various sub-systems of the solar system, together with certain terrestrial systems like pendula and freely falling bodies.

One might envision this person as simply stopping with the set I_0 – as simply having the theory of mathematical physics $\langle H, I_0 \rangle$. But, on the other hand, he might be convinced that H can be made to apply to other physical systems that are like those in I_0 in some respects. That is, he might take his initial success in applying H to I_0 as an indication that he could meet with similar success in attempting to apply H to physical systems that are similar to those in I_0. He might think that I_0 can be expanded by adding other physical systems of the same kind and that H could still be successfully applied to this extended range of intended applications. One way to characterize these expectations on the part of one who has successfully applied H to I_0 is this. He actually has the theory of mathematical physics $\langle H, I \rangle$ and describes the set of intended applications I by listing paradigm examples to members of I – the members of I_0. That is, the members of I_0 are paradigm examples of the kind of physical systems that he expects to successfully account for with the core H. They are such paradigm examples because he has, in fact, already succeeded in using H to account for their behavior.

This view of the genesis of theories of mathematical physics is very similar to a view of the genesis of scientific theories in general that has been offered by Kuhn [21]. Indeed, the view of theories of mathematical physics suggested here may, quite naturally, be regarded as simply a special case of Kuhn's thesis. I regard both the view suggested here and Kuhn's thesis as substantive, historical claims about how theories of mathematical physics (or scientific theories in general) come into being, or how people come to have them. People could, conceivably, come to have theories in quite a different way. That they do not, is a claim to be argued for by looking at the historical records. I shall not, at this point, undertake to provide such arguments – even for the claim restricted to theories of mathematical physics. Rather, I shall confine my attention to some consequences of the restricted claim. Those readers interested in a defense of the claim may consult Kuhn's work. It should only be noted that there is a possibility that this view might be more defensible for the special case of theories of mathematical physics than for theories in general.

Now, supposing that this view of the genesis of theories of mathematical physics is generally true, one can note several interesting features of the historical development of such theories. First, people who have the theory of mathematical physics $\langle H, I \rangle$ will tolerate a certain amount of vagueness as to what is in the characteristic range of intended applications. They will only be able to "point to" certain physical systems that are surely in I – the range of the initial successful application of H. They will not be able to give necessary and sufficient conditions for membership in I. We have already encountered an example of this situation in our discussion of classical particle mechanics. There we were able to get some idea of what was to count as a system of "physical" particles by listing some examples of things that "must" be such systems, but we are unable to specifically delineate the class of systems of "physical" particles. At that point, we acknowledged this as a lacuna in our logical reconstruction of classical particle mechanics. In the light of the view now being considered, this is an essential limitation on the enterprise of logical reconstruction. If the characteristic range of intended applications in theories of mathematical physics is always described by listing as paradigm examples the range of the initial successful application of the core, then one can never hope to arrive at a completely precise rendering of the empirical content of the theory. At best, one can hope to arrive at a precise formulation of the claim $I_t \in N_{\mathscr{E}_t}$, – the content of the theory restricted to individuals "explicitly" believed to be members of I.

That vagueness must be tolerated in descriptions of the characteristic range of intended applications follows immediately from this view of the genesis of theories of mathematical physics. That the limits of the characteristic range may be decided arbitrarily in some cases does not. It may well be that, in the case of particular theories, there is a certain amount of arbitrariness. We have already mentioned the possibility that so much arbitrariness surrounds decisions about what is in I that it may *always* be possible to "save" the theory when it encounters recalcitrant data by ruling the troublesome physical systems out of I arbitrarily. Let us also suppose that there is sufficient arbitrariness to allow that, roughly speaking, anything to which H can be made to apply is arbitrarily counted as a member of I. That is, let us suppose that there is so much arbitrariness in what is to count as a member of I that it is always possible to decide questions about what is in I "in favor" of the theory – either to save it in the face of recalcitrant data, or to enlarge the set of individuals known to belong to its characteristic range of intended applications.

If, in the situation just described, we were to decide to systematically decide all the doubtful cases in a way which "gives the theory the benefit of the doubt", then one might claim that the listing of the paradigm individuals does not really determine I. What does determine the membership of I is essentially the property of being, together with the members of I_0, a part of the range of a successful application of the core H. We have effectively removed all the vagueness about the content of I by systematically deciding the cases which permit arbitrariness "in favor" of the theory. Because there are so many of these cases, this amounts to saying that an individual is to count as a member of I if and only if, when it is added to I_0, H can be successfully applied to the enlarged set. Roughly, membership in the characteristic range of applications is determined solely by what can be added to I_0 without destroying the applicability of H to the resulting set of physical systems. There are no *a priori* expectations about the sort of things to which the theory should be applicable. A failure of the theory to live up to some expectations about where it ought to apply could never be a reason to give up the theory, except in the limiting case of the paradigm set of applications. The applicability of the formalism H determines, to a large extent, but not exclusively, the characteristic range of intended applications of the theory. The only restriction on this set I besides the applicability of H to it is that I_0 must be a sub-set of it. Very roughly speaking, the theory determines its own range of applications; it is intended to apply to just exactly what it, in fact, applies to – nothing more nothing less. The only restriction on the theory's determining its own range of applications is that the initial paradigm applications must be included in it. Again, roughly, we have hit upon some physical systems – the members of I_0 – to which H can be successfully applied in an "interesting" way. This, in itself, is a highly significant property of these systems. We are interested in discovering all the physical systems that have this property. These systems comprise the characteristic range of intended applications of the theory in question.

I know of no philosopher who explicitly holds the view that this is the correct way to regard the characteristic range of intended applications in theories of mathematical physics. However, the view is not entirely unattractive. The chief attractiveness of the view is that it accounts for the alleged tenacity with which theories of mathematical physics are retained in the face of recalcitrant data. Conversely, the chief objection to the view is that it appears to be all too successful in accounting for this. It might appear that, on this view, recalcitrant data could never cause one to give up a theory. This is not quite true. To see precisely what the situation is, let us consider separately the claims that the condition we are considering is both necessary and sufficient for membership in I.

First consider the claim that it is a sufficient condition for membership in I. Roughly, any physical system that can be added to I_0 producing a new set to which H may also be successfully applied is "automatically" counted as a member of I, regardless of whether it shares any other "interesting" properties with members of I_0. For example, one might take success in explaining the behavior of gases by kinetic theory as *conclusive* justification for the claim that gases are systems of particles 'conclusive' in the sense that *no* subsequently discovered dissimilarities between gaseous systems and the paradigm examples of particle systems would cause us to give up this claim. It is important to note that situations could arise in which maintaining this sufficient condition would be incompatible with having the theory. In the simplest case, there might be two physical systems, each of which could be singly added to I_0 and H successfully applied to the enlarged set, and yet both together could not be added to I_0 and H successfully applied to the result. Believing this to be the case is incompatible with the belief that some particular application of H to I is successful (D62-1). Intuitively, what is going on here is this. There are two incompatible ways to extend the range of applications of the theory beyond the paradigm applications. The theory might develop beyond the initial success in either of two mutually exclusive directions. Since our sufficient condition for membership in I allows us to make no distinction between these two possible expansions, we are forced to give up the theory. I know of no real-life example of a situation in which people have been forced to give up a theory for this reason. The apparent absence of such examples could mean that people never actually maintain such a sufficient condition. There is always some way to decide which possible expansion is the "right" one on the basis of which new system is more like the members of I_0. On the other hand, the absence of examples could merely mean that such problematic cases have not, in fact, arisen.

Now, consider the claim that it is a necessary condition for membership in I. Roughly, no physical system will be counted as a member of I unless it can be added to I_0 and H successfully applied to the enlarged set, no matter how much like the members of I_0 this new system is. For example, had there been repeated failure to account for, say, Brownian motion, within the framework of classical particle mechanics, this might have provoked physicists to simply disregard the readily apparent similarities between these systems and the paradigm examples of particle systems and to deny that they were particle systems. Actual examples of such *blatant* arbitrariness are hard to find, but one might take denying that light is corpuscular to be a less blatant instance of the same phenomenon. At any rate, it is this necessary condition that makes it appear that the view being considered gives much too strong an account to the steadfastness with which theories may be retained in the face of recalcitrant data. Essentially, it allows us to arbitrarily disregard all putative counterexamples to the theory, no matter how "intuitively" like the paradigm applications they are.

Though I will not provide any empirical evidence for this claim, I believe that the situation just described is not an accurate picture of any real-life physical theory, though it could conceivably be. Rather, I think the typical case lies somewhere between the two extreme situations we have considered: vagueness, but no arbitrariness about the content of I; and vagueness removed by always arbitrarily giving the theory the benefit of the doubt. Typically, I believe, the description of I by listing the paradigm initial applications in I_0 will admit of a certain amount of arbitrary "clarification", but not enough to *always* give the theory the benefit of the doubt. One might expect that most hard decisions about what is to be included in I would be resolved by further empirical investigation of the individuals in question. But if this failed to yield decisive results, if the decision did have to be made arbitrarily, then it would be made in such a way as to give the theory the benefit of the doubt – to "save" it or to increase its scope. That is, there will be some cases in which empirical investigation of "borderline" cases could ultimately force a person to give up the theory, while there will be others where this empirical investigation will be inconclusive. In these latter cases, the decision about what is in I may be made arbitrarily and it is natural to expect that it will be made in a way that allows the theory to be retained and its scope enhanced. The upshot of this is that, even when the description of I by listing paradigm examples admits of some arbitrariness, it is plausible to think this description as determining a unique characteristic range of intended applications. This is because we decide the cases where there is room for arbitrariness in a systematic way. In effect, we let the theory decide them for us.

At this point, we may profitably reconsider a previously noted (p. 267) difficulty with our account of what it is for a person to have a theory to mathematical physics (D62). We noted that this account provides no guarantee that all persons who have the theory $\langle H, I \rangle$ at the same time believe the same, or almost the same statements associated with the theory. We are now in a position to indicate how Kuhn's thesis, or our version of it for theories of mathematical physics, provides a means of remedying this defect. The idea is simple. It is certainly plausible to think that the initial successful application of the core of the theory is essentially the same for all those who have the theory. Different people who have the theory at a later time in its development *may* believe different statements. They may be more or less clever in seeing ways to extend the theory, and more or less successful in convincing their colleagues what evidence supports the claims they make with the theory. There is some room for disagreement, at least short term disagreement, among the practitioners of the same theory, even in the face of the same data. It would be implausible to require unanimity among the persons who have the theory. Yet, it might be that more unanimity ought to be required than simply agreement on the initial successful application. I confess that my intuitions on this point are rather weak. However, it is clear that Kuhn's thesis strongly suggests that we should modify our notion of what it is to have a theory of mathematical physics so as *at least*, to require that everyone who has the theory has it "because of" the same initial success.

Indeed, a little reflection reveals that Kuhn's thesis suggests far more than this. It suggests that our identity criteria for theories need to be modified. Suppose our formulations of Kuhn's thesis for theories of mathematical physics is generally true and two people satisfy the requirements of (D62) for the pair $\langle H, I \rangle$. If they do not have appropriate beliefs about the same initial successful application of the core H, there is a strong tendency to say that they really have different theories. The theories are intuitively different in that attempts to extend the scope of the theory beyond its initial success will be of an essentially different character. They will begin trying to enlarge a different initial set of physics systems and trying to refine a different expansion of H. This might be true, even though the extension of the set I determined by both sets of initial successful applications just happened to be the same. This suggests that we should include the initial successful application of the theory in our identity criteria for theories of mathematical physics. This suggestion is made precise in (D63) where we define 'a theory of mathematical physics in the sense of Kuhn'.

(D63) x is a *theory of mathematical physics in the sense of Kuhn* if and only if there exists an H, I, I_0, and $\mathscr{E}_0$ such that:
(1) $x = \langle H, I, I_0, \mathscr{E}_0 \rangle$;
(2) $H = \langle M_0, N_0, r, M, C \rangle$ is a core for a theory of mathematical physics;
(3) $\mathscr{E}_0$ is an expansion of H;

(4) $I_0 \subseteq I \subseteq N_0$;
(5) (D61–4);
(6) (D61–5);
(7) (D61–6).

In the quadruple $\langle H, I, I_0, \mathscr{E}_0 \rangle$, I_0 and $\mathscr{E}_0$ are to be interpreted in such a way that $I_0 \in N_{\mathscr{E}_0}$ is the claim of the initial successful application of the theory in question. The only formal properties we require of these entities are specified in (D63–3) and (D63–4). That they are to be interpreted in the way just mentioned only appears when we come to say what it is to have a theory of mathematical physics in the sense of Kuhn. This we do in (D64):

(D64) If p is a person and $\langle H, I, I_0, \mathscr{E}_0 \rangle$ is a theory of mathematical physics in the sense of Kuhn, then *p has $\langle H, I, I_0, \mathscr{E}_0 \rangle$ at time t* if and only if:
(1) (D62–1);
(2) If $\mathscr{E}_t$ is an expansion of H such that, for all expansions $\mathscr{E}$ of H such that p believes at t that $I \in N_{\mathscr{E}}$, $N_{\mathscr{E}_t} \subseteq N_{\mathscr{E}}$ and p believes at t that $I \in N_{\mathscr{E}_t}$ then:
  (a) $N_{\mathscr{E}_t} \subseteq N_{\mathscr{E}_0}$;
  (b) (D62–2–a);
  (c) (D62–2–b);
(3) For p, the members of P_0 are paradigm examples of members of I.

The first thing to note about (D64) is that we are using the notion of x's being a paradigm example of y for person p' as an undefined primitive. We have already said something, on an intuitive level, about this notion. To strive for more precision about it would involve us in deep philosophical issues far beyond the scope of this work. For our purposes, it is sufficient to note that (D64–3) entails that, for all times t, p believes that $I_0 \subseteq I$. This, together with (D64–2–a) (plus the assumption that p believes all logical consequences of his beliefs) entails that p believes that $I_0 \in N_{\mathscr{E}_0}$. That is, everybody who has the theory $\langle H, I, I_0, \mathscr{E}_0 \rangle$ believes that $I_0 \in N_{\mathscr{E}_0}$. Intuitively (D64–2–a) requires that all attempts by anybody who has the theory to apply H to I be compatible with the attempt represented by the expansion $\mathscr{E}_0$. In this sense the expansion $\mathscr{E}_0$ can be regarded as determining, at least partially, the entire development of the theory. In a similar way, the members of I_0 will always be members of the largest list of individuals believed by persons who have the theory to be members of I. In this sense, the scope of the theory can be viewed as successive enlargements of I_0. Note that (D64) does not require that every person who has the theory, at some time while he had the theory, believe nothing stronger than that $I_0 \in N_{\mathscr{E}_0}$, or that $I \in N_{\mathscr{E}_0}$. Typically, of course, those who "discover" the theory will initially have beliefs no stronger than this. But, most of those who have the theory will come to have it at a mature stage in its development. They will come to have the same beliefs that the discoverers had initially, but they will, in general, come to have much stronger beliefs as well. One might expect that there would be a substantial sharing of the beliefs held by those who have the theory at a mature stage in its development. However, the present account does not attempt to provide a precise account of this.

We may summarize the results of this discussion, roughly, in the following way. The view of what it is to be a theory of mathematical physics and what it is to have such a theory that is expressed by (D61) and (D62) claims that all those who have a theory share *two* things: a core mathematical formalism, and a commitment to use that formalism in dealing with the same class of physical systems. This view does not require that there be any unanimity whatsoever among the people who have the theory as to the particular way that they use the common formalism to deal with the common range of intended applications. The view expressed by (D63) and (D64) requires that, in addition to the core formalism and the characteristic range of intended applications, all persons who have the theory share the same "starting point" in attempting to use the core formalism to say as much as possible about the characteristic range of intended applications. They may not all actually start from this point – only the discoverers of the theory do this – but they all apply the formalism in a way that is partially determined by the initial successful application. That is, the initial paradigm applications are always included in the list of things they believe to be applications and their ways of postulating special laws are always compatible with the initial success of the theory. This is a very minimal amount of unanimity in the way that the core formalism is applied in dealing with the characteristic range of intended applications. Intuitively, it seems clear that much more unanimity is present among those who have real-life theories of mathematical physics. One might expect that a great part of this could be accounted for by the fact that all those who have the theory share the same broad principles of inductive logic. That is, once the minimal conditions of unanimity were satisfied, the common principles of induction and exposure to roughly the same data would assure, in the long run, a significant amount of unanimity. A thorough investigation of this possibility would take us into areas of confirmation theory and inductive logic that are beyond the scope of this work.

Thus far, we have given a rather extensive account of how the persons who have a theory of mathematical physics might, from time to time, believe different statements that are characteristically associated with this theory. We have looked at different ways that their beliefs might change while they still had the same theory. Using Kuhn's terminology, one might say that we have given an account of the practice of "normal science" within the framework of a particular theory of mathematical physics. We have not, so far, said very much about how people come to have theories of mathematical physics, or about how they cease to have one theory and come to have another dealing with roughly the same range of phenomena. Again in Kuhn's terminology, we have said very little about "scientific revolutions" as they occur in mathematical physics. I confess, at the outset, that this is a subject about which I find it extremely difficult to say anything that is both precise and interesting. Nevertheless, the view of the logical structure of theories of mathematical physics I have been defending does appear to have some consequences relevant to such questions and I shall try to make these explicit.

To begin, our account of what it is to be a theory of mathematical physics allows us two cases. First, there is the case where a theory of mathematical physics comes into being dealing with a class of phenomena which has never before been dealt with by any theory that could properly be called 'a theory of mathematical physics'. The genesis of classical particle mechanics and classical thermodynamics appear to be examples of this. It is difficult to see how the earlier theories dealing with roughly the same phenomena could be reconstructed to yield anything like what we have called 'a theory of mathematical physics' – at least not a theory that makes use of the device of theoretical functions. Kepler's description of planetary motion might be regarded as a theory of mathematical physics, but a very rudimentary one containing only non-theoretical functions. On the other hand, the theory of relativity and quantum mechanics provide examples of theories which came into being to replace earlier theories of mathematical physics which were found to be inadequate. Let us consider the first case.

There are two significant aspects to situations in which theories of mathematical physics come into being where there have been no such theories before. The first aspect – the quantification of the data – has been rather widely discussed. It is obvious that no theory of mathematical physics can be constructed to deal with a class of phenomena until the phenomena have been described in terms of something like numerical functions. That is, some way must be found to "represent" the qualitative data quantitatively. A great part of the work of Newton, Galileo, and earlier medieval physicists may be viewed as efforts toward providing a quantitative description of phenomena involving moving bodies – a kinematics. Roughly speaking, these attempts may be seen as seeking relational properties of the individuals involved in the phenomena that are both intuitively significant and satisfy conditions necessary to permit their being "represented" by mathematical structures like, for example, numerical extensive systems (see p. 18). A great deal can be said about the conditions these relations must satisfy in order to permit quantification. Very little has been said about which relations are "interesting" and which are not. Whether one has quantified a phenomenon in an interesting way is determined both by the intuitive familiarity of the resulting concept and its fruitfulness in facilitating the construction of a successful theory. The concept of acceleration was intuitively rather remote from familiar, pre-classical mechanics ways of describing motion – for example, by describing the geometrical configuration of the body's path. Yet, it was not so remote as to appear to have nothing to do with the more familiar concepts *and* it could be incorporated in a successful theory.

This aspect of constructing *ab initio* a theory of mathematical physics can be viewed as constructing a part of the core to the theory – the set of partial possible models N_0 – in such a way that the phenomena to be accounted for can be regarded as models for this abstract mathematical structure. The second significant aspect of the genesis of a theory of mathematical physics is the construction of the rest of the core. Where do the theoretical functions "come from"? How is their relation to the non-theoretical functions discovered? Why do we impose the constraints that we do on the values of these functions? There is one pat answer to all these questions: we simply pick theoretical functions, constrain their values and relate them to non-theoretical functions in *any* way that "works". That is, the *only* important aim in constructing the theory is that claims made with the theory are true and, perhaps, that the formalism is as simple as possible. I believe this simple answer is inadequate. It is inadequate in that it fails to account for the fact that the claims of the theory are expected to, in some sense, 'explain' the observed facts about the phenomena in question. There is something beyond merely a tractable and accurate description of the data that is demanded of a theory. In choosing the way the remaining mathematical structure is to be constructed the theorist is guided by more than simply a desire to systematize accurately. He wants to make the data intuitively "intelligible" in some way that goes beyond this. The theory must, in some way, allow us to "understand" better the phenomena it accounts for.

I am unable to give a complete account of the conditions a theory of mathematical physics must satisfy to assure that its claims are genuinely explanatory. Just as in efforts to provide an account of scientific explanation in general (see, for example [15]), necessary conditions for something's being an explanation are not hard to come by, but sufficient conditions elude us. Indeed, the difficulty in the special case of theories of mathematical physics can, rather naturally, be viewed as a special case of the difficulty encountered in providing sufficient conditions for scientific explanation, in general. In the general case, the difficulty arises when one tries to distinguish between genuinely explanatory, "lawlike" generalities and *ad hoc* generalities which "just happen" to agree with the data available. In the case of theories of mathematical physics, the difficulty arises when one tries to distinguish between cores which have genuinely explanatory theoretical functions and those which do not. In the general case, it has been argued that the distinction in question can only be drawn relative to the historical context in which the explanation is offered ([13], pp. 84 ff.). Very roughly, what counts as a lawlike generality at a particular time is determined by what generalities have been successfully maintained in the past. There are (perhaps insurmountable) difficulties encountered in filling in the details of this account. However, the same general insight appears to be applicable to theories of mathematical physics.

A crucial feature of theories of mathematical physics that "generate" genuinely explanatory claims appears to be roughly this. The theoretical functions and the constraints on their values can be recognized as being "something like" pre-theoretical concepts that are relevant to the phenomena in question. For example, the mass-function in classical particle mechanics can be related in an obvious way to a pre-theoretically familiar property of ordinary, medium sized physical objects – namely, the property of offering some resistance to being set in motion. Likewise, the force function can be related in an obvious way to the pre-theoretical notion of how much "push and pull" it takes to set some medium-sized physical object in motion and keep it moving. One might say, very roughly, that the concepts of mass and force in classical particle mechanics are simply a quantitatively precise "reconstruction" of these intuitive, pre-theoretical concepts. The claims made using the theory of mathematical physics containing these concepts derive their explanatory force from this connection with intuitively familiar notions. This provides a kind of historical continuity in our thinking about phenomena involving moving bodies. One is inclined to say, after seeing what classical particle mechanics has to say about some particular instance of motion, "Ah yes, that's just what I would have expected".

It must be emphasized that this is only a suggestion about where one might look for an account of the explanatory force of theories of mathematical physics. Following it up presents many obvious difficulties. For example, it is not clearly necessary that *all* theoretical functions be connected with pre-theoretical concepts if the theory is to be explanatory. The concept of entropy in classical thermodynamics had no obvious connection with any pre-theoretic property of heat engines when this theory was discovered. The connection with "randomness" was only subsequently discovered. Perhaps more important, the notion of a function appearing in a theory of mathematical physics being "connected" with or "derived from" some pre-theoretical concept is simply not a very precise one. But, even without attempting to make this notion more precise, one can see intuitively that questions are sure to arise about how close the connection must be and just what properties of the theoretical functions are most influenced by this connection. Again, I think it is possible to go some small way toward sketching an answer to this question.

First, it appears to me that a number of the properties of the theoretical functions like, for example, continuity and differentiability are simply going to be required for reasons of mathematical convenience. It is unlikely that any features of the pre-theoretic properties connected with these functions will demand these. Indeed, it appears more likely that we might be willing to "fudge", or "idealize" the pre-theoretical notions a bit to assure that the corresponding theoretical functions have these properties. (This remark applies, as well, to the relational properties that underlie the non-theoretical functions in theories of mathematical physics.) Second, it appears that the relations among the theoretical functions and the non-theoretical functions which characterize the theory – the basic mathematical structure M – will not be very strongly influenced by the pre-theoretical background. At least, if one can regard classical particle mechanics as typical, this appears to be the case. In the "pre-theory" the property of resistance to being set in motion was never clearly distinguished from the property of weight; the velocity of a body, rather than its acceleration, was sometimes claimed to be proportional to the force acting on it. One may view the statement of the second law as an attempt at *legislating* a coherent way of relating the concepts that were currently being employed to deal with motion. In this way, the basic laws that appear in the theory of mathematical physics represent an attempt at providing a single, coherent way of using certain concepts where one *might not* have existed before. For this reason, one should not expect that *all* the intuitive, pre-theoretical features of these concepts will be mirrored in the functions in the theory that correspond to them. Indeed, if the theory of mathematical physics is successful, one might expect that everyday usage of the pre-theoretical concepts would be modified to bring it into harmony with usage suggested by the properties of the corresponding functions in the theory.

The considerations just mentioned suggest that, though the mathematical physicist may typically regard himself as simply clarifying and making precise already existing concepts with the help of a mathematical formalism, he in fact has a considerable amount of latitude in the way that he does this. Existing concepts do not remain unscathed when they are taken up, made quantitative, and incorporated into a theory of mathematical physics as theoretical functions. However, this is not the whole story. Though the relations between theoretical and non-theoretical functions that appear in the basic mathematical formalism of the theory may be chosen with some latitude, the same kind of latitude may not be present when we come to apply this formalism. Features of the intuitive, pre-theoretic analogues of the theoretical functions may severely limit its applications. For example, prior to Newton the weight of a body was regarded as an intrinsic property of the body and not conceptually separated from the property of resisting being set in motion. Thus, when the latter property appears as the mass function in classical particle mechanics we are strongly inclined to view a body's mass value as an intrinsic property of the body. Similarly, weight was regarded as being proportional to the quantity of "matter" in an object – that is, it was believed to be extensive with respect to concatenation, and this leads us to expect that the mass-function in classical particle mechanics will also be extensive with respect to concatenation. In the case of forces, the pre-theoretic concept of a force required that it be "produced" by some physical object and took the capacity to produce certain forces as an enduring property of physical objects. These features of the pre-theoretic concepts of force and mass do not get "built into" the basic mathematical structure of classical particle mechanics. Rather, according to our account of the matter, they appear as constraints on the application of that structure. In the case of classical particle mechanics, at least, the nature of these constraints on the application of the basic mathematical structure appear to be much more closely determined by the pre-theoretical concepts than is the basic mathematical structure itself.

According to our account of the way theoretical functions operate, there is some reason to think that the situation exemplified by classical particle mechanics is typical. First, it is natural to expect that the pre-theoretical background will have considerably more influence on the properties of the non-theoretical functions than the theoretical functions. The reason is this. The pre-theoretical properties associated with the non-theoretical functions must be used to determine the values of these functions in a rather direct way, while the same is not true for theoretical functions. A bit more precisely, the non-theoretical functions must be "representations" of pre-theoretical relational properties (or slightly idealized modifications of these) to which we have empirical access. Moreover, we have to have a relatively strong grip on the truth conditions of statements involving these properties antecedently to determining any values for non-theoretical functions. We are thus not at liberty to impose a great deal of modification on these concepts. We may "clean them up" slightly to assure that the non-theoretical functions have some "nice" mathematical properties, but we can not dabble with intuitively significant changes in their truth conditions. In contrast, our account of theoretical functions does not require that we have any "direct" access to the values of these functions. Though the explanatory force of the theoretical functions surely comes from their association with some pre-theoretical properties, we do not have to have, *antecedently*, a very firm grip on the truth conditions of statements involving these properties. This is because we are never compelled to appeal directly to such statements in determining the values of these functions. This means that we can afford to legislate a bit about these truth conditions.

We can be a bit arbitrary in the way we choose to connect up the theoretical functions with the non-theoretical functions and let the implications of this connection for the pre-theoretical concepts associated with the theoretical functions "come out in the wash".

Our view also allows us to understand why one might expect that the influence of the "pre-theory" on the theoretical functions appears primarily in the constraints rather than in the basic mathematical formalism or in the special theoretical laws that might be postulated. Suppose that connection between the theoretical and non-theoretical functions in the basic mathematical structure has very little, or even no, empirical content. As we have noted, the second law taken alone, in the absence of constraints and special force laws, has no empirical content (see Chapter VI). If we then impose constraints on the theoretical function values that are suggested by the pre-theoretical concepts associated with them, it will be possible to cling to these constraints tenaciously even though particular claims made using them turn out to be falsified (see above). This is because, in this situation, the real empirical content of the claims of the theory is determined by the theoretical laws that are postulated. These laws may always be sacrificed to save the constraints. In this situation, one might say that "essential" features of the concepts associated with the theoretical functions enter in two ways. First, in the basic mathematical formalism an essential connection is made between theoretical and non-theoretical functions. This is not entirely arbitrary. It may be dictated by strong intuitive pre-conceptions, but it need not be. Second, in the constraints on the application of this basic formalism some essential properties of the individual theoretical functions are exscinded. Now, it may well be that, in a situation of "conceptual flux" which existed before the initial success of the theory in question there was much more intuitive agreement about "essential" properties of various individual concepts than about how they all "fit together". In such a situation, it would be natural for these essential features to appear in the constraints, while the connection between the concepts appearing in the basic mathematical formalism was the result of an apparently arbitrary choice, perhaps from among several intuitively acceptable alternatives.

In the situation just described, it is important to understand what features of the theory are significant in determining the values of theoretical functions for particular individuals and how these values might change as the theory develops. According to the view that identifies theories of mathematical physics as ordered pairs $\langle H, I \rangle$, the possible values of a particular theoretical function for a particular individual in a domain of a member of I are completely (though not, in general, uniquely) determined by a particular expansion of H, $\mathscr{E}_t$. For different expansions, one would generally find different, perhaps even incompatible, possibilities for this value. In this respect, the extension of the concepts associated with the theoretical functions might be said to change as the theory develops. However, in the case where the theory develops linearly with a series of successively more restrictive expansions of H, the successive possibilities for a given theoretical function value will all be compatible and the value will become successively more uniquely determined. In this case, it does not seem intuitively appropriate to say that the extension of these concepts is changing. Rather, it appears intuitively that we are simply finding out more and more about this concept. It should also be noted that our knowledge about the range of possibilities for a particular theoretical function value will, typically, be strongly dependent on our knowledge about what is in I. The more members of I we have explicit information about, the more we are able to say about the range of possible theoretical function values for some particular individual. In this sense one *might* say that the extension of theoretical concepts changes as the theory develops. But it would clearly be more accurate to say that our knowledge of this extension changes. However, in situations where I is described by listing paradigm individuals *and* where doubtful cases are arbitrarily decided "in favor of" the theory it may seem more plausible to say that the extension of these concepts changes. Modifying the identity conditions for theories in the manner attributed to Kuhn does not significantly change this situation. The only difference is that now all the expansions of H must be restrictions of $\mathscr{E}_{t_0}$ and we always know that $I_0 \subseteq I$.

Let us now consider the second kind of situation in which a new theory of mathematical physics might come into being – the case where the new theory replaces an earlier theory of mathematical physics. To begin, we should recall our account of how one who has a given theory of mathematical physics might come to be dissatisfied with it and eventually give it up. The crucial feature to be recalled is this. According to our account, it is highly unlikely that encountering recalcitrant data will ever *compel* a person who has a theory of mathematical physics to give it up. In the typical cases, where the characteristic range of intended applications and some parts of its members are described intentionally, there are always going to be other beliefs that one might give up in the face of recalcitrant data in preference to giving up the beliefs that are essential to having the theory. Accepting Kuhn's thesis changes this situation only slightly. Now the discovery of new data about the paradigm set of intended applications could falsify the claim of the initial "successful" application of the core, and thus compel one to give up the theory. But even here, having the theory is relatively impervious to recalcitrant data.

Though the possibility exists of tenaciously maintaining the theory, even in the face of repeated failure to successfully apply H to I, it is not a situation that can be comfortably accepted. In such a situation, one might suspect that serious doubts would arise in the minds of those who have the theory – doubts about the ultimate prospects of finding a successful application. Such doubts might naturally cause them to consider the possibility of doing better with another theory. But, it appears doubtful that they would go so far as to actually give up the theory *unless* they had, actually at hand, a theory whose prospects looked better. One explanation for this sort of conservatism is not difficult to find. Even though no application of H has been found that is compatible with everything known about all the individuals actually known to be members of I, it may well be that there are successful applications of H to significant sub-sets of I. The fact that classical particle mechanics gives an adequate account of the motions of medium-sized physical objects moving at familiar velocities motivates one to retain this theory even when it runs into difficulties in dealing with electrically charged bodies moving at high velocities. At least it motivates one to retain it until a real alternative is available. Until some new mathematical structure is discovered that is just as successful as the old, where it has been successful, and, in addition, gives some reason to expect that it will be more successful in dealing with the members of I that cause difficulties for the old theory, it is likely that those who have the old theory will continue "plugging away" – trying to apply it successfully.

What can one say about the new theories that arise to replace theories of mathematical physics that find themselves in such difficult straits? It appears pretty clear that, if these new theories are going to have any explanatory force at all, they can not be completely devoid of connections with the previously existing theory. In much the same way as when theories of mathematical physics arise where there has been no such theory before, the functions appearing in the theory must have some touch with intuitively familiar concepts. In the present case, the existing theory of mathematical physics is the natural source for the concepts that provide this kind of "historical continuity" in our theorizing. Indeed, in this case one might hope to give a somewhat more precise account to this continuity. One might attempt to characterize the relation between the new theory and the old theory by the following requirement. The new theory must be such that the old theory reduces to (a special case) of the new theory. It is generally claimed that something like this is true for classical particle mechanics and special relativity theory as well as for "classical" special relativity and relativistic quantum mechanics. I do not know how to argue that this *must* be true in general. It may be that other types of "conceptual connections" besides reduction would suffice to endow the new theory with the familiarity needed for explanatory force. On the other hand, a compelling feature of the reduction relation is that it provides some means of accounting for why we were so successful in applying the old theory to some situations, and yet failed in attempts to extend it beyond these.

If we accept the reduction requirement as at least a partially adequate account of how a new theory must be related to the old it replaces, then the account of the reduction relation for theories of mathematical physics provided in the previous chapter sheds some light on this situation. If we are out to require that there be some connection between *all* the concepts of the old and new theory, it appears that strong reduction is clearly what we want. Indeed we may even, intuitively, want to demand more. But the main point to be made here can be made without becoming more precise about this. The point is this. The functions that appear in the new theory should be thought of as being associated with "different" concepts than the functions to which they correspond in the old theory. That is, the functions in the new theory appear in a different mathematical structure – they stand in different mathematical relations to one another; they admit of different possibilities of determining their values – than the corresponding functions in the old theory. If one wants to be precise about the role of a function in the new theory, the thing to consider is how it fits into the mathematical formalism and whether it is used as a theoretical or non-theoretical function. For example, the mass function in classical particle mechanics and the mass function in special relativity theory are to be regarded as two theoretical functions appearing in distinct theories of mathematical physics. One should not expect that means of determining the values of one *necessarily* have anything to do with determining the values of the other. Of course, it is an interesting fact that classical particle mechanics stands in a reduction relation to special relativity and that the mass functions in the theories correspond in this reduction relation. But this should not obscure the fact that these functions have different formal properties and, in this sense, they are associated with different concepts. One might look at the situation this way. Just as our pre-Newtonian notion of mass (a somewhat confused one) was modified when it was "taken up" into classical particle mechanics, so the classical notion of mass is modified when it is "taken up" into special relativity. The only difference is that, in the latter case, we are able to be quite precise about what modifications have occurred by comparing the mathematical structure of the two theories in question and looking at the reduction relation holding between them.

This completes what I have to say about a "cradle to grave" account of theories of mathematical physics. In capsule summary, the view of theories of mathematical physics that I have sketched may be thought of as a kind of compromise between an instrumentalist view of theories and a descriptive, or statement view. Essentially associated with every theory of mathematical physics is some mathematical structure. This structure is a tool, an instrument. It is a tool that people use in dealing with certain chunks of their experience. But they use it in a particular way. They use it to describe, perhaps to explain, this experience. At any rate, they use it to make statements about certain phenomena they encounter. These statements are empirical statements. They are vulnerable to the vicissitudes of experience in the same way that more mundane empirical statements are. But confidence in the usefulness of the mathematical formalism as a tool in making true statements about this class of phenomena is not so vulnerable to experience. On Kuhn's view, at least, one becomes convinced of the usefulness of a particular mathematical structure by actually succeeding in applying it successfully to some limited range of phenomena. This success raises the expectation that the same structure might be successfully applied to phenomena that are, in some sense, like those that appeared in the initial successful application. This leads us to try. A bit of success encourages more ambitious attempts to include more within the scope to the theory. But a few failures do not necessarily indicate disaster. So long as some of our applications remain intact as we expand our information, we still have confidence in the usefulness of the mathematical structure. It is only when it repeatedly fails to live up to its expectations that we consider chucking it aside for something better. But just as a broken oar is better than none, we don't chuck it aside until we have something better.

BIBLIOGRAPHY

[1] Adams, Ernest W., 'Axiomatic Foundations of Rigid Body Mechanics', Unpublished Ph.D. dissertation, Stanford University, 1955.
[2] – , 'The Foundations of Rigid Body Mechanics and the Derivation of its Laws from Those of Particle Mechanics', in *The Axiomatic Method* (ed. by Henkin, Suppes, Tarski), North-Holland, Amsterdam, 1959, pp. 250–265.
[3] Beth, E. W., 'On Padoa's Method in the Theory of Definition', *Indagationes Mathematicae* **15** (1953) 330–339.
[4] Campbell, Norman R., *Foundations of Science*, Dover, New York, 1957.
[5] Carathéodory. 'Untersuchungen über die Grundlagen der Thermodynamik', *Math. Ann.* **67** (1909) 355.
[6] Carnap, Rudolf, *Philosophical Foundations of Physics*, Basic Books Inc., New York, 1966.
[7] Craig, William, 'On Axiomatization Within a System', *Journal of Symbolic Logic* **18** (1953) 30–32.
[8] – , 'Replacement of Auxiliary Expressions', *Philosophical Review* **65** (1956) 38–55.
[9] Craig, William and Vaught, R. L., 'Finite Axiomatizability Using Additional Predicates', *Journal of Symbolic Logic* **23** (1958) 289–308.
[10] Duhem, Pierre, *The Aim and Structure of Physical Theory*, Atheneum, New York, 1962.
[11] Falk, G. and Jung, H., 'Axiomatik der Thermodynamik', in *Handbuch der Physik* III/2 (1959) pp. 119–175.
[12] Giles, R., *Mathematical Foundations of Thermodynamics*, Macmillan, New York, 1964.
[13] Goodman, Nelson, *Fact, Fiction and Forecast*, Bobbs-Merrill, New York, 1965.
[14] Hamel, G., 'Die Axiome der Mechanik', in *Handbuch der Physik* **5**, pp. 1–42.
[15] Hempel, Carl G., *Aspects of Scientific Explanation and Other Essays in the Philosophy of Science*, Free Press, New York, 1965.
[16] – , 'The Theoretician's Dilemma', *Minnesota Studies in the Philosophy of Science*, University of Minnesota Press, Minneapolis, **II** (1958) 37–98.
[17] – , 'Fundamentals of Concept Formation in Empirical Science', University of Chicago Press, Chicago, **II**, No. 7 (1952).
[18] Hertz, Heinrich, *The Principles of Mechanics* (translated by D. E. Jones and J. T. Walley), London 1899.
[19] Jamison, Benton N., 'An Axiomatic Treatment of Lagrange's Equations', Unpublished M.S. thesis, Stanford University, 1956.
[20] Joos, Georg, *Theoretical Physics*, Hafner, New York, 1950.
[21] Kuhn, Thomas S., *The Structure of Scientific Revolutions*, University of Chicago Press, Chicago, 1962.
[22] Mach, Ernst, *The Science of Mechanics*, Open Court, La Salle, Ill., 1960.
[23] – , *History and Root of the Principle of Conservation of Energy*, Chicago 1944.
[24] Mackey, George W., *The Mathematical Foundations of Quantum Mechanics*, W. A. Benjamin, Inc., New York, 1963.
[25] Mates, Benson, *Elementary Logic*, Oxford University Press, New York, 1965.

BIBLIOGRAPHY

[26] McKinsey, J. C. C., Sugar, A. C., and Suppes, P. C., 'Axiomatic Foundations of Classical Particle Mechanics', *Journal of Rational Mechanics and Analysis* **II** (1953) 253–272.
[27] McKinsey, J. C. C. and Suppes, P. C., 'On the Notion of Invariance in Classical Mechanics', *British Journal for the Philosophy of Science* **V** (1955) 290–302.
[28] Montague, Richard, 'Deterministic Theories', in *Decisions, Values and Groups* (edited by Washburne), Pergamon Press, New York, 1957, pp. 325–370.
[29] Nagel, Ernest, *The Structure of Science*, Harcourt, Brace & World, New York, 1967.
[30] Narlikar, V. V., 'The Concept and Determination of Mass in Newtonian Mechanics', *Philosophical Magazine* **XXVII**, No. 7 (1939) 33–36.
[31] Pendse, C. G., 'A Note on the Definition and Determination of Mass in Newtonian Mechanics', *Philosophical Magazine* **XXIV**, No. 7 (1937) 1012–1022.
[32] – , 'A Further Note on the Definition and Determination of Mass in Newtonian Mechanics', *Philosophical Magazine* **XXVII** (1939) 51–61.
[33] – , 'On Mass and Force in Newtonian Mechanics', *Philosophical Magazine* **XXIX** (1940) 477–484.
[34] Putnam, H., 'What Theories Are Not', in *Logic, Methodology and Philosophy of Science: Proceedings of the 1960 International Congress*, Stanford University Press, Stanford, Cal., 1962, pp. 240–251.
[35] Ramsey, Frank P., 'Theories', in *The Foundations of Mathematics*, Littlefield, Adams & Co., Patterson, New Jersey, 1960, pp. 212–236.
[36] Scheffler, Israel, *The Anatomy of Inquiry*, Alfred A. Knopf, New York, 1963.
[37] Simon, Herbert A., 'Axioms of Newtonian Mechanics', *Philosophical Magazine* **XXXVI**, No. 7 (1947), 888–905.
[38] – , 'The Axiomatization of Classical Mechanics', *Philosophy of Science* **XXI**, No. 4 (1954), 340–343.
[39] – , 'Definable Terms and Primitives in Axiom Systems', in *The Axiomatic Method* (ed. by Henkin, Suppes, Tarski), North-Holland, Amsterdam, 1959, 433–453.
[40] – , 'The Axiomatization of Physical Theories', *Philosophy of Science*, 1970.
[41] Suppes, Patrick C., *Introduction to Logic*, Van Nostrand, New York, 1957.
[42] – , 'A Comparison of the Meaning and Uses of Models in Mathematics and the Empirical Sciences', *Synthese* **XII**, No. 2/3 (1960) 287–301.
[43] – , 'Models of Data', in *Logic Methodology and Philosophy of Science: Proceedings of the 1960 International Congress*, Stanford University Press, Stanford, Cal., 1962, pp. 252–261.
[44] Suppes, Patrick C. and Zinnes, J. L., 'Basic Measurement Theory', in *Handbook of Mathematical Psychology*, vol. I, Wiley, New York, 1963, pp. 1–76.
[45] Toulmin, Stephen, *The Philosophy of Science: An Introduction*, Harper & Row, New York, 1960.
[46] Truesdell, Clifford A., 'Foundations of Continuum Mechanics', in *Delaware Seminar in the Foundations of Physics*, Springer, New York, 1967, pp. 35–48.

INDEX

Adams, E. W.   8, 9, 217, 234
application equivalent   187
application of theories   27, 184
applied core,   181
   correctly   182
   successfully   182
axiomatized deductive theory,   7
   relation to scientific theory   7
axiomatic systems   5

Beth's definability theorem   59

is a CPM   115
is a CRBM   239
Cathéodory   8
Cavendish balance   62
connected with   255
constrained,
   only theoretically   172
   only non-theoretically   172
   vacuously   172
   heterogeneously   173
   effectively vacuously   173
constraints   66, 170
coordinating definitions   7
core   171
Craig eliminability   55
Craig, W.   52

is a DNCPM   140
D-extendible   52
D-Ramsey eliminable   52
Duhem, P.   70, 90

is an E_y 42
empirical statement   23
effect dominates   187
effect equivalent   188
expanded core,
   theoretically   179, 180
expansion   180
an extension of y 42

extensive scale system   86
extensive system,   18
   numerical   18

F-relation   209
Fletcher trolley   61
formal equivalence,
   weak sense   194
   strong sense   194
formal identity of theories   184
formal language   6

is a GNCPM   141
Galilean transformations   149
Galilean invariant force laws   150
Giles, R.   8
group theory   9

is an HNCPM   141
Hamel, G.   8
having a theory   266, 294
Hempel, C.   53
holism   70, 89

is an INCPM   141
initially dominates   187
initially equivalent   187
instrumentalism   1
intended application of a theory   66
interpretation   6

Kuhn, T.   70, 90, 288

law,   179
   theoretical   179
   non-theoretical   179
logical consequence   7

is an MPM   131
McKinsey, J.   11, 113
Mach, E.   59
Mackey, G.   8
measurement, theory of   18

measurement, fundamental 19
measurement, derived 20
mechanics,
  generalized 207
  Lagrangian generalized 208
  particle 114, 131
model 7, 10
Montague, R. 8

is an NCPM 129
Newtonian mechanics 16, 45, 129, 139, 140
null expansion 180

is a PM 114
is a PK 118
is a P' 47
is a P_0 41
Padoa's principle 39
particle kinematics 118
Pendse, C. G. 51
possible model 17
possible partial model 41
preferred theory 205

is an RBK 239
is an RBM 235
Ramsey eliminable 49
Ramsey eliminable in the strong sense 49
Ramsey, F. 46
Ramsey sentence 53

Ramsey solution to the problem of theoretical terms 46
reduction,
  close 223, 224, 229
  strong 225, 229
restriction 47

is an S 11
is an S' 47
is an S_0 17
set theoretic predicate 9
Simon, H. 8, 51, 133
Sugar, A. 11
Suppes, P. 8, 9, 11

theory of mathematical physics, 179–181, 183, 260
  a frame for a 165, 166
  in the sense of Kuhn 293
  n-ary matrix for a 162
theoretical terms, the problem of 38
theoretically effect-dominates 188
theoretically effect-equivalent 188
θ-dependent 31
θ-theoretical 33
Toulmin, S. 1
Truesdell, C. 8

is a UNCPM 139

Vaught, R. 52